Amorphous Food and Pharmaceutical Systems

Edited by

Harry Levine
Nabisco R&D, Kraft Foods, East Hanover, New Jersey, USA

ROYAL SOCIETY OF CHEMISTRY

The proceedings of a conference, sponsored jointly by the BioUpdate Foundation and the Biotechnology Group of the RSC, entitled The Amorphous State – A Critical Review, held at Churchill College, Cambridge on the 15–17 May 2001.

Special Publication No. 281

ISBN 0-85404-866-9

A catalogue record for this book is available from the British Library

Published by The Royal Society of Chemistry
Thomas Graham House, Science Park, Milton Road, Cambridge CB4 0WF, UK

Registered Charity Number 207890

For further information see our web site at www.rsc.org

Typeset by Vision Typesetting, Manchester, UK
Printed and bound by Athenaeum Press Ltd, Gateshead, Tyne and Wear, UK

Preface

The Amorphous Aqueous State – Some Personal Reminiscences

Water and its mysteries have dominated and still continue to dominate my professional life, which has now spanned some 50 years. Curiously, I now find that the first 25 years were devoted to studies of dilute aqueous solutions *in* water, and the second 25 years to studies of very dilute solutions *of* water. The interest in 'residual water' arose during a marketing brainstorm that my then-employer, Unilever Ltd, was in the process of conducting. The aim was to persuade consumers about the wholesomeness of frozen food products. A new word was coined and added to the marketspeak vocabulary: 'frozeness'. This was seen as a desirable attribute, because it was associated with 'freshness'.

At about the same time, the middle 1970s, Unilever developed an interest in plant cell and tissue culture. Because I 'knew all about water', my colleagues and I were charged with the scientific backup for 'frozenness', and also the development of suitable cryopreservation methods for the maintenance of plant embryos, destined for eventual growth into palm trees. We soon found that all this was easier said than done. Thus began the long and tortuous path that has eventually led, by a random path, to the 2001 Cambridge conference, proceedings of which are collected in this book.

None of us had prior hands-on experience of freezing or cryobiology. So, like most physically trained scientists, we began our pilgrimage with a study of the effects produced by freezing model systems, in our case water-soluble polymers. Calorimetry seemed to be a useful method to monitor such effects, and a somewhat dilapidated Perkin Elmer DSC-2 was available. With help of friends in the Engineering Division, it was soon given an overhaul and converted to make it suitable for studies at subzero temperatures. The initial chart recorder traces of cooling and heating runs provided more questions than answers. As is frequently recorded by others in the scientific literature, so we, too, rediscovered the wheel several times over. Thus, we discovered undercooling, nucleation, and eutectic phase behaviour. What puzzled us was the universal appearance of a discontinuity in the heat output curve in the neighbourhood of $-30\,°C$, irrespective of the solution under study. Eventually, we were driven to the conclusion that we were observing glass transitions in the frozen solutions. We called it T_g', but without quite understanding its significance. Others have tried since then to give it different names, but that has only confused the issue. T_g' has fortunately become embedded in the literature, and by now, we all know what we are talking about, or do we?

The universality of T_g' no longer needs emphasising; it now forms the basis of several process technologies, mainly in the food and pharmaceutical industries. In 1974, however, we were intrigued by the amorphous material that remained after water had been removed by freezing. What might be its ultrastructure, and why did it still contain water? This all led to some sophisticated electron microscopic studies by Patrick Echlin and Helen Skaer, both at the University of Cambridge. In 1977, we jointly published our first T_g' paper.[1] At that time, my personal future was still shrouded in the mists of uncertainty, but the collaboration with my Cambridge colleagues continued for many years and led to some firm friendships. Had a fortune teller told me at the time what my future held for me, I would not have believed it.

After my departure from Unilever, I was fortunate to be able to continue the 'glass' studies at the Department of Plant Sciences in Cambridge, without the need for monthly project reports and budget forecasts. By that time, I had become fascinated by the concept of undercooled water as a means of preserving live cells and tissues. It actually worked, but the procedures were too complex to lend themselves to commercial exploitation. It did, however, give me the opportunity of adding Pierre Douzou, of cryobiochemistry fame, to my circle of friends. In the world of Plant Sciences, we must have been unique, because much of our experimental material consisted of mammalian erythrocytes. When lysed and centrifuged, the outcome did, however, resemble the beetroot juice that was spattered around the laboratory walls. During the undercooling work, I was brought face to face with the fact that the effects produced by freezing, on the one hand, and low temperature, on the other, have nothing in common; low temperature preserves and stabilises, but freezing kills! Indeed, I began to realise that on this planet, cold (freezing) is the most widespread threat to life.

That brought us back to glasses and glass transitions, perhaps even *in vivo*. My colleagues and I became interested in the physics and chemistry of natural survival mechanisms, in freeze-tolerance and freeze-resistance phenomena, and by a roundabout route, that brought us to supersaturated solutions and glasses of polyhydroxy compounds. I went for help to the glass experts in the Materials Science Department. They listened politely to my story, but when it came to water-soluble glasses, they looked at me as though I was talking metaphysics. It was only when I reminded them of sugar candy and candy floss that a look of recognition returned to their faces. Since then, they, too, have become obsessed with the materials science of aqueous solid solutions.

This all happened at around the time when I received my first visit from Harry Levine, which eventually led to a wonderful friendship with him and Louise

Slade. They 'bought into' my stories, and jointly we developed the new branch of technology, until it was ready to be presented. They then threw themselves with vigour into measuring hundreds of T_g values, writing papers and preaching the gospel of glass transitions to the food processing industry, with amazing success. I later tried the same approach with the pharmaceutical industry, but was not nearly as successful. Even today, the myths of water activity and bound water are hard to kill!

In the meantime, we came to realise that freeze- and drought-tolerance, as exhibited by many species, probably relies on *in vivo* vitrification mechanisms and is promoted by the biosynthesis of lyoprotectants of different chemical origins, with PHCs predominating. Trehalose received extensive press coverage and was even claimed, mistakenly, by some to be unique as protectant against desiccation. It was only one step to suggest that similar mechanisms might be applied to the *in vitro* stabilisation of labile molecules, of supramolecular structures, and perhaps even of intact cells and tissues. Initial experiments proved to be encouraging, and some pharmaceutical companies began to take an interest. This persuaded us to file for patents, but the University was reluctant to assist in such activities. That is how Pafra Biopreservation came into existence, a startup (or upstart) enterprise, located in the Cambridge Science Park. Forced to exist under stringent financial controls, our small group was yet able to develop the stabilisation technology to the point that we were runners-up (after Marconi) in the Prince of Wales Award for Technology competition, and also received several government awards and grants.

During the twelve years of its existence, our Cambridge laboratory was able to welcome and host twelve scientists, from the Netherlands, USA, Japan and Russia, from graduate student level to professors on sabbatical leave. My colleagues and I had to work hard to keep our heads above water, financially speaking. In the meantime, the visitors were able to advance our collective understanding of *in vivo* ice nucleation, of glassy carbohydrates, and of phenomena relating to nucleation and crystallisation in such glasses, their hydrates, and their ability to stabilise proteins. Those were exciting times for us all.

At one stage, my fellow directors at Pafra Board of Directors put it to me that we were good at spending money, but what might we do to earn some money. It was this suggestion that got us into freeze-drying. It dawned on us that this capital-, labour- and energy-intensive process was universally used in the pharmaceutical industry, but that there appeared to be little understanding of the technology. It was strictly a trial-and-error operation, too often with expensive errors. Even worse, the process did not receive any mention in chemical engineering texts, and there appeared to be no freeze-drying research in any university engineering departments anywhere in the world. With our accumulated knowledge of water, drying, carbohydrates, glasses and stability, we set about to develop a freeze-drying consulting service to industry. Tony Auffret, our Technical Manager, was mainly responsible for creating an enviable reputation for this enterprise. The development had another beneficial spinoff; it put us in touch with Mike Pikal, then in far-away Indianapolis, surely the undisputed King of Freeze-Drying. Thus, another friendship was formed and cemented. By the time

Pafra Biopreservation was sold in 1997, our clientele (or patients?) included 18 of the world's 20 largest pharmaceutical companies, in addition to many smaller ones. To us, it was a case of David and Goliath, and we were constantly astounded by how megaPharma, where millions of dollars are spent annually on R&D, could be so ignorant of a technology right at the heart of their operations. Our archives bear witness to the number of 'hospital cases' we received for treatment. While Harry and Louise were touring the globe, visiting bakeries wherever they went, teaching the polymer/material science approach to food processing, such as baking cookies and crackers, so we did a similar job for the pharmaceutical industry.

At some stage, Harry, Louise and I reached the conclusion that the science and technology so basic to our respective industrial interests needed tidying up. There were too many holes, too many unanswered questions, and there appeared to be few well-directed research approaches. We set about the construction of a highly subjective short list of 'experts' who were familiar with the outstanding problems and actively engaged in relevant research. And so it was that the first Amorph conference was put together in 1995. It was completely sponsored by industry, which enabled us to bring together 35 invited 'experts' at Girton College, Cambridge. The format was novel, because no participant was permitted to speak for longer than 5 minutes; it was to be a true Discussion Conference. Louise kept a record, apparently of every word that was spoken, and the proceedings were written up in the form of an informal report, which is attached (see Appendix I) to this Preface.

This book constitutes the record of the 2001 follow-up (see Appendix II) to the 1995 discussion conference. During the intervening years, the list of 'experts' has grown. More scientists have become fascinated by the puzzles of water-soluble amorphous systems, their properties and their applications. Important contributions by the 'newcomers' feature in this book, alongside those of the old-timers. The reader is left to judge whether all the problems and questions highlighted in 1995 have been resolved. If not, then what else is required?

Although the significance of water-based amorphous states has become more widely recognised, there is plenty of tutorial work left, and I hope still to be able to make a contribution. My most recently acquired friends at Inhale Therapeutic Systems Inc. allow me annually to 'indoctrinate' newly employed scientists, but also to discuss with their experienced colleagues matters relating to drying, stability and amorphisation. There are also still Intellectual Property issues, associated with our former patents, that rumble on in various law courts and require attention.

The BioUpdate Foundation, which I helped to found, in association with yet another friend, Andre Schram, provides post-experience courses on various aspects of biotechnology. The amorphous state forms an important part of the courses on protein stability. Our freeze-drying course is also an evergreen, and continues to attract participants from many European pharma companies, whenever and wherever it is presented.

In summary, I have been fortunate to get to know, and often to befriend, so many scientists in so many countries. It is said that a rolling stone gathers no moss. This rolling stone has gathered plenty, both in dilute solutions and, more recently, in aqueous glasses.

Reference

1. F. Franks, M.H. Asquith, C.C. Hammond, H.B. Skaer and P. Echlin, *J. Microsc.*, 1977, **110**, 223.

Felix Franks
London, March 5, 2002

Appendix I: Summary Report of the Discussion Symposium on Chemistry and Application Technology of Amorphous Carbohydrates

Girton College, Cambridge, UK, April 4–6 1995. Symposium Organizers: Felix Franks and Harry Levine, Symposium Manager: BioUpdate Foundation. Report compiled by Felix Franks, from notes supplied by Louise Slade.

The Premise

The physical properties of amorphous carbohydrates in the anhydrous state or at a low moisture content play an important role in the processing and product quality of cereal-based and various other foods and the stabilization of pharmaceuticals and biotechnological products (*e.g.* as excipients in freeze-drying). There is an increasing awareness that, in all such applications, thermomechanical properties partly determine the choice of suitable formulations. Despite their increasing importance in food and pharmaceutical process technology, the chemistry of such amorphous sugars is substantially unexplored. Formulations and recipes are usually arrived at on a hit-or-miss basis with little basic understanding of the reasons for success or failure.

The Objective

The Symposium was convened to discuss and define the relevant problems, rank them in some order of importance and suggest effective experimental, theoretical and computational approaches for their study.

The organizers of the Symposium express their gratitude for the generous support by the sponsoring companies.

Participation

Participants included 30 invited scientists with known interest and expertise in the subject, and an equal number of observers, nominated by the sponsoring companies.

Report

This report is not intended to be a printed version of the full Symposium proceedings. It is compiled in note form as a summary of *significant aspects* of the discussions. In its layout it conforms approximately to the format of the Symposium and should be read in conjunction with the Symposium programme and the list of participants.

The discussions relating to each session have been 'tidied up' and are summarized in precis form, according to subject matter, rather than in the chronological order in which they were introduced during the session. Contributors to each discussion topic are indicated but remarks made during the discussions are not attributed to individuals. Participants are reminded that the contents of the agenda document and of this report are privileged information and must not be quoted or referred to without the explicit permission of the individual contributors, whose identities can be obtained from the BioUpdate Foundation.

Follow-up

The principals of the BioUpdate Foundation are now considering suitable follow-up actions to what was considered to be (by the majority of participants) a most productive and novel exchange of ideas.

Topic 1 – Relationship Between Molecular Structure and Glass Transition

Contributors: Le Meste, Ablett, Randall, Slade, Huang, Angell, Franks, Cesaro, Brady, McInnes, Zografi, Bizot, Levine, Karel, Foster.

Relationships (if any) between molecular structure, interactions and their temperature dependences; differences between entangling and nonentangling systems; *i.e.* are there molecular and network T_g values. Are some experimental methods more sensitive to one or the other? How is the structure of a biopolymer related to its T_g? Currently T_g needs to be measured, cannot be predicted, *e.g.* from structure and/or interactions.

How can structural features of amorphous carbohydrates be measured? A better definition of 'solid' is required in relation to amorphous phases.

Can parameters of importance in glassy carbohydrates be predicted; *e.g.* the 'universal' constants in the WLF equation. Why does ΔC_p of vitrification decrease with molecular weight for a series of oligomers?

The 'heretical' view was expressed that a liquid formed immediately after completion of melting is NOT an equilibrium state. How does its viscosity/temperature relationship differ from a supercooled liquid?

Is it possible to measure isomerization rates in a sugar melt? Different tautomers might possess different T_g values. T_g depends on annealing temperature; find 12 °C differences, particularly high where other thermal events at higher temperatures are possible. Refer to fructose behaviour; similar behaviour might occur with galactose and ribose.

Recent NMR studies suggest that the H-bonded network structure of a β-furanose might be easily disrupted by a 'foreign' stereoisomer and that a β-pyranose melt should be more viscous than the corresponding β-furanose. This implies that a freshly prepared melt will increase in viscosity during relaxation. Furanose–pyranose conversions typically have an activation energy of $10\,kJ\,mol^{-1}$, so that the process could be trapped.

Glycerol (three carbons) and sorbitol (six carbons) exist as single conformers, but fructose exists as a mixture of many possible conformers. Presumably conformers should be miscible? Immiscible amorphous phases might coexist. A pure pyranose crystal could give rise to isomer mixtures or discrete phases on fusion, certainly on a 5–10 nm scale. Hence could have time-dependent entropy and viscosity changes. For instance, the NMR spectrum of fructose held at 120 °C reveals the appearance of different isomers with time; this might lead to a depression of T_g. There was agreement that the conformer composition of sugars is temperature and concentration dependent.

Sorbitol and mannitol also have preferred conformations, at least in solution; they are solvent-dependent. In the crystal, the two polyols adopt different conformations (planar zig-zag *vs* 'sickle'). Nothing is known about the fused state. Speculation that α and β sugars may well have diffusion coefficients that differ by 10%.

Many sugar sub-states exist with measurably different energies (boats, chairs). They would affect entropic contributions to glass transitions. How much entropy is trapped in a glass, compared to the entropy of fusion? Compare fructose and sucrose: apparently more 'trapped' entropy in fructose. Relevance to T_m/T_g and 'fragility' concept, because T_m location depends on the entropy of fusion. Also consider the contribution of tautomeric mixtures in this context.

Contrast sucrose with lactose, raffinose and trehalose, all at low moisture contents: Gordon Taylor equation fits, except for sucrose, where the effect of water on T_g is larger than calculated. How is the 'structure' of the dry sugar related to its T_g? Do internal hydrogen bonds play a role? Possibly, but rotations of C–C bonds do occur.

The vitrification potential of salts was mentioned. Mg gluconate is a particularly good glass former, with $T_g = 80\,°C$; an amorphous 50:50 mixture of sucrose and Mg gluconate is stable. Other glass-forming salts include Na gluconate and Na citrate.

Iso-maltose is claimed to be more flexible and has lower T_g than maltose. Other evidence to the contrary. The hydrodynamic volume is important.

T_m/T_g is used as an indicator of 'fragility' of fluids. Thus, the ratio is 2.0 for water (strong) and 1.0 for the most fragile liquids. Actually should use T_b/T_g as the true relation, because both temperatures refer to the liquid phase.

The question of 'unique' sugars was raised. As regards fragility, stachyose

$\gg$ maltose; implies high W_g, (as well as T_g and high mol. wt) has a role to play. Trehalose is more fragile than expected.

Are there predictive relationships between structure and fragility? How can fragility be measured by a single method? Perhaps tan δ, measured at a single frequency, say 10 MHz with respect to its value at T_g. Possibly Brillouin scattering?

Topic 2 – Chemical Reactivity of Solid Sugars

Contributors: Karel, Zografi, Hatley, Angell, Huang, Foster, Ablett, Franks.

Reactions do occur below T_g. Often (not always) rate depends on $(T - T_g)$. Freeze-dried materials have porous matrix which collapses above T_g. Loss of volatiles or oxygen uptake can then exhibit reduced rates above T_g. Crystallization events can also affect chemical reactions: when anhydrous sugar crystallises, the matrix is diluted, but when sugar crystallises as a hydrate, the matrix may be diluted or concentrated. The role of crystallization enhanced reactions in food stability was mentioned.

Consider two components + water; one a good glass former (*e.g.* trehalose), the other is an 'additive', but should be considered as a reactant, rather than simply as a plasticizer.

Most reactions require an initial proton transfer step, usually involving water, but sugars can also play that role. Also raised the question of the meaning of pH in a system with 2% moisture.

Should be possible to identify T_g more reliably in complex systems containing proteins, *e.g.* where partly superimposed transitions in tertiary structure (denaturation) can occur. Could have complex behaviour where possibility of tautomerism exists. Fructose shows heat flow discontinuities at 240 and 320 K. Is the upper transition related to an isomerization?

Can WLF equation account for chemical reactions (Maillard)? The rate is said to depend on ΔT and moisture content, but ΔT is itself a function of moisture content. Scepticism expressed about the validity of WLF kinetics in such situations. Chemical reaction rate depends more on diffusion (mobility), perhaps ONLY on translational diffusion. Water acts primarily as plasticizer, but also as reactant (hydrolysis). WLF constants are NOT 'universal' constants. It was suggested that WLF or similar kinetic models are just as universal as the Arrhenius model which is itself a special case.

Long discussion on Maillard reactions; they can occur in glasses, *e.g.* lactose + insulin yield ketoamine products. Most studies refer to prenucleated systems. Crystallization requires both nucleation and growth below T_g. PHB (polyhydroxybutyrate) crystallises *in vivo* ($T_g = 10\,^{\circ}C$) but is found to be amorphous at room temperature after extraction.

Topic 3 – Chemistry and Biochemistry in Supersaturated Carbohydrate Mixtures

Contributors: Hatley, Foster, Karel, DeLuca, Franks, Shalaev.

Discussion of the survival of microorganisms and viruses in glassy matrices, also mention of chemical reactions, *e.g.* enzymatic and acid inversion of sucrose in glasses.

Photochromic material in a glassy matrix requires 2 s at 60 °C to change colour, involves movement of two five-membered rings. Possibly conformational changes can occur in proteins, even in the vitreous state. Reference to Klibanov's work on enzyme–substrate complex in a solvent from which the substrate can be removed by washing, but the protein cannot depress T_g. The protein 'remembers' its conformation when in the presence of the substrate. Suggestion that, within the (limited) resolution of FTIR, protein conformation can be maintained during drying, but slow aggregation can occur during subsequent storage.

In the oxidation of NADH: Arrhenius kinetics apply both above and below T_g, but large decrease in E_A at T_g. Could be due to multistep reaction, each with its own E_A.

Monoclonal antibodies (MCA) lose activity by aggregation after 'conventional' freeze-drying and storage for 60 days at 35 °C. Sucrose and maltose can protect, but in combination, the sugars are more potent than would be predicted from their individual effects. Source of aggregation: one MCA complex dissociates, followed by irreversible misassociation of subunits. If the protection by sugars requires sugar 'bridges', then two types of OH spacings are required to explain the observed effects. Question about the nature of hydrogen bonding patterns between sugar molecules.

The question of intramolecular reactions within glasses was raised, *e.g.* deamidation of aspartate. Not enough data are available. Suggestion that even intramolecular rearrangements or condensations require an intial proton transfer step, usually from solvent. Can a sugar in the amorphous matrix act as proton donor/acceptor in such steps?

The question of protein cold denaturation during freeze-drying was raised which might be responsible for the loss of 'quality'.

Topic 4 – Physical Processes (*e.g.* Crystallization) of, and within Amorphous Carbohydrates; Kinetics; Effects of Residual Moisture
And
Topic 5 – Solid Solutions Involving Carbohydrates

Contributors: Reid, Levine, Zografi, McInnes, Ring, Hemminga, Pikal, Huang, Blanshard, Randall, Roos, Shalaev, Franks, Foster, Angell, Slade, Flink, Mathlouti, Brady.

A general plea was made for better definitions/descriptions of 'amorphous', 'amorphous structure', 'supercooled/supersaturated'. Any difference between quenched liquid and dry milled crystal? What is considered to be the size limit of a 'crystal'? How many unit cells? Freeze-dried amorphous samples left in the freeze-drier at 60 °C for 2 h, 85 °C for 4 h or 120 °C for 1 h (re)crystallize without any indication of melting.

How is (incipient) crystallinity detected. The detection of a crystalline phase requires dimensions of several unit cells, say 10 nm. X-ray diffraction becomes unreliable as a quantitative estimate for degrees of crystallization below 10%. NMR is more reliable in such cases; it detects 'crystallinity' by the existence of specific bond orientations

The technology of creating amorphous materials by milling crystals was discussed; requires particle size 1–6 μm. T_g values of milled and cold quenched materials are identical; reference also made to 'cotton candy' technology.

Crystals subjected to pressure lose their characteristic Raman spectra; when the pressure is released some substances remains amorphous, others revert to crystallinity. The same results have been found with the effects of irradiation on crystals. The results might form the basis of categorising materials as 'good' or 'bad' glass formers.

Crystallization kinetics from amorphous phases need study. According to the literature, no crystallization occurs below T_g, but this is hardly the case. The addition of even small amounts of PVP has a major effect on the crystallization of sugars, although T_g is not markedly affected. A 'magic' inhibition of sucrose crystallization is observed in 7:1 mixtures of sucrose:fructose. Raffinose, trehalose and lactose are also effective crystallization inhibitors. A sorption mechanisms in solution, *i.e.* by poisoning, was acceptable, but how do they act in anhydrous systems?

A case of the drug indomethacin was mentioned: in a formulated product of T_g 45 °C it crystallises in three weeks at room temperature, and (more slowly) even at 20 °C. It can be studied by DSC down to $(T_g - T) = 15$ °C. A fit of the VTF equation and extrapolation suggest a relaxation time of 10 s at T_g. At $(T_g - T) = 50$ °C, crystallization can be prevented for one year. If storage under such conditions is not practical, PVP can be added to raise T_g. For 20% PVP no crystallization occurs in three weeks, even at 4 °C above T_g.

The effect of additives on the crystal forms obtained is of interest; examples were presented of very different effects produced by the addition of glucose and fructose on the crystal habit of sucrose. Quite apart from crystal growth effects, the question was put how crystal nucleation is affected by 'foreign' sugars. Nucleation history also governs how amorphous materials are produced, whether from the melt, from solution by freeze-drying or evaporation, or by spray-dried powders.

The crystallization of mannitol is of particular importance. Homogeneous amorphous mannitol, obtained by spray drying and containing 0.5% water, has a T_g of 36 °C. At room temperature it crystallizes completely within two weeks but without loss of water. Similar results are found with mannitol/glycine mixtures ($T_g = 50$ °C). Monitoring by X-ray diffraction over several weeks shows gradual polymorphic changes. Crystallization at $-20\,°C$ was also reported for mixtures with $T_g = 100\,°C$.

Evidence for method of preparation effects on the ability to detect T_g: Maltodextrin/protein mixtures (0.5% moisture) reveal no T_g at room temperature, but T_g becomes visible after storage at $-90\,°C$ and subsequent heating.

What can be said about the relative stabilities of systems protein-water-X, where X is sucrose or ficoll? T_g (ficoll) $\gg 132\,°C$, T_g (sucrose) $= 70\,°C$. Thus, for same T_g, e.g. $45\,°C$, require more sucrose in formulation. Also free volume of polymer is larger.

Discussion of relative densities of sucrose *vs* sucrose/PVP and sucrose/ficoll. Note also in ternary systems have separation of polymer rich and sucrose rich phases. Mobility of probes in ternary mixtures of water/glucose/NaCl find retardation by NaCl. Two amorphous phases not uncommon in protein/sugar/water mixtures; observe two distinct glass temperatures.

Phase separation also observed in starch/fructose/water systems. Thus, 20% fructose is completely miscible with starch/water, but at higher fructose concentrations observe two separate T_g profiles.

The complex phase behaviour of water/sugar/salt mixtures was highlighted with reference to the solid/liquid state diagram of the water/sucrose/NaCl system. During cooling, depending on the initial composition, especially the sucrose:NaCl ratio, ice crystallization takes place first, but $NaCl.2H_2O$ crystallization can also be induced. Two distinct stages of freeze concentration can now be identified: after the completion of primary ice crystallization and after the subsequent completion of secondary ice and NaCl crystallization. The point of *maximum* freeze concentration after the crystallization of ice and NaCl can be regarded as a quasi-eutectic point in the ternary state diagram. The dependences of glass and softening temperatures on the sucrose:NaCl ratio can be represented in two dimensions and is of some practical importance in freeze-drying operations. If NaCl can be induced to crystallize, it lends mechanical rigidity to the cake during the sublimation of ice. It then becomes possible to perform primary drying above T_g' (or T_s), without danger of product collapse.

Another aspect of ternary mixtures in relation to freeze-drying was mentioned: the use of mass transfer agents. *Tert*-butanol (TBA) accelerates the sublimation of ice from frozen sugar solutions, especially at some specific TBA:sugar ratios. Thus, the freeze-drying of lactose or sucrose is time consuming because of the low collapse temperatures. The addition of TBA permits primary drying above the nominal collapse temperature. Concern was voiced about difficulties experienced with the removal of TBA; some preparations retain 4.5% after drying. TBA appears to be retained (encapsulated) within the sugar.

The question was asked whether solutions, cooled at different rates, but to the same T_g' would exhibit different ice growth rates. The general opinion was in the affirmative.

Question of the existence of sucrose hydrate(s) was raised again: no agreement! 1949 X-ray evidence for hydrate(s) may (not) be reliable/reproducible? Observation of anhydrous crystal at $>5\,°C$, but crystal hydrate of different appearance at $<5\,°C$, with different refractive index. Suggestion to wash out crystals with ethanol at low temperature. Other suggestions: use temperature gradients (zone refining) to detect and grow crystals.

Discussion of protein denaturation in ternary and quaternary systems with low moisture contents; *e.g.* T(denat) in maltose/water glass is depressed up to 25–35% moisture, but no further depression beyond that. Aggregation immediately follows denaturation. Distinction must be drawn between denaturation caused by unfolding and chemical inactivation which may occur without unfolding.

In water/sugar/protein systems observe preferential hydration of protein. As water becomes limiting, the sugar is dehydrated before the protein. Extensive literature on lysozyme/LDH and the effects of polyols and water on the drying stability.

Glassy films containing 50% ovalbumin in fructose, glucose, sucrose are rehydrated and monitored by infra-red; the effects of the three sugars on the IR spectrum are very different. Is anything known about the hydrogen bonding patterns between sugars/polyols and proteins, as water is removed? Not much. Insulin–lactose adducts prevent insulin aggregation during drying; reason unknown.

The question of the significance and determination of residual moisture content generated much discussion. In a partially crystalline mixture, the 'residual moisture content' *must be referred to the water content per unit mass of amorphous phase* which is not always easy to determine. The question arose if there is a universal relationship between water activity and critical moisture content, or does it depend on the particular composition. A useful definition of critical moisture content is the water content that is able to depress T_g to room temperature.

There exists an extensive literature on hydration and hydration numbers of sugars; what is its relevance? Hydration number is an operational definition, therefore the method of measurement must be defined. The discussion generated much heat. It was suggested that measurements of a_w provide an indication of the deviation from ideal solution behaviour which, in turn, depends on the hydration number. This was contradicted: deviations from ideal behaviour are ascribed to solute–solute interactions which, in the case of sugars, are of a repulsive nature.

Molecular dynamics simulations suggest 2.6 water mol per sugar OH in xylose, but how does this correspond with the reported $n_h = 5$ for sucrose? Presumably hydration might be defined in terms of the distances of nearest neighbour water molecules from sugar OH groups by stipulating acceptable hydrogen bond lengths. It used to be said that 3 mol of water per sugar mol causes collapse of freeze-dried sugars, but that referred to room temperature. The discussion was left unresolved, because (a) there was no agreement about the exact definition of 'hydration' and (b) its temperature dependence.

Topic 6 – Dynamics, Especially of Small Molecules within Amorphous Carbohydrates

Contributors: Levine, Angell, Zografi, Karel, Ablett, Hemminga.

The diffusion of molecules making up the matrix and those of a 'diffusion probe' must be clearly distinguished. For most chemical reactions diffusion is required and the diffusing species must have a molecular weight of the same order as the molecules comprising the glassy matrix. If diffusion is required, then it should be possible to manipulate the kinetics by choosing a strong/fragile matrix, because diffusion persists well below T_g in a fragile matrix.

Residual motions are possible in glasses, *e.g.* CO_2 diffuses rapidly through PTFE below its T_g. Also observe mobility of ions in glasses. Despite high viscosity of a 'solid' glass, ESR detects fast motions. How do such motions depend on the superstructure of the 'solid'?

In reactions that do occur below T_g, the rate often depends on $(T_g - T)$. In freeze-dried materials have porous matrix which collapses above T_g. Loss of volatiles or oxygen uptake can then exhibit reduced rates above T_g. Crystallization events can also affect chemical reactions: when anydrous sugar crystallizes, the matrix is diluted, but when sugar crystallizes as a hydrate, the matrix may be diluted or concentrated. The role of crystallization enhanced reactions in food stability was emphasized.

It was generally accepted that the main effect of residual moisture in an amorphous sugar is one of plasticization, *i.e.* a depression of T_g. However, there is(are) a further, specific effect(s) of water at low moisture levels, unrelated to $(T - T_g)$, especially when $T_{exp} \ll T_g$. It was also stated that annealing plays a part in changing the influence of water on the properties of a low moisture system. This may be due to secondary transitions (β-transitions) during annealing. Participants were reminded that in soft plastic materials, T_g controls the softness, but β-transitions control the impact strength. The same was true for the rather different effects of water on the properties of starch and gluten. In this connection it was mentioned that for starch, a 35% water content corresponds to one water molecule per sugar OH group.

Topic 7 – Model Systems and Analytical Methods

Contributors: Roos, Karel, Mathlouti, Pikal, Bardat, Huang, Ablett, Hemminga, Randall, DeLuca, Skrabanja, Shalaev, Franks, Foster, Mathlouti, Duddu, Reid, Slade, McInnes, Angell, Levine, Izzard.

Participants searched for a 'single best' method for the study of solid carbohydrates, in particular for the determination of T_g. The following techniques were discussed: solid state NMR, DSC, dielectric permittivity, mechanical spectroscopy, microscopy, X-ray diffraction, infrared spectrophotometry, viscosity. The consensus was that a combination of techniques would prove suitable for the elucidation of intractable problems posed by amorphous materials. Thus, thermal and mechanical techniques provide macroscopic views, while certain spectroscopic techniques yield results on a molecular scale, but optical methods may require transparent samples. ESR requires the addition of probes, or the attachment of labels to molecules under study; that may affect the properties to be investigated. There is a need for a rapid in-line method to determine T_g.

It was suggested that pure sugars are rarely available; trace metals give rise to decomposition on melting and experimental artefacts. In particular, trace metals can enhance oxidative degradation. Mention was also made of spurious 'enzyme activity' in trehalose solutions, probably due to the incomplete removal of enzymes during purification.

Thermomechanical Analysis

In practice a mechanical method (DMTA) should be most useful; it is sensitive and easy to interpret, *i.e.* elastic modulus and loss. A statement that DMTA loss peaks correspond exactly with the 'true' T_g was questioned; perhaps tan δ corresponds to the softening temperature T_s, rather than to T_g. This view was generally accepted. Mechanical methods are inappropriate for frozen systems where ice melting may be the predominant factor, but is of no practical significance. The discussion turned to the difficulties of applying mechanical methods to powdered materials. Special cells have been constructed, or the sample can be pelleted, but experience shows that reproducibility is difficult to achieve. How can moisture loss/gain be controlled during DMTA measurement? Powder is kept between sheets of plastic of known permeability and a pressure of 4 kbar is applied and the distance of penetration measured. The method is used for coffee and milk powders. It was suggested that the pharmaceutical industry relies too heavily on DSC and is unaware of the successes achieved with DMTA.

Nuclear Magnetic Resonance/Electron Spin Resonance

Nuclear magnetic relaxation may well be the most potent method; can separately study motions of sugar and water, *e.g.* by ^{17}O enrichment, can also separate rotational and translational diffusive motions and also specific groups rotations, such as CH_2OH in sugars. A warning was voiced concerning 'over-enthusiastic' interpretations of experimental data, based on insuffient information that can be found in the literature. In complex protein/carbohydrate/water systems at low moisture contents, NMR will pick up mainly the protein protons. With increasing moisture content, the water protons provide the major contribution to the spectrum. Computer fitting of data by a single exponential may possibly be adequate at very low moisture contents, but is unsatisfactory in most cases and provides misleading information.

The focus turned on ESR which was suggested as a suitable technique for monitoring rotational motion (translation does not affect the signal). Can use simple spin labels of the same dimensions as sugar molecules, hence fitting into the sugar matrix. Signal depends on molecular shape and free volume. Correlation times for solids (include glass) are of the order of ms to 0.1 ms; for liquids 0.1 ms to 1 ps. For malto-oligomers find simple exponential relaxation rates. Dispersion of relaxation rates in more complex systems (gelatin, gluten) have also been observed. Matches of ESR with NMR results were reported. Flexibility (relaxation rate) differs for different bonds; this leads to dispersion. If a single relaxation time is measured, it must relate to the probe and not to the sugar matrix. Questions were asked whether in higher oligomers the possibility of partial helical structures can give rise to hydrophobic effects. Other questions related to the ESR probe mobility/temperature slope below T_g (*i.e.* activation energy). Apparently members of a homologous series have the same E_A. If so, then how far can the similarity within a homologous series be pushed?

Results now available for several disaccharide–water systems over large ranges of composition and temperature, although NMR cannot be applied very successfully to glassy states. On the other hand, measurements are possible at very high pressures and have been performed on sucrose solutions up to 0.5 kbar. In dilute solutions sugar motions are uncorrelated, *i.e.* sugar molecules rotate as independent units, and the results correlate well with DSC. With increasing concentration sugar and water motions become coupled, presumably due to sugar–(water)–sugar links. At even higher concentrations, under conditions of vitrification, sugar and water motions become uncoupled; differences in correlation times can reach eleven orders of magnitude at low temperatures! Also rotational and translational diffusive motions in highly concentrated sugar solutions are uncoupled. The NMR results on water do not match DSC data (inasmuch as DSC results can be related to specific chemical species in a mixture). Liquid-like rotational diffusion of residual water (in isolated microdroplets?) can be observed down to 125 K!

X-ray Diffraction

Although X-ray diffraction is recognized as a unique diagnostic method for the estimation of crystallinity and the identity of crystallising species, there was also agreement that the method cannot be applied quantitatively at degrees of crystallization below *ca.* 10% or, more importantly, above 90%. Best results obtained for anhydrous crystals, because all the residual moisture is in the amorphous phase. The usefulness of the technique is limited for mixtures with a low amorphous content. It was pointed out that crystallinity of starch can be measured by the heat of solution; the results match those obtained by X-ray diffraction.

Dielectric Relaxation

Dielectric methods able to monitor properties of amorphous solids and very viscous fluids over a wide range of frequencies. Mention was made that dielectric thermal analysis of freeze-dried sucrose indicates two frequency-dependent peaks, believed to measure cooperative motions of whole sucrose molecules, rather than group rotations. The need for reliable deconvolution methods to separate motions of small molecules from those of a glassy matrix was emphasized.

Infrared/Raman Spectroscopy

Limited information on infrared applications; mainly for proteins. Any such data can only refer to protein secondary structure. Raman spectra are more sensitive to tertiary structures. Many conformational changes would be 'invisible' to infrared. Reliable or interpretable information on dry proteins in KBr pellets; infrared not very useful for wet samples. Little useful infrared information on sugars in solution; even less for dry, amorphous sugars. Quench melted sucrose spectrum resembles dilute solution, but supersaturated solution spectrum resembles crystal. Raman spectra extremely complex, no normal mode analysis yet available. Comparison of crystal, glass and solution spectra for alditols surprisingly reveals main differences in the C–H stretch region and not in the C–O or O–H regions.

Differential Scanning Calorimetry

DSC is popular, apparently easy to perform; this statement led to a protracted discussion of the usefulness/application/interpretation of DSC. A need to agree on how to define T_g', in order to discuss relaxation times. For a canonical glass transition, the onset of the C_p discontinuity correponds to a relaxation time $\tau = 100$ s; at the midpoint of the transition, $\tau = 10$ s. If DSC results are to be compared to NMR data, it is necessary to know the order of magnitude of τ. Most useful to quote onset, midpoint and endpoint of DSC signal. To avoid enthalpic relaxation effects (hysteresis), C_p should be measured at constant frequency, *e.g.* 10 Hz, possibly separate real and imaginary parts with the aid of modulated DSC (see below), but also at constant heating rate to reveal enthalpy relaxation effects. The measurements are then complementary and resemble the dielectric relaxation results ε' and ε''. Can also compare mechanical 10 s modulus with a DSC scan rate that gives a T_g' corresponding to a 10 s relaxation time.

The scope and potential of DSC was discussed at some length. Single scans, encompassing the whole temperature range of interest, often provide only limited information. The power of DSC can be considerably enhanced by interrupted scans, annealing programmes, thermal cycling and temperature-step DSC, and also by isothermal (time) scanning. The caveat was made that in such experiments the sample should always be allowed to reach a steady state (within the normal period of observation) at each temperature increment during heating or cooling processes. At one point it was suggested that a review of DSC-based methods that could be applied to studies of amorphous carbohydrates would be useful.

The influence of sample preparation/history on the appearance of the DSC scan was discussed. Shifts of up to 5 °C in the measured T_g' have been observed in samples cooled at different rates

The potential of modulated DSC (MDSC) was briefly discussed, but the consensus of those with experience was that the technique had still to establish itself, that the available software was suspect and that interpretations of experimental data was not as straightforward as the manufacturers claimed. Possible uses discussed included the distinction of instrumental artefacts from real thermal transitions. The example was cited of multitransition processes where the individual thermal events can be better resolved by MDSC than is possible with standard DSC. Those of the participants who had applied the technique did not profess to be 'experts' and the advent of MDSC was treated with guarded optimism.

Although melting is a continuous process, the kinetics are not constant, since the rate of melting increases with temperature. The melting of small crystals might show up as a discrete step in a DSC trace. It was agreed that large changes in surface area might affect the appearance of DSC traces.

The 'problem' of enthalpy relaxation (spurious endotherms in DSC traces) received mention, because it can give rise to misleading interpretations. On the other hand, a study of enthalpy relaxation with time can be extremely useful, because it can provide estimates of relaxation times (shelf lives) of products (*e.g.* potato starch) stored under different conditions.

Several participants reported that in frozen, freeze-dried and ground bulk sugar preparations (*i.e.* in disperse systems) two discontinuities in the heat flow curves are observed. In homogeneous quenched or air-dried preparations (*e.g.* films), only one discontinuity is detected. Of the two steps, the lower one is reversible and reproducible, the upper one is irreversible and depends on details of sample preparation. Their correct interpretation gave rise to much debate. The point was made that heat flow discontinuities do not necessarily correspond to glass transitions. The higher temperature transition has been explained in terms of the beginning of ice melting (but it also occurs in systems that contain no ice). It was questioned why ice melting, which is a first order process, should be associated with a discontinuity in C_p. An alternative explanation was advanced: the lower temperature transition is the 'true' glass transition, but viscous flow just above T_g is too slow to be detected by DSC in real time. The upper temperature transition is observed by conventional DSC when the heat absorption, associated with a physical or chemical change, reaches the order of 0.1 mW. This usually occurs when the viscosity of the glass has been reduced by a factor of $10^{6,7}$. The observed heat flow change is thus indicative of a change in the sample configuration (collapse) and a subsequent change in the thermal conductivity of the system, rather than a change in C_p. Its details depend on previous sample history, scanning rate, chemical/physical composition, *etc.* This is the effect that has come to be described as T_g' in current literature. In practice, it is the important effect for the processing (*e.g.* drying) of materials, because it monitors collapse, but it is not a glass transition. In support of this interpretation it was said that this transition was the only one to be detected by DTMA which directly measures deformation *within the period of observation*. There was, however, no general agreement on the nature of T_g'. Dissent was voiced mainly by those who do not routinely observe the transition at the lower temperature, or whose measurements are confined to frozen systems. It was also suggested that, on complete annealing, this transition is also reversible. In contrast it was said that 'complete annealing' was synonymous with crystallization.

The question of nomenclature was raised several times; it is continuing to cause confusion. The collapse temperature (T_c) has been around for many years. T_g' (1980s) has gone into the folklore of amorphous materials as *the* glass transition, but T_m' has also entered the literature (1990s), as has also the softening (sticky) temperature T_s (1995), although the concept has been around in powder technology for many years. Whatever the transition is called, there was agreement that for frozen sucrose solutions it lies at $-32\,°C$ and that it is of importance for the setting of correct freeze-drying process parameters. Other temperatures mentioned include T_g'', believed to be the 'real' glass temperature by some, and referred to loosely as a β-relaxation (unspecified) by others. Mention was made that in freezing systems, the true point of maximum freeze concentration can only be achieved with the aid of prolonged annealing, rather than by single DSC runs at cooling/heating rates of 10 K min^{-1} or faster. The transition at the lower temperature has by some workers been designated as T_g'' but has not yet been given much attention. The suggestion was put that the lower of the two transitions should logically be designated T_g', but it was questioned whether this could ever be achieved in practice, in view of the popularity that the process commonly characterized as T_g' now enjoys in the literature.

For protein systems, in the absence of sugars, T_g is usually difficult or impossible to detect. This emphasises that DSC is inappropriate in some cases, where the glass transition is not accompanied by a discontinuity in the specific heat. The explanation was given in terms of strong and fragile fluids. On folding, a protein loses many degrees of configurational freedom and hence becomes a strong fluid for which the glass transition is not accompanied by a measurable ΔC_p. The addition of a sugar reveals the T_g (of the mixture). By altering the protein:sugar ratio and extrapolation, a T_g for the pure protein can be obtained. Alternatively, by denaturing a protein, the material is converted into a fragile fluid (increased number of degrees of freedom) and T_g can then usually be observed by DSC. Mechanical measurements always reveal softening transitions for proteins (referred to as T_g', see above).

Computer Simulation

The usefulness of computer modelling (Monte Carlo, Molecular Dynamics) was briefly discussed. The systems which are currently amenable to study include single sugar molecules in 'computer water', *i.e.* infinitely dilute; the approach then provides structural and dynamic information about hydration and water exchange rates. Anomerization rates can also be determined for such solutions. It appears that hydration details for anomers are identical, except in the vicinity of the anomeric OH group. It seemes reasonable that an increase in the sugar concentration leads to a change in the anomeric ratio. It was suggested that the replacement of the ring oxygen by sulfur would enable tautomers to be studied individually.

In the case of disaccharides, trehalose differs from its isomers, in that water appears to penetrate into the 'mouth' of the disaccharide gap. It must be remembered, however, that trehalose differs in structure from its more common isomers, being a 1,1-linked sugar.

Rotational and translational time correlation functions for water and isolated sugar molecules in solution can be modelled. However, the modelling of diffusion-related processes in viscous systems is out of the question because of the long time scales involved. It will be many years before melting and crystallization process become amenable to computer simulation.

Topic 8 – Amorphous Carbohydrates as Biostabilisers

Contributors: Zografi, Pikal, Hemminga, Angell, Randall, Franks, Foster, Roos, Bardat, DeLuca, Reid, Hatley, Karel, Flink.

Amorphous carbohydrates find widespread application as stabilizers (excipients) of labile biologicals. There is evidence for their function as *in vivo* protectants against desiccation injury, *e.g.* in seeds, bacterial spores, overwintering insects. These aspects were not touched upon.

Most discussions revolved around practical points. When amorphous sucrose, trehalose and raffinose are exposed to water vapour, they take up water and eventually crystallize in the order sucrose before trehalose before raffinose. The reason is that sucrose crystallizes in the anhydrous state, the other two sugars as $3H_2O$ and $5H_2O$ hydrates respectively.

The role of thermodynamic stability *vis-a-vis* chemical stability of proteins was mentioned. The so-called water replacement mechanisms came in for criticism: if water stabilizes, then why freeze-dry at all? It was suggested that ΔG_{denat} can be calculated from data at high temperature and extrapolated back to room temperature. The conclusion was that, at room temperature, the protein was in a state of cold denaturation. The reasoning was not altogether clear to some participants. For instance the point was made that it is not easy to extract the contribution of the protein to ΔG, because the thermodynamic function relates to the whole system, the composition of which changes during drying. Also, in measurements of $\Delta G(T)$, the dry system passes through a glass transition. Below T_g, equilibrium thermodynamics hardly apply. The conventional solution equilibrium data for $\Delta G(T)$ cannot be applied to dry systems.

Residual enzyme activity after drying in the presence of sucrose and dextran was discussed in detail. Surveying data for catalase, LDH, MDH, ADH, GDH and PFK produced little in the way of systematic trends, except that sucrose was always the better protectant. One reason might be impurities in dextran. The reason for inactivation was given as dissociation of quaternary structures, rather than chemical reactions.

The question was asked why so much reliance is placed on freeze-drying if proteins can be stabilized effectively and simply in solution at subzero temperatures (undercooled). There was unanimous agreement that carbohydrate stabilizers that are effective at high temperatures also stabilize proteins at low temperatures.

What might be the phase behaviour in protein/salt/sugar/water systems during drying? Is there a possibility of phase separation, such as is found in solutions, *e.g.* in systems in which the excipient crystallises during drying? Would two distinct values of T_g' be measurable? No data are available. For human dried growth hormone (hGH) preparations alone, or with hydroxyethyl starch, both with 0.5% moisture, $T_{denat} = 195\,°C$. In the presence of trehalose, $T_{denat} = 155\,°C$. It appears that T_{denat} scales with T_g. Initial losses of protein activity were reported for preparations without added sucrose or with very low levels ($< 0.5\%$) of sucrose, where similar levels of ficoll produced stable products. Large pH decreases have been observed for phosphate-buffered proteins during freeze-drying; they can be prevented by the addition of sufficient sucrose ($> 1\%$).

Cyclodextrins enhance the solubility and stability of some drugs. It was noted that β-cyclodextrin (ex Sigma) contains 15% water, equivalent to 11.5 mol water per mol. Several comments were made about the improvement in freeze-drying conferred by modified cyclodextrins, but the mechanism of complex formation is not understood. It was reported that no T_g can be detected for any dried hydroxypropyl-β-cyclodextrin, although the frozen solution yield a good measurement for T_g' ($-14\,°C$).

The need for a better understanding of polysaccharide–peptide interactions at low moisture contents was emphasized, with special reference to drug delivery problems. Substances such as biodegradable starch derivatives, alginic acid, hyaluronic acid and chitosan have the advantages of being soluble, easily chemically modifiable, and of offering drug protection until the point of release. A monovinyl HES derivative is useful for trapping peptides; it can be emulsified in an organic solvent and cross linked with acrylamide and sterilized. The degree of crosslinking governs the degree of later swelling and the rate of peptide release. Doubts were expressed about FDA approval and double bond formation during free radical-induced cross linking.

The possibility of sterilizing glasses was discussed, with special reference to the diffusion of radicals and gases (ethylene oxide). There exists an extensive literature on β- and γ-irradiation. Both are more effective at higher moisture contents, but the mechanisms are not accounted for in terms of T_g.

There is general need for long-term storage stability data on pharmaceutical products, say for 6 years. Reliable accelerated storage methods are urgently needed. Should tests be carried out at high temperatures or high RH? Other matters of concern include the nature, and potential toxicity of any degradation products, even at trace levels. In some drying processes, failure appears quickly, but unpredictably; this is probably due mainly to in-process causes. Moisture sorption from incorrectly dried stoppers is also a case for concern.

The survival of *S. cerevisiae*, stored frozen at $-80\,°C$ is improved by most sugars (except fructose long-term), although above $-25\,°C$ glucose is no longer effective. The efficacy is measured by CO_2 production. Suggestions were made for a comparison of lactose and maltose, both with T_g' values of $-37\,°C$.

Several mentions were made about the survival of microorganisms, seeds and viruses in glassy matrices, but time limited the proper discussion of this important topic. In the case of seeds, T_g and moisture content appear to be good indicators of viability.

Evaluations and Suggestions for Follow-up Events

Participants were asked for their views as regards follow-up activities to the Workshop. Possible alternatives were suggested as closed discussion meetings like the Girton Workshop or open meetings of a more conventional type, with student participation encouraged, and opportunities for oral/poster presentations. The present format was favoured by the majority, but with pleas for fewer topics on the agenda. Frustration was expressed at the limited time allowed to the presenters of problems (5 minutes) by the moderators. A better grouping of the topics might make it possible to have 10 minute introductions. Well prepared visual aids would assist the presentations and discussions. The desirability of poster presentations was mentioned by several participants.

The question whether participation should be limited to either food or pharmaceutically oriented scientists ('stale bread or monoclonal antibodies') was answered in the negative. Participants gained from listening to and mixing with experts in other fields, but with identical problems.

Observers were asked for their reactions; several of the comments included the word 'fantastic'. It was a valuable experience and enabled most of the participants/observers to hear at first hand about the frontiers of the scientific and technological knowledge in the areas discussed. A suggestion was made that polymer scientists should be included, but participants were reminded that the Workshop had been intended to address problems of amorphous carbohydrates, and not amorphous solids generally.

A suggestion that instrument manufacturers should have been invited drew the reply that such invitations had been extended but had not been taken up by the companies concerned. The Workshop ended with a vote of thanks to the organisers and a request for any additional comments to be sent to Louise Slade before the end of April.

Appendix II: The Amorphous State – A Critical Review, Churchill College, Cambridge, May 15–17, 2001. Joint Organizers: Felix Franks and Harry Levine

During the past decade, the important role played by amorphous water-soluble substances (solid aqueous solutions) has been increasingly recognised within the food and pharmaceutical industries. A small 'by invitation' expert conference was convened in Cambridge during 1995, to analyse the then state of knowledge and, in a conference report [Appendix I], to identify important gaps that required further investigation.

This follow-up conference, sponsored jointly by the BioUpdate Foundation and the Biotechnology Group of the Royal Society of Chemistry, aimed at a critical review of progress achieved during the intervening years. It also provided an open forum for the presentation of original research and the posing of further questions. The 1995 report served as an *aide memoire*, against which progress could be measured.

The conference was divided into eight sessions, addressing the following topics:

- Structure and its significance in the application technology of amorphous materials
- Glassy state dynamics and its significance for stabilization of labile bioproducts
- Theories of unstable aqueous systems: how can they help the technologist?
- Progress in food processing and storage
- Rational pharmaceutical formulation of amorphous products
- Chemistry in solid amorphous matrices
- Residual water, its measurement and its effects on product stability
- Novel experimental approaches to studies of amorphous aqueous systems

Each session was introduced by a summary review lecture, charting progress over the past decade, and was followed by a limited number of contributed papers, supplemented further by a selected number of posters.

The organizers were keen to encourage attendance of a healthy mixture of 'experts' and younger colleagues, in particular graduate students actively engaged in research related to the properties and applications of amorphous materials. To this end, they established a fund for the award of bursaries to enable young scientists and technologists to attend the conference and to present their research in poster and oral forms.

The organizers therefore wish to express their gratitude to the following for generous contributions to the Bursary Fund:

Pfizer Global Research and Development, Sandwich, Kent, UK
Nabisco, East Hanover, NJ, USA
Inhale Therapeutic Systems Inc., San Carlos, CA, USA
Pillsbury Co., Minneapolis, MN, USA
Cargill Inc., Wayzata, MN, USA
Nestle Research Centre, Lausanne, Switzerland
The Royal Society of Chemistry (Biotechnology Group), London, UK
The BioUpdate Foundation, London, UK and Amsterdam, The Netherlands

It was gratifying for the organizers to be able to report that the fund was able to award 12 full bursaries to postdoctoral researchers and graduate students from several countries, even as far away as New Zealand!

In addition to the corporate contributions, the organizers also expressed their gratitude for individual donations from several sources.

Several plenary speakers expressed their wish to see some form of publication following the conference. The Royal Society of Chemistry Publication Department agreed to produce a volume, which would contain the collected keynote presentations, as well as chapters submitted by those participants who wanted to contribute. Harry Levine volunteered to act as editor of this book.

Contents

Theories of Unstable Aqueous Systems: How Can They Help the Technologist?

Progress in Food Processing and Storage

Rational Pharmaceutical Formulation of Amorphous Products

Chemistry in Solid Amorphous Matrices

Residual Water, its Measurement, and its Effects on Product Stability

Novel Experimental Approaches to Studies of Amorphous Aqueous Systems

List of Contributors

Adhikari, Benu, Department of Chemical Engineering, University of Queensland, St. Lucia, QLD 4072, Australia

Auffret,* Tony, Pharmaceutical Sciences, Pfizer Global R&D, Sandwich, UK

Austin, T.K., Preformulation Team, Pharmaceutical and Analytical R&D, AstraZeneca R&D Charnwood, Bakewell Road, Loughborough, Leicestershire, LE11 5RH, England

Bhandari,* Bhesh R., Food Science & Technology Group, School of Land and Food Sciences, University of Queensland, Gatton, QLD 4345, Australia

Braga da Cruz, I., Escola Superior de Biotecnologia, Universidade Catolica Portuguesa, Porto, Portugal

Brownsey, Geoff, Institute of Food Research, Colney Lane, Norwich NR4 7UA, UK

Buera,* M.P., Departamento de Industrias, Facultad de Ciencias Exactas y Naturales, University of Buenos Aires, 1428 Buenos Aires, Argentina

Caffin, N., Food Science & Technology Group, School of Land and Food Sciences, University of Queensland, Gatton, QLD 4345, Australia

Colonna, Paul, Institut National de la Recherche Agronomique, BP 71627, 44316 Nantes Cedex, France

Corti, H.R., Unidad de Actividad Quimica, Comision Nacional de Energia Atomica, Pcia. Buenos Aires, Argentina

D'Arcy, B., Food Science & Technology Group, School of Land and Food Sciences, University of Queensland, Gatton, QLD 4345, Australia

Debenedetti,* Pablo G., Department of Chemical Engineering, Princeton University, Princeton, NJ, USA

Errington, Jeffrey R., Department of Chemical Engineering, Princeton University, Princeton, NJ, USA

Franks,* Felix, BioUpdate Foundation, 25 The Fountains, 229 Ballards Lane, London N3 1NL, UK

Goff,* H.D., Dept. of Food Science, University of Guelph, Guelph, ON, NiG 2W1, Canada

Gunning, Yvonne M., Food Materials Science Division, Institute of Food Research, Norwich Research Park, Colney Lane, Norwich NR4 7UA, UK

Hagiwara, Tomoaki, Department of Food Science and Technology, Tokyo University of Fisheries, 4-5-7 Konan, Minato-ku, Tokyo 108-8477, Japan

Halley, P., Department of Chemical Engineering, University of Queensland, St. Lucia, QLD 4072, Australia

Howes, Tony, Department of Chemical Engineering, University of Queensland, St. Lucia, QLD 4072, Australia

Igarashi, Toshio, X-Ray Research Laboratory, Rigaku Corp., 9-12, Matsubara-cho, 3-chome, Akishima-shi, Tokyo, 196–8666 Japan

Izutsu,* Ken-ichi, National Institute of Health Sciences, Tokyo, Japan

Kajiwara,* Kazuhito, Department of Biosciences, Teikyo University of Science and Technology, 2525 Yatsusawa Uenohara-machi, Kitatsuru-gun, Yamanashi, 409-0193 Japan

Katagiri, Chihiro, Institute of Low Temperature Sciences, Hokkaido University, Sapporo 060-0819, Japan

Kawai,* Kiyoshi, Department of Food Science and Technology, Tokyo University of Fisheries, 4-5-7 Konan, Minato-ku, Tokyo 108-8477, Japan

Kim, Yu Jin, Department of Food Science and Technology, Tokyo University of Fisheries, 4-5-7 Konan, Minato-ku, Tokyo 108-8477, Japan

Kishi, Akira, X-Ray Research Laboratory, Rigaku Corp., 9-12, Matsubara-cho, 3-chome, Akishima-shi, Tokyo, 196-8666 Japan

Kojima, Shigeo, National Institute of Health Sciences, Tokyo, Japan

Kou,* Yang, Department of Food Science, University of Massachusetts, Amherst, MA 01003, USA

Lalloue, Benedicte, Food Materials Science Division, Institute of Food Research, Norwich Research Park, Colney Lane, Norwich NR4 7UA, UK

Lechuga-Ballesteros,* David, Inhale Therapeutic Systems Inc., San Carlos, California, USA

Levine,* Harry, Nabisco R&D, Kraft Foods, 200 DeForest Av., East Hanover, New Jersey 07936, USA

Longinotti, M.P., Unidad de Actividad Quimica, Comision Nacional de Energia Atomica, Pcia. Buenos Aires, Argentina

Lourdin,* Denis, Institut National de la Recherche Agronomique, BP 71627, 44316 Nantes Cedex, France

MacInnes,* William M., Nestle Research Center, Vers-chez-les-Blanc, CH-1000 Lausanne 26, Switzerland

Malcata, F. Xavier, Escola Superior de Biotecnologia, Universidade Catolica Portuguesa, Porto, Portugal

Mazzobre, M.F., Departamento de Industrias, Facultad de Ciencias Exactas y Naturales, University of Buenos Aires, 1428 Buenos Aires, Argentina

Miller, Danforth P., Inhale Therapeutic Systems Inc., San Carlos, California, USA

Montoya, K., Dept. of Food Science, University of Guelph, Guelph, ON, NiG 2W1, Canada

Motegi, Akihito, Department of Biosciences, Teikyo University of Science and

Technology, 2525 Yatsusawa Uenohara-machi, Kitatsuru-gun, Yamanashi, 409-0193 Japan

Munekawa, Sigeru, X-Ray Research Laboratory, Rigaku Corp., 9-12, Matsubara-cho, 3-chome, Akishima-shi, Tokyo, 196-8666 Japan

Murase,* Norio, Department of Biotechnology, School of Science and Engineering, Tokyo Denki University, Hiki-gun, Saitama 350-0394, Japan

Noel, Timothy R., Food Materials Science Division, Institute of Food Research, Norwich Research Park, Colney Lane, Norwich NR4 7UA, UK

Oliveira, Jorge C., University College Cork, Cork, Ireland

O'Sullivan, D., Preformulation Team, Pharmaceutical and Analytical R&D, AstraZeneca R&D Charnwood, Bakewell Road, Loughborough, Leicestershire, LE11 5RH, England

Parker,* Roger, Food Materials Science Division, Institute of Food Research, Norwich Research Park, Colney Lane, Norwich NR4 7UA, UK

Pikal,* Michael J., School of Pharmacy, University of Conneticut, Storrs, CT, USA

Pilosof,* Ana M.R., Departamento de Industrias, Facultad de Ciencias Exactas y Naturales, University of Buenos Aires, 1428 Buenos Aires, Argentina

Pradistsuwana, Chidphong, Department of Food Technology, Chulalongkorn University, Thailand

Reid,* David S., University of California, Food Science & Technology, One Shields Av., Davis, CA 95616 USA

Ring, Steve G., Food Materials Science Division, Institute of Food Research, Norwich Research Park, Colney Lane, Norwich NR4 7UA, UK

Ross, Edward W., OTD, U.S. Army Natick Soldier Systems Center, Natick, MA 01760, USA

Ruike, Masatoshi, Department of Biotechnology, School of Science and Engineering, Tokyo Denki University, Hiki-gun, Saitama 350-0394, Japan

Sahagian, M.E., Dept. of Food Science, University of Guelph, Guelph, ON, NiG 2W1, Canada

Schiraldi,* Alberto, DISTAM, Universita di Milano, Via Celoria 2, 20133 Italy

Schoonman,* A., Nestle Research Center, Vers-chez-les-Blanc, CH-1000 Lausanne 26, Switzerland

Shalaev,* Evgenyi, Groton Laboratories, MS-8156-004, Global R&D, Pfizer Inc., Groton, CT 06340 USA

Slade,* Louise, Nabisco R&D, Kraft Foods, 200 DeForest Av., East Hanover, New Jersey 07936, USA

Sopade, P.A., Food Science & Technology Group, School of Land and Food Sciences, University of Queensland, Gatton, QLD 4345, Australia

Steele,* G., Preformulation Team, Pharmaceutical and Analytical R&D, AstraZeneca R&D Charnwood, Bakewell Road, Loughborough, Leicestershire, LE11 5RH, England

Sugie, Masashi, Department of Biosciences, Teikyo University of Science and Technology, 2525 Yatsusawa Uenohara-machi, Kitatsuru-gun, Yamanashi, 409-0193 Japan

Suzuki,* Toru, Department of Food Science and Technology, Tokyo University of Fisheries, 4-5-7 Konan, Minato-ku, Tokyo 108-8477, Japan

Takahashi, Hiroshi, Faculty of Engineering, Gunma University, Maebashi-shi 371-8510, Japan

Takai, Rikuo, Department of Food Science and Technology, Tokyo University of Fisheries, 4-5-7 Konan, Minato-ku, Tokyo 108-8477, Japan

Taub, Irwin A., OTD, U.S. Army Natick Soldier Systems Center, Natick, MA 01760, USA

Truong,* Vinh, Food Science and Technology, School of Land and Food Sciences, University of Queensland, Gatton, QLD 4345, Australia

Ubbink, J.B., Nestle Research Center, Vers-chez-les-Blanc, CH-1000 Lausanne 26, Switzerland

Watzke, H.J., Nestle Research Center, Vers-chez-les-Blanc, CH-1000 Lausanne 26, Switzerland

Yoshioka, Sumie, National Institute of Health Sciences, Setagaya-ku, Tokyo 158-8501, Japan

Zhang, Jiang, Inhale Therapeutic Systems Inc., San Carlos, California, USA

Zografi, George, School of Pharmacy, University of Wisconsin-Madison, 777 Highland Av., Madison, WI 53705 USA

Zylberman, Vanesa, Departamento de Industrias, Facultad de Ciencias Exactas y Naturales, University of Buenos Aires, 1428 Buenos Aires, Argentina

*Principal contributor

THE AMORPHOUS STATE – A CRITICAL REVIEW
Churchill College, Cambridge, May 15 – 17, 2001

	TUESDAY	WEDNESDAY	THURSDAY
08.30	Registration	Theories of unstable aqueous systems: how can they help the technologist? Pablo Debenedetti	Residual water: its measurement and effects on product stability David Lechuga-Ballesteros
09.00	Introduction	Unstable systems: Descamps, Taylor	Residual water: Surana(2)
09.15	Panel discussion – Progress since 1990 (Chair: Harry Levine)		
10.30	Coffee	Coffee	Coffee
11.00	'Structure' in amorphous solids: the amorphous/crystalline continuum Evgenyi Shalaev	Progress in food processing and storage, based on amorphous product technology Louise Slade	Use, misuse, abuse of experimental approaches to studies of amorphous aqueous systems David Reid
11.45	Structure: Truong, Kou	Food process technology: Sopade, Goff	Novel techniques: Schiraldi, Surana(1)
13.00	**Lunch**	**Lunch**	**Lunch**
14.00	Glassy state dynamics – significance for biostabilisation; role of carbohydrates Roger Parker	Rational pharmaceutical formulation: can regulators be educated? Tony Auffret	End of Conference
14.45	Dynamics: Lourdin, Surana(3), Schoonman	Pharmaceutics: Toner, Kett	
16.00	Tea	Tea	
16.30	Inspection of posters	Discussion of poster presentations	
19.00	**Dinner**	**1900 Reception and Course Dinner in an 'Old Cambridge' venue – Girton College**	
Evening	Chemistry in solid amorphous matrices: implications for biostabilization Michael Pikal		

List of Amorph 2001 Participants

Vinh Truong
University of Queensland
School of Land & Food Sciences
Food Science & Technology
Gatton, QLD 4345
Australia
Tel. 617 07 54601224
Fax 617 54601171
e-mail TV@jst.uq.edu.au

Carolina Schebor
University of Buenos Aires
Faculty of Exact & Natural Sciences
Departamento de Industrias
Ciudad Universitaria
1428 Buenos Aires
Argentina
Tel. 54 114576 3366
Fax 54 114576 3366
e-mail cschebor@di.fcen.uba.ar

Ana Pilosof
University of Buenos Aires
Faculty of Exact & Natural Sciences
Departamento de Industrias
Ciudad Universitaria
1428 Buenos Aires
Argentina
Tel. +54 11 4576 3374
Fax +54 11 4576 3366
e-mail apilosof@di.fcen.uba.ar

Michael Lynch
Technologie Servier
Anlytical Division
27 rue Eugène Vignat
BP 1749
45007 Orleans Cedex 1
France
Tel. +33 2 38 81 60 00
Fax +33 2 38 54 01 31
e-mail Michael.Lynch@fr.netgrs.com

Alberto Schiraldi
Professor of Physical Chemistry
University of Milan
DISTAM
Via Celoria 2
20133 Milano
Italy
Tel. +39 02 70602063
Fax +39 02 70638625
e-mail alberto.schiraldi@unimi.it

Kazuhito Kajiwara
Teikyo University of Science &
Technology
Department of Biosciences
2525 Yatsusawa Uenohara-machi
Kitatsuru-gun
Yamanashi 409-0193
Japan
Tel. +81 554 63 4411
Fax +81 554 63 4431
e-mail Kajiwara@ntu.ac.jp

Norio Murase
Tokyo Denki University
Department of Biotechnology
College of Science & Engineering
Hatoyama, Hiki-gun
Saitama 350-0394
Japan
Tel. +81 492 96 2911
Fax +81 492 96 5162
e-mail nmurase@b.dendai.ac.jp

Miang Hoong Lim
University of Otago
Food Science Department
PO Box 56
Dunedin
New Zealand
Tel. 64 3 479 7953
Fax 64 3 479 7567
E-mail Miang.lim@stonebow.otago.ac.nz

William Michael MacInnes
Nestle, Nestec SA
Nestlé Research Centre
Process Research Team
Vers-Chez-Les Blanc
CH-1000 Lausanne 26
Switzerland
Tel. +41 21 785 8747
Fax +41 21 785 8554
e-mail william.macinnes@rdls.nestle.com

Suched Samuhasanegtoo
University of Nottingham
Food Sciences
Sutton Bonington Campus
Loughborough, Leicestershire
LE12 5RD
UK
Tel. 0115 951 6198
e-mail suched@yahoo.com

Peter Adeoye Sopade
University of Queensland
Dept. of Chemical Engineering
St Lucia
Brisbane QLD 4072
Australia
Tel. +61 7 3365 3931
Fax +61 7 3365 4199
e-mail p.sopade@cheque.uq.edu.au

Pilar Buera
University of Buenos Aires
Faculty of Exact & Natural Sciences
Departamento de Industrias
Ciudad Universitaria
1428 Buenos Aires
Argentina
e-mail pilar@di.fcen.uba.ar

Douglas Goff
University of Guelph
Department of Food Science
Guelph, ON NIG 2WL
Canada
Tel. 519 824 4120
Fax 519 824 6631
e-mail dgoff@noguelph.ca

Denis Lourdin
INRA
Rue de la Gérandière
BP 71627
44316 Nantes Cedex 3
France
Tel. +33 2 4067 5147
Fax +33 2 4067 5066
e-mail lourdin@nantes.inra.fr

Chidphong Pradistwuwana
P/a Tokyo University of Fisheries
Dept. Food Technology
4-5-7, Konan, Minato-ku,
Tokyo 108-8477
Japan

Bhesh Bhandari
University of Quennsland
School of Land & Food Sciences
Food Science & Technology
Gatton, 4345 QLD
Australia
e-mail BB@fst.uq.edu.au

Maria Florencia Mazzobre
University of Buenos Aires
Faculty of Exact & Natural Sciences
Departamento de Industrias
Ciudad Universitaria
1428 Buenos Aires
Argentina
Fax +54 11 4576 3366
e-mail florm@di.fcem.uba.ar

Philippe Letellier
Technologie Servier
Analytical Pre-Development and
Post-AMM Department
27 rue Eugène Vignat
BP 1749
45007 Orleans Cedex 1
France
Tel. +32 2 38 81 60 00
Fax +33 2 38 81 61 77
e-mail philippe.letellier@fr.netgrs.com

Marc Descamps
University Lille1
Dept. Physics
LDSMM
Bat P5
59655 Villeneuve d'Ascq Cedex
France
e-mail Marc.Descamps@uni-lille1.fr

Ken-ichi Izutsu
National Institute of Health Sciences
Drug Division
1-18-1 Kamigoga
Setagaya, 158-8501
Tokyo
Japan
Fax 81 3 3707 6950
e-mail izutsu@nihs.go.jf

Kiyoshi Kawai
Tokyo University of Fisheries
Dept. of Food Science & Techn.
4-5-7 Konan, Minato-ku
Tokyo 108-8477
Japan
Tel. +81 3 5463 0623
Fax +81 3 5463 0585
e-mail fm00513@cc.tokyo-u-fish.ac.jp

Jens Liesebach
University of Otago
Food Science Department
PO Box 56
Dunedin
New Zealand
e-mail Jens_Liesebach@gmx.de

Lynne Taylor
Research Scientist
AstraZeneca
Solid State Analysis
Mölndal, S431 83
Sweden
Tel. +46 31 7761282
Fax +46 31 7763835
e-mail: Lynne.Taylor@astrazeneca.com

Paru Sellappan
Nestec SA
Food Science & Process Research
Nestlé Research Center
Vers-chez-les-Blanc
CH-1000 Lausanne 26
Switzerland
Tel. +41 21 785 8175
Fax +41 21 785 8554
e-mail parvathi.sellappan@rdls.nestle.com

Robert Alcock
Quadrant Healthcare
Pharmaceutical Sciences
1 Mere Way
Ruddington
Nottingham NG11 6JS
UK
Tel. 0115 974 7474
Fax 0115 974 8494
e-mail AlcockR@quadrant.co.uk

Toru Suzuki
Tokyo University of Fisheries
Dept. Food Science and Technology
4-5-7, Konan, Minato-ku
Tokyo 108-8477
Japan
Tel. +81 3 5463 0623
Fax +81 3 5463 0585
e-mail toru@tokyo-u-fish.ac.jp

Janet McFetridge
University of Otago
Food Science Department
PO Box 56
Dunedin
New Zealand
Tel. +64 3 4797800
Fax +64 3 4797567
e-mail mcfja061@student.otago.ac.nz

Isabel Braga da Cruz
Nestlé/Escola Superior de
Biotechnologia
Nestlé Research Center
Process Research Team
Vers-Chez-Les-Blanc
CH-1000 Lausanne 26
Switzerland
Tel. +41 21 785 83 68
Fax +41 21 785 85 54
e-mail g102.laboratory@rdls.nestle.com

Annemarie Schoonman
Nestec SA
Food Science and Process Research
Nestlé Research Center
Vers-chez-les-Blanc
CH-1000 Lausanne 26
Switzerland
Tel. +41 21 7859229
Fax +41 21 7858554
e-mail
Annemarie.Schoonman@rdls.nestle.com

Eric Langner
Quadrant Healthcare Research
1 Mere Way
Ruddington
Nottingham NG11 6JS
UK
Tel. 0115 9747474
Fax 0115 9748494
e-mail langnee@quadrant.co.uk

Felix Franks
BioUpdate Foundation
25 The Fountains
229 Ballards Lane
London N3 1NL
UK
Tel. 020 8922 1686
e-mail bioup@dial.pipex.com

Roger Parker
Institute of Food Research
Food Material Science
Norwich Research Park
Colney Lane
Norwich NRA 7UA
UK
Tel. 01603 255284
Fax 01603 507723
e-mail roger.parker@bbsrc.ac.uk

David O'Sullivan
AstraZeneca
R&D Charnwood
Preformulation and Biopharmaceutics
Bakewell Road
Loughborough
Leics. LE11 5RH
UK
Tel. 01509 645 113
Fax 01509 645 546
e-mail david.o'sullivan1@astrazeneca.com

Tony Auffret
Pfizer Ltd
Central research (D049)
Pharmaceutical Sciences
Ramsgate Road
Sandwich
Kent CT13 9NJ
UK
Tel. +44 1304 641162
Fax +44 1304 653909
e-mail Tony-Auffret@sandwich.pfizer.com

Pablo G Debenedetti
Princeton University
Dept. of Chemical Engineering
Princeton, NJ 08544
USA
Tel. +1 609 258 5480
Fax +1 609 258 0211
e-mail pdebene@priceton.edu

Mehmet Toner
Harvard Medical School
Massachusetts General Hospital,
Shriners Hospital for Children
51 Blossom Street
Boston MA 02114
USA
Fax +1 617 371 4950
e-mail mtoner@sbi.org

Raj Suryanarayanan
University of Minnesota
Depart. of Pharmaceutics,
College of Pharmacy
308 Harvard Street, SE 9 -127 WDH
Minneapolis, MN 55455
USA
Tel. +1 612 624 9626
Fax +1 612 626 2125
e-mail suryo001@tc.umn.edu

Louise Slade
Nabisco
Corp. Res.
PO Box 1944
East Hanover NJ 07936
USA
e-mail SladeL@nabisco.com

Alex Fowler
University of Massachusetts@Harvard
Mechanical Engineering
Dartmouth
North Dartmouth, MA 02747
USA
Tel. +1 508 999 8542
Fax +1 508 999 8881
e-mail afowler@umassd.edu

David S Reid
University of California
Food Science & Technology
One Shields Avenue
Davis, California 95616-8571
USA
Tel. +1 530 752 8448
Fax +1 530 752 4759
e-mail dsreid@ucdavis.edu

Julian Blair
Quadrant Healthcare Research
1 Mere Way
Ruddington
Nottingham NG11 6JS
UK
Tel. 0115 974 7474
Fax 0115 974 8494
e-mail blairj@quadrant.co.uk

Steve Ablett
Unilever Research
Measurement Science
Colworth Laboratory
Sharnbrook
Bedford MK44 1LQ
UK
Tel. 01234 222491
Fax 01234 222599
e-mail steve.ablett@unilever.com

Imad A Farhat
University of Nottingham
Division of Food Sciences
School of Biosciences
Sutton Bonington Campus
Loughborough, Leics. LE12 5RD
UK
Tel. 0115 9516134
Fax 0115 9516142
e-mail imad.farhat@nottingham.ac.uk

Richard Buscall
ICI Strategic Technology Group
PO Box 90 Wilton
Middlesbrough
Cleveland TS90 8JE
UK
Tel. 01642 437613
Fax 01642 436306
e-mail richard_buscall@ici.com

Neha Patel
Vectura Ltd
Dept. Research & Development
University of Bath Campus
Claverton Down
Bath BA2 7AY
UK
Tel. 01225 826161
Fax 01225 826942
e-mail neha.patel@vectura.co.uk

Trevor Gard
Quadrant Healthcare Sciences
Pharmaceutical Sciences
1 Mere Way
Ruddington
Nottingham NG11 6JS
UK
Tel. 0115 974 7474
Fax 0115 974 8494
e-mail gardt@quadrant.co.uk

Martin Izzard
Unilever Research
Colworth House
Sharnbrook
Bedford
UK
Tel. 01234 222815
Fax 01234 222007
e-mail martin.izzard@unilever.com

William Macnaughtan
University of Nottingham
Division of Food Sciences
School of Biosciences
Sutton Bonington Campus
Loughborough, Leics LE12 5RD
UK
Tel. 0115 951 6112
Fax 0115 951 6142
e-mail
bil.macnaughtan@nottingham.ac.uk

Vanessa Marie
University of Nottingham
Food Science
School of Biological Science
College Road
Sutton Bonington
Loughborough LE12 5RD
UK
e-mail SCXVM@nottingham.ac.uk

Nigel Griffith-Skinner
GlaxoSmithKline
Department Physical Properties
Bldg 427
Temple Hill
Dartford Kent
DA1 5AH
UK
e-mail ngs47213@gsk.co.uk

Jeffrey Errington
Princeton University
A-215 Engineering Quadrangle
Dept.of Chemical Engineering
Princeton, NJ 08544
USA
Tel. + 609 258 5413
Fax + 609 258 0211
e-mail jerring@princeton.edu

David Lechuga
Inhale Therapeutic Systems, Inc.
Pharmaceutical Research
150 Industrial Road
San Carlos
CA 94070-6256
USA
Tel. + 1 650 631 3254
Fax + 1 650 631 3150
e-mail dlechuga@ihale.com

Harry Levine
Nabisco
Corp. Res
PO Box 1944
East Hanover NJ 07936
USA
Tel. 973 503 3923
Fax 973 503 2364
e-mail LevineH@nabisco.com

Victor Huang
Pillsbury Company
Department R&D
330 University Ave SE
Minneapolis, MN 55414
USA
Tel. 612 3308286
Fax 612 3308064
E-mail Vhuang@pillsbury.com

Yang Kou
University of Massachusetts
Dept. Food Science
4 University Park
Waltham MA 02453
USA
Tel. + 1 508 233 4443
Fax + 1 508 233 5200
e-mail y-kou@usa.net

Evgenyi Y Shalaev
Groton Laboratories
Global R&D Pfizer
Dept. Pharmaceutical R&D
8156-04, Pfizer
Eastern Point Rd
CT 06340 USA
Tel. 860 441 5909
Fax 860 441 0467
e-mail
Evgenyi_Y_Shalaev@groton.pfizer.com

Rabul Surana
University of Minnesota
Department of Pharmaceutics,
College of Pharmacy
308 harvard Street, SE 9-125 WDH
Minneapolis, MN 55455
USA
Tel. 612 624 7968
Fax 612 626 2125
e-mail sura0007@tc.umn.edu

Brian A Pethica
Princeton University
School of Engineering and App. Science
Dept. of Chemical Engineering
220C, Engineering Quadrangle
Princeton, New Jersey, 08544-5263
USA
Tel. 609 258 1670
Fax 609 258 0211
e-mail bpethica@princeton.edu

Pavinee Chinachoti
University of Massachusetts
Dept. Food Science
Chenoweth Lab.
Amherst, MA 01003
USA
Tel. +1 413 545 1025
Fax +1 413 545 1262
e-mail pavinee@foodsci.umass.edu

George L Collins
TherMold Partners L.P.
652 Glenbrook Road
Stamford CT 06906
USA
Tel. 203 977 8161
Fax 203 977 8237
e-mail gcollins@thermoldlp.com

Vicky Kett
School of Pharmacy
Queen's University of Belfast
97 Lisburn Road
Belfast BT9 7BL
Northern Ireland
e-mail v.kett@qub.ac.uk
Tel. 028 9027200
Fax 028 90272028

Michael Fitzgerald
Pfizer
Pharmaceutical Pharm. R&D
PGRD
Sandwich, Kent CT13 9NJ
UK
Tel. 01304 643928
Fax 01304 653909
e-mail
Michael_Fitzgerald@sandwich.pfizer.com

L D Hall
University of Cambridge
Herschel Smith Laboratory
for Medicinal Chemistry
Forvie Site Robinson Way
Cambridge CB2 2PZ
UK
Tel. 01223 336805
Fax 01223 336748
e-mail ldh11@hslmc.cam.ac.uk

Michael J Pikal
School of Pharmacy
University of Connecticut
372 Fairfield Road
POB U-92
Storrs, CT 06269-2092
USA
Tel. +1 860 486 3202
Fax +1 860 486 4998
e-mail pikal@uconnvm.uconn.edu

Rajesh Patel
Glaxo SmithKline Pharmaceuticals
Pharmaceutical Development
New Frontiers Science Park (South
Third Avenue
Harlow Essex CM19 5AW
UK
Tel. 01279 622000
Fax. 01279 643910
e-mail Rajesh_2_Patel@sbphrd.com

Michael S Otterburn
Cargill Inc.
Speciality Food Ingredients
15407 McGinty Road West
Wayzata
MN 55391 - 2399
USA
Tel. +1 952 742 6454
Fax +1 952 742 4330
e-mail michael_otterburn@cargill.com

Andrew Edward Bayly
Procter & Gamble
Newcastle Technical Centre
Whitley Road
Longbenton
Tyne & Wear NZ 12 9TS
UK
Tel. 0191 279 1627
Fax 0191 279 2757
e-mail bayly.ae@pg.com

Introduction – Progress in Amorphous Food and Pharmaceutical Systems

Harry Levine

NABISCO R&D, KRAFT FOODS, EAST HANOVER, NEW JERSEY 07936, USA

When the conference convener, Felix Franks, asked me to kick-off the opening session by chairing an introductory panel discussion on 'progress since 1990', I thought it might be interesting (to all the conferees, not just to me) to begin that discussion by presenting a chronology of key milestones on the subject, from my own personal and admittedly biased perspective. That chronology is shown in Table 1.

For me, the story began in 1979 (at General Foods [GF], where I had started working as a polymer chemist in 1976, following a 2-year post-doctoral fellowship) when I happened to find and read a paper by Felix Franks and co-workers[1] on biological cryoprotection. In that seminal paper, the concept of T_g', a characteristic subzero glass transition temperature for a maximally freeze-concentrated aqueous solution, was introduced for the first time. As illustrated in Figure 1, T_g' was shown on a differential scanning calorimetry thermogram and a solute-water state diagram, and was identified as $\approx T_r$, the recrystallization temperature, for water-compatible polymers such as poly(vinyl pyrrolidone) and hydroxyethyl starch. It occurred to me that Franks' new conceptual approach to biological cryoprotection might be applicable to the storage stabilization of ice-containing frozen foods. I thought I should try to meet Prof. Franks and discuss the matter with him.

In early 1980, during a multi-purpose business trip to Europe, I had an opportunity to visit Cambridge University in England and meet Felix Franks for the first time. That was a turning point in my life, both professionally and personally. When I returned to GF, I arranged to invite Felix to visit our Technical Center in Tarrytown, NY and present a guest seminar, which was co-hosted by two of my young colleagues in Central Research, Tim Schenz and Louise Slade, along with myself. As a consequence of that first visit to Tarrytown, Felix became a paid consultant to General Foods from 1980–87, as well as a scientific mentor (and lifelong friend) to Louise and me. [Twenty two years later, I still enjoy telling people that Felix was the person responsible for bringing Louise and me together, first professionally and then personally.]

1

Table 1 *A chronology – from a personal perspective*

1979	Seminal paper on 'biological cryoprotection' found and read – Franks *et al.*, *J. Microscopy*, 1977.[1] T_g' ($\approx T_r$) shown for the first time [Figure 1] on a DSC thermogram and solute–water state diagram
1980	Levine met Franks at Cambridge, then Levine, Slade and Tim Schenz hosted Franks at a General Foods, Tarrytown seminar. Franks became GF consultant (1980–87) and mentor to Levine and Slade
1980–82	Other historical papers found and read – White & Cakebread (1966)[2] – glassy state in foods; Luyet, MacKenzie, Rasmussen (1939–77)[3–6] – biological cryoprotection; Parks *et al.* (1928–34)[7,8] – glucose glasses; van den Berg (1981 thesis)[9] – glass transition in starch
1982	Slade and Levine attended Franks' Cambridge Conference on Biological Cryoprotection – met, *e.g.* Angell and the Crowes
1985	Slade attended Franks' Cambridge Conference on 'Water Activity?' and gave ground-breaking lecture[10] on 'Food Polymer Science' approach 'Beyond "Water Activity"…'
1987	Levine and Slade began teaching their Food Polymer Science short course [Figure 2 – magazine cover]
1989	Levine and Slade organized ACS 'Water Relationships in Foods' symposium → edited 1991 Plenum book[11] (>600 citations to date)
1991	Slade and Levine's CRC Crit. Revs. monograph 'Beyond Water Activity…'[12] published → >470 citations to date
1992	Blanshard's ACTIF program at Nottingham University (1989–92) → 1992 Easter School → 1993 'Glassy States in Foods' book[13]
1994	'What are Harry and Louise talking about?' [media Figure 3]
1995	Franks and Levine organized 'AMORPH 1' conference at Cambridge – 'Discussion Symposium on Chemistry and Application Technology of Amorphous Carbohydrates' → Summary report on current status [Highlights – Table 2]
2001	Franks' list of 'outstanding questions' on Solid Aqueous Solutions of Carbohydrates[14] [Table 3] →
5/15/01	'AMORPH 2001' at Cambridge – 'The Amorphous State – A Critical Review'

During the period from mid-1980 through 1982, while Louise and Tim and I were busy building a very large experimental database of DSC-measured T_g' values for hundreds of food ingredients, materials and products, I also found and read many other historical papers on glasses, glassy states and the glass transition in food and biological systems. They included: White and Cakebread's 1966 paper on the glassy state in sugar-containing foods;[2] an extensive series of works by Luyet, Rasmussen and MacKenzie, published from 1939 to 1977, on frozen aqueous solutions of small carbohydrates with application to biological cryoprotection;[3–6] an even earlier series of papers by Parks *et al.* (1928–34) on the physical chemistry of glucose glasses;[7,8] and a 1981 doctoral thesis by van den Berg on the glass transition in starch–water systems.[9]

In 1982, Louise and I managed, after much difficulty with corporate prohibitions, to obtain GF approval to attend our first scientific conference at Cambridge's Girton College, on the subject of biological cryoprotection and organized by Felix. There, we met for the first time such well-known workers in this field as Austen Angell and the Crowes, John and Lois, whose many papers we

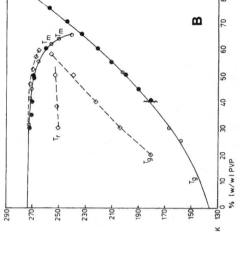

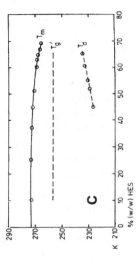

A. (a) DSC heating scan for quench cooled 50% PVP solution, showing the thermal transitions mentioned in the text; (b) DSC heating scan for a slow frozen 50% PVP solution; (c) DSC cooling scan for a 50% PVP solution which had previously been heated to 360 K in the calorimeter. Scanning directions are shown by horizontal arrows; scanning rate: 10 K min^{-1}.

B. Solid–liquid state diagram for water-PVP. Drawn out lines and $\odot$ refer to $\bar{M}_n = 44{,}000$; broken lines and $\diamond$ refer to $\bar{M}_n = 700{,}000$; the data of MacKenzie & Rasmussen (1972) are shown as filled-in circles, and their lower limit of vitrification at 40% is shown on the T_g curve.

C. Solid–liquid state diagram for water-HES showing melting points, T_m, and tentative values for the devitrification, T_d, of quenched samples. T_g is the glass transition of the concentrated solution ($\sim 70\%$) which remains after slow freezing and the separation of ice.

Figure 1 *Seminal work by Franks et al.*[1] *on* T_g'

subsequently started to add to our reading collection. To everyone attending that conference, except Felix (and his wonderful wife, Hedy), Louise and I were totally unknown novices, but we absorbed like sponges every bit of new information to which we were exposed.

Already by 1985, Louise and I had Felix to thank for paving the way for us with many more well-placed introductions to scientific notables in the field of water in foods [*e.g.* John Blanshard, Marc Karel, Ted Labuza, Martine Le Meste, Peter Lillford, David Reid, Denise Simatos, Cornelius van den Berg] and with invitations to present talks on our research at various international technical meetings in Europe. We were beginning to know people in the field and become known by some of them for our GF work on 'cryostabilization technology' and its industrial applications to frozen foods. In 1985, Louise was allowed to attend another conference organized by Felix at Cambridge, this one on 'Water Activity…?', where she gave a ground-breaking invited presentation[10] on our 'food polymer science' approach 'beyond "water activity"…'. At home, I learned afterwards from various sources that her talk caused quite a stir among many of the participants at that conference.

In 1987, Louise and I were invited to develop and begin teaching externally a 2-day-long industrial short course on our food polymer science approach to moisture management and water relationships in foods. [Figure 2 shows the cover of a trade magazine from that time, on which was featured a hand (mine)-drawn state diagram reproduced from our first course manual.] That first course offering (of 26 to date, attended by > 700 participants) coincided with Louise's and my decision to leave General Foods and together join a new Fundamental Science department being created in Corporate Research at Nabisco.

There at Nabisco, working under an enlightened team of R&D managers, Louise and I were enthusiastically encouraged, for a number of good-for-the-company reasons, to continue and expand our external scientific activities in the areas of publishing papers, presenting talks, and teaching our short course, as well as organizing conferences and symposia. Among the latter was a major international symposium in 1989 on Water Relationships in Foods, which we were invited to organize on behalf of the American Chemical Society's Agriculture and Food Division. The proceedings from that symposium were published in 1991 as a book of the same title,[11] which we edited for Plenum Press. To date, that book has accumulated a total of > 600 citations.

In 1991, we finally also published our 'Beyond Water Activity…' monograph[12] in CRC Critical Reviews, a paper that had been 'in preparation', at Felix's request, since the time of Louise's presentation[10] at Cambridge in 1985. While Felix showed remarkable patience, I got so tired of having to tell him 'better late than never' and 'good things are worth waiting for'. In the end, Louise and I have been most gratified to see that review become a 'citation classic', with > 470 citations to date.

1992 marked a milestone in the ACTIF research program [Amorphous and Crystalline Transitions in Foods] at Nottingham University in England. The original program – a consortium among Nottingham (led by John Blanshard)

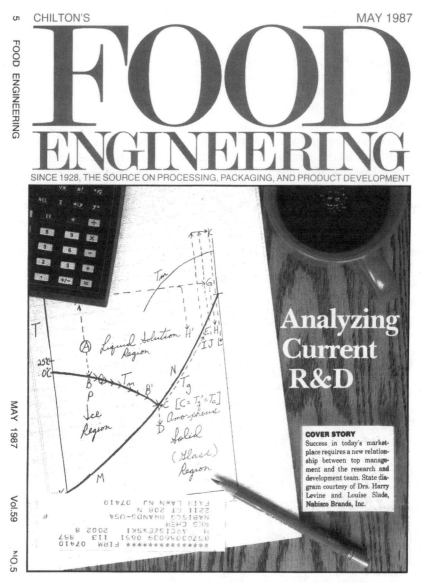

CHILTON'S MAY 1987

FOOD ENGINEERING

SINCE 1928, THE SOURCE ON PROCESSING, PACKAGING, AND PRODUCT DEVELOPMENT

Analyzing Current R&D

COVER STORY
Success in today's market-place requires a new relation-ship between top manage-ment and the research and development team. State dia-gram courtesy of Drs. Harry Levine and Louise Slade, Nabisco Brands, Inc.

Figure 2 *A state diagram featured on the cover of a trade magazine in 1987*

and a number of major European industrial sponsors (led by Unilever and Peter Lillford) – was designed to be a three-year program beginning in 1989. By the time of its conclusion in 1992 (but prior to follow-up programs called ACTIF II and InterACTIF), it had already become extremely productive and widely known throughout the field. To mark the occasion, Blanshard and Lillford convened one of Nottingham's well-known 'Easter School' conferences, this one on the topic of 'The Glassy State in Foods', the proceedings of which were published as a major book in 1993.[13]

And now for a touch of humor [at least, friends of ours thought it was hilarious]. In 1994, during the political campaign season in the US, there was a series of TV advertisements featuring a married couple named 'Harry and Louise', who discussed in excruciating detail the pros or cons (I forget which) of a Clinton (Hilary!) health care proposal of that time. Those several commercials ran so frequently and for so long that they generated various derogatory comments and reactions in the media of the day, two pieces of which are shown in Figure 3. Since by this time, we had been teaching our short course for eight years (and even had it videotaped once or twice), had published many voluminous review papers, and never seemed to be able to deliver an invited presentation without going well over our allotted time, we were unhappily accustomed to hearing people often say 'what are Harry and Louise talking about?'. But seeing that written in large bold print in a newspaper was something else again.

In 1995, Felix and I organized the first 'AMORPH' conference at Cambridge, a 'Discussion Symposium on the Chemistry and Application Technology of Amorphous Carbohydrates'. The previously unpublished summary report from that conference, compiled by Felix, is reproduced here in its entirety in Felix's Preface. For my Introduction to the AMORPH 2001 conference, I summarized

What are Harry and Louise talking about?

Aren't you glad you don't have friends like Harry and Louise? Does anybody?
This is supposed to be an ordinary couple who apparently spend

Robert Reno

Figure 3 *Humorous media pieces from 1994 on 'Harry and Louise'*

highlights from Felix's 1995 summary report, which are shown in Table 2.

Among our preparations for AMORPH 2001 at Cambridge, Felix produced a list of 'outstanding questions' on Solid Aqueous Solutions of Carbohydrates,[14] which are reproduced here in Table 3. Those questions were intended to form the basis for the presentations (plenary lectures and invited contributions), posters and discussion sessions during AMORPH 2001 on 'The Amorphous State – A Critical Review'. The rest of this book comprises the written proceedings from that conference, which I've had the honor and pleasure of editing, for publication by the Royal Society of Chemistry.

Table 2 *Highlights from 'AMORPH 1' Summary Report*

Topic 1 – Relationship between Molecular Structure and Glass Transition

Angell's 'strong/fragile' classification scheme for glass-formers, along with T_m/T_g ratio, may be useful in distinguishing behaviors among various common mono- and disaccharide sugars.

Topic 2 – Chemical Reactivity of Solid Sugars

Some chemical reactions, especially those that do not rely on translational diffusion, can and do occur below T_g. Chemical reaction rates (*e.g.* for Maillard browning), both below and above T_g, depend on mobility of reactants that may include water (*e.g.* hydrolysis).

Topic 3 – Chemistry and Biochemistry in Supersaturated Carbohydrate Mixtures

Accounts of survival of microorganisms, seeds, and viruses in glassy matrices, and of varying storage stability of enzymes, proteins and drugs, after drying in 'protecting' sugar glasses, were discussed.

Topic 4 – Physical Processes (*e.g.* Crystallization) of and within Amorphous Carbohydrates; Kinetics; Effects of Residual Moisture
and
Topic 5 – Solid Solutions involving Carbohydrates

According to crystallization kinetics theory, crystallization does not occur below T_g. But very slow crystallization at $T < T_g$ has been reported, *e.g.* for the drug indomethacin. However, at 50 °C below T_g, indefinite stability against crystallization can be achieved.

Topic 6 – Dynamics, especially of Small Molecules within Amorphous Carbohydrates

Translational diffusion of small molecules (with low T_g), even including water, at $T < T_g$ of an amorphous carbohydrate matrix of higher T_g (due to higher MW) can occur, but will be much slower than the diffusion of the same small molecule in a bulk liquid state.

Topic 7 – Model Systems and Analytical Methods

After a search for a 'single best' method for determination of T_g for solid carbohydrates, the consensus was in favor of a combination of techniques, which could include DMTA, DSC (or MDSC), DETA, NMR, ESR, dielectric permittivity, X-ray diffraction, IR/Raman spectroscopy, microscopy, and computer simulation.

Topic 8 – Amorphous Carbohydrates as Biostabilizers

Amorphous carbohydrates find widespread application as stabilizers (excipients) of labile biologicals. The glassy solid state can be a necessary but not sufficient condition for stability. Stabilization with trehalose can be more successful than with sucrose, which can be due to the higher T_g and lower chemical reactivity of trehalose.

[Note added in 2001 – trehalose is now FDA-approved and commercially available (but expensive) for use in foods in the United States.]

Table 3 *Solid aqueous solutions of carbohydrates – outstanding questions*

- Supersaturated carbohydrate solutions
- Solvent effects on intramolecular mobility
- Molecular structure and thermomechanical properties
- Relationship between crystal structure and glass 'structure'
- Methods for studying hydrogen bond topology
- Crystallization of oligosaccharides from aqueous glasses
- Paradoxes posed by sugar hydrates: the raffinose mystery
- Dynamics within aqueous glasses: uncoupling of molecular motions
- Relaxation rates and fragility: analysis of DSC scans
- Complexities of ternary (multicomponent) systems
- Sugar glasses as stabilizing principles: ecology and technology

---

References

1. F. Franks, M.H. Asquith, C.C. Hammond, H.B. Skaer and P. Echlin, *J. Microscopy*, 1977, **110**, 223.
2. G.W. White and S.H. Cakebread, *J. Food Technol.*, 1966, **1**, 73.
3. B. Luyet, *J. Phys. Chem.*, 1939, **43**, 881.
4. D. Rasmussen and B. Luyet, *Biodynamica*, 1969, **10**, 319.
5. A.P. MacKenzie and D.H. Rasmussen, *Water Structure at the Water–Polymer Interface*, H.H.G. Jellinek (ed.), Plenum Press, New York, 1972, 146.
6. A.P. MacKenzie, *Phil. Trans. Roy. Soc. Lond. B*, 1977, **278**, 167.
7. G.S. Parks, H.M. Huffman and F.R. Cattoir, *J. Phys. Chem.*, 1928, **32**, 1366.
8. G.S. Parks, L.E. Barton, M.E. Spaght and J.W. Richardson, *Physics*, 1934, **5**, 193.
9. C. van den Berg. *Vapour Sorption Equilibria and Other Water–Starch Interactions: A Physico-Chemical Approach*, PhD thesis, Agricultural University, Wageningen, Netherlands.
10. L. Slade and H. Levine, Faraday Division, Royal Society of Chemistry, Industrial Physical Chemistry Group, Conference on Concept of Water Activity, Cambridge, UK, 1985.
11. H. Levine and L. Slade (eds.), *Water Relationships in Foods*, Plenum Press, New York, 1991.
12. L. Slade and H. Levine, *Crit. Revs. Food Sci. Nutr.*, 1991, **30**, 115.
13. J.M.V. Blanshard and P.J. Lillford (eds.), *The Glassy State in Foods*, Nottingham University Press, Loughborough, UK, 1993.
14. F. Franks, personal communication, 2001.

Structure and its Significance in the Application Technology of Amorphous Materials

The Concept of 'Structure' in Amorphous Solids from the Perspective of the Pharmaceutical Sciences

Evgenyi Shalaev[1] and George Zografi[2]

[1]GROTON LABORATORIES, MS-8156-004, GLOBAL R&D, PFIZER INC., GROTON, CT 06340
[2]SCHOOL OF PHARMACY, UNIVERSITY OF WISCONSIN-MADISON, 777 HIGHLAND AV., MADISON, WI 53705

1 Introduction

Partially or completely amorphous pharmaceutical materials are often created during the manufacture of dosage forms, when an initially highly crystalline bulk drug is processed by standard methods, such as milling, compaction, wet granulation, freeze-drying and spray-drying. The importance of amorphous materials in pharmaceutical research and development has been highlighted in a number of reviews.[1–3] Formation of a disordered solid may be intentional, when higher solubility and bioavailability are desired, or it may be accidental, during the processing of crystalline materials. In both cases, the physical and chemical stability of such a disordered solid drug must be considered to have possibly been compromised. Because of the very strict regulatory requirements for maintaining the stability of pharmaceuticals, an understanding of amorphous/crystal relationships is an essential part of the drug development process.

In the pharmaceutically related literature on amorphous solids, the main focus has been on studies of the relationships between molecular mobility and chemical and physical stability.[4,5] In particular, many efforts have been devoted to identifying a 'critical' temperature, below which an amorphous material will be completely stable on practical time scales.[6] Candidates for such a critical stability temperature include the glass transition temperature (T_g),[6] a 'chemical reaction glass transition temperature' and the Kauzmann temperature $(T_K$ or $T_0)$ (see refs. 4–7 for examples). It has been recognized, however, that although molecular mobility is important, it is not the only factor that determines the stability of amorphous solids. For example, it has been suggested recently that effects (e.g. polarity) of the medium can be important in controlling the chemical stability of amorphous solids.[8] In particular, the role of 'solid-state pH' in amorphous systems is attracting increasing attention in the pharmaceutical literature.[9–11]

Another critical factor that is expected to have a significant impact on the stability of amorphous solids is the structural nature of molecules in the solid state. Indeed, there are well-established relationships between crystal structure and chemical reactivity in crystalline solids.[12] However, studies on amorphous structure/stability relationships are still under-represented in the pharmaceutical literature. In the present paper, we review different aspects of the structural features of amorphous solids, in relation to their chemical and physical stability. The paper includes five following sections. In the first, we ask the following question, 'is there more than one amorphous state of the same material?' In particular, the issue of 'polyamorphism' is considered. Both true polyamorphism (*i.e.* involving a phase transition between two amorphous phases) and relaxation 'polyamorphism', where there is a continuous space of dynamic amorphous states, are discussed. The second section is devoted to the microscopic heterogeneity of amorphous materials, and in particular, how this might relate to chemical reactivity. The third part of this review considers the (dis)similarity between local structures of amorphous and crystalline states of the same material. Several examples, such as indomethacin and polyhydroxy compounds, are considered and possible relationships between differences in the local structures of amorphous solids *vs* crystals and crystallization potential of amorphous solids are highlighted. The topic of crystalline mesophases is considered in the fourth section. Such mesophases include liquid crystals, plastic crystals/orientationally disordered crystals and conformationally disordered crystals, all of which combine the properties of both crystalline (*i.e.* long-range order) and amorphous (*i.e.* glass transition) materials and are considered to be intermediate between crystalline and amorphous states. The last part of this review deals with the glass transition of material in small or confined spaces. Examples of 'confined glasses' in pharmaceutical systems include slightly disordered crystalline materials, liposome/drug complexes, and frozen aqueous solutions of carbohydrates or proteins.

2 Polyamorphism

It is well-established that the properties of amorphous glasses can be different for the same material, depending on the previous history of treatment and time of storage. Thus, from a kinetic perspective, we can speak of an amorphous glass as existing in different states with different glass properties, *e.g.* T_g, relaxation time. For those interested in the stability of amorphous states, the differentiation of such states is of great interest. Such differences can be depicted in the form of a Gibbs free energy plot, as shown in Figure 1. In contrast to behavior based on kinetic properties, which is described by curves 'glass 2' and 'glass 2″', we can also encounter true polyamorphic systems, where a first-order phase transition occurs between two amorphous phases in the supercooled or equilibrium liquid state, as depicted by curves 'liquid 1' and 'liquid 2' (Figure 1). In addition, Figure 1 illustrates that liquid/liquid phase transitions may be hindered, if T_{g1} happens to be higher than the temperature of liquid/liquid phase transition, T_{ll}. Such phase transitions between two liquids with different structures and densities have

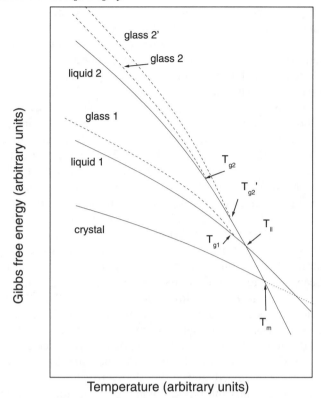

Figure 1 *Schematic Gibbs free energy curves for a hypothetical single-component system that exhibits liquid/liquid phase transition. Liquid 1 and liquid 2 describe two liquids that exhibit a first-order phase transition (thermodynamic polyamorphism). Glass 2 and glass 2' represent relaxation 'polyamorphism'. The difference between curves 'liquid' and 'glass' is exaggerated for clarity*

been predicted theoretically.[13-16] A first-order transition between two liquid phases of the same composition, for example, has been described by a potential energy hypersurface with megabasins that show local minima of potential energy in a glassy state.[16] Angell has suggested that 'polyamorphs' differ in their strength, with a low-density form acting as a 'strong glass-former' and a high-density form as a 'fragile glass-former'.[16] Experimental examples of materials showing such phase transitions between amorphous phases with different densities include water,[17] silica,[18] carbon,[19] and the melt of Y_2O_3–Al_2O_3.[20]

For organic materials, there are fewer documented examples of polyamorphism than for inorganic substances. For several polymers, *e.g.* atactic poly(styrene) (at-PS), atactic poly(methylmethacrytate) (at-PMMA), atactic poly(propylene oxide) (at-PPO), isotactic poly(methylmethacrylate) (it-PMMA) and isotactic poly(vinyl chloride) (it-PVC), a liquid/liquid phase transition has been observed at T between T_g and the melting temperature, T_m.[21] This liquid/liquid phase transition usually occurs at a temperature approximately equal to $1.2T_g$ (in Kelvin). The transition is considered to be highly cooperative (involving the entire polymer chain), corresponding to a transition from a polymer melt with local order to a completely disordered polymer melt. A special case of polyamorphism in biological polymers might be associated with native *vs* denatured proteins, which can form glassy states with very different packing of the structural elements, *i.e.* coils or helices.[16]

Triphenyl phosphite has been suggested as a small organic molecule that undergoes a liquid/liquid phase transition from a supercooled liquid to a new amorphous 'glacial phase' at temperatures of 227 to 213 K, which are between T_g (176 K) and T_m (295 K).[22] This 'glacial phase' is X-ray amorphous, has a higher viscosity, longer spin-lattice relaxation times and a higher density than does its regular amorphous form. In addition, non-isothermal crystallization of the 'glacial phase' occurs at a higher temperature, with a lower enthalpy than that observed with an ordinary supercooled liquid. Evidence that the formation of the glacial phase from the supercooled liquid involves a first-order transition is supported by the observations of an induction period and the appearance of opacity (a characteristic of phase boundaries), which accompany the conversion. It has also been suggested that the apparently amorphous 'glacial phase' may in fact be a crystalline phase with a lower degree of order than that found in a three-dimensional crystal, *e.g.* a liquid crystalline phase.[22] To prove if this is so would require additional experiments using, for example, small-angle X-ray diffraction. If the 'glacial phase' of triphenyl phosphite is indeed a liquid crystalline phase, it would be an example of a group of solid states known as 'crystalline mesophases',[23] to be discussed later, rather than a true liquid 'polyamorph'.

Another example of a possible liquid/liquid phase transition in the amorphous state is the thermal transition observed for fructose at 100 °C by Slade and Levine.[6] A similar thermal event for fructose at ~ 65 °C has been observed when it was mixed with a phospholipid.[24] Presently, it is not clear whether these thermal events observed for fructose correspond to first-order liquid/liquid transformations and, hence, true polyamorphism. However, if this is so, it would be the first example of polyamorphism in a system of practical interest in the food and pharmaceutical sciences.

There is a lack of discussion in the literature on the possible nature of polyamorphism in small organic molecules. A possible reason for liquid/liquid polyamorphism in such systems may be the distinctly different states of molecular conformation which can occur. If this is the case, polyamorphism in general may be characteristic of many glassy carbohydrates under certain sets of conditions, because carbohydrates are known to be able to exist in different conformations.

With regard to the second type of 'polyamorphism', involving kinetic states of amorphous systems, there are numerous examples that arise from different types of processing, such as milling, freeze-drying or spray-drying. For example, it has been shown that freeze-dried and spray-dried cefamandole nafate have very different X-ray diffraction patterns.[25] While both materials exhibit typical amorphous diffraction patterns, *i.e.* a broad halo without sharp crystalline peaks, the spray-dried material has a weak and relatively narrow line at 5 degrees 2Θ, suggesting that it is more ordered than the freeze-dried material. This suggestion is supported by the fact that the heat of dissolution for the spray-dried sample was closer to that of the crystalline material. Moreover, when the freeze-dried material was annealed below T_g, both its X-ray diffraction pattern and its heat of solution were shifted closer to that of the spray-dried material. Therefore, according to these results for this system, we would conclude that amorphous forms arise in different kinetic states, rather than in thermodynamic states, when the process used to produce them is altered and that this represents the result of thermal and processing history, rather than two distinct liquid phases.

Another example of process-induced differences is amorphous tri-*O*-methyl-β-cyclodextrin produced by milling and quenching of the melt by rapid cooling.[26] In these systems, first of all, the enthalpy relaxation rate was shown to be faster for the milled sample. Indeed, the milled material, which tended to relax more quickly, also showed a stronger tendency to crystallize upon standing. One other possible reason for greater crystallization rates in the milled samples, however, may have been the presence of residual seed crystals not present in the quench-cooled melt. Caution in concluding that different processes always lead to different amorphous states in the glass can be prompted from the observation that the T_g and heat capacity values for the milled sample and the quench-cooled melt of tri-*O*-methyl-β-cyclodextrin were quite similar, despite differences in enthalpy relaxation rates.[26] Also, two amorphous forms of raffinose, produced by freeze-drying and by dehydration of its crystalline pentahydrate,[27] exhibited the same T_g and similar water sorption profiles and rates of recrystallization when equilibrated at relative humidities greater than 45%.

One of the first examples of a possible influence of thermal history and processing on chemical stability of pharmaceuticals in the amorphous state was that of the antibiotic moxalactam.[28] Here, two samples prepared by freeze-drying were subjected to identical conditions except that one underwent secondary drying at 40 °C, while the other was so treated at 60 °C for the same length of time. Both samples had the same residual water content, but were characterized by different rates of decarboxylation. The rate of decarboxylation was slower for the sample dried at the higher temperature, presumably because it was in a more relaxed state and hence in a state of greater order.

2.1 Significance in the Pharmaceutical Sciences

To date, the existence of any true polyamorphism in pharmaceutical materials has not been established, thus calling for more careful examination of amorphous drugs and excipients in this regard. Unambiguous proof of such polyamorphism would be the observation of a discontinuity in the Gibbs free energy curve for two amorphous forms, as shown schematically in Figure 1. If such polyamorphism is detected and understood, it may be possible to take advantage of a particular 'liquid' polymorph in the formulation process. Clearly, of more current significance is the very strong possibility that the kinetic properties of an amorphous material can be affected by the process used to produce it. Consequently, characterization of an amorphous material must involve careful attention to thermal history, as well as to the production process to be used.

3 Heterogeneity in Amorphous Solids

Amorphous solids that contain single components have been shown to exhibit dynamic properties that affect molecular mobility and, hence, the translational and rotational motions that affect their physical and chemical stability. Strong evidence exists to suggest that in most cases of interest, the relaxation times for such materials, particularly in the vicinity of T_g, exist as a distribution of relaxation times, most likely due to spatial heterogeneity. This is reflected in the non-exponential behavior for the rate of structural relaxation, X(t), as expressed in the well-known Kohlrausch–Williams–Watts (KWW) equation:

$$X(t) = \exp\left(-\frac{t}{\tau_{KWW}}\right)^{\beta_{KWW}} \tag{1}$$

where X(t) is a property of the material, τ_{KWW} is the average relaxation time and β_{KWW} measures the extent of non-exponentiality and the distribution of relaxation times. Values of β_{KWW} vary from 0 to 1, with 1 representing a single-exponential relaxation process. Typically, many organic amorphous materials have values of β from 0.3 to 0.8,[29,30] indicating a broad distribution of relaxation times. An alternative view, which attributes non-exponentiality to homogeneous relaxation that is intrinsically non-exponential due to certain types of cooperativity, has also been proposed,[31] but with less experimental support.

Recent experiments and simulations have provided strong support for the role of spatial heterogeneity in the vicinity of T_g.[32] Such support, for example, has been obtained in single molecule experiments[33] that allow a direct answer concerning the distribution of relaxation times of exponentially relaxing elements. The rotational motion of Rhodamine G molecules in a film of poly(methylmethacrylate) was monitored using transient fluorescence intensity. Experiments were performed at 5 to 15 K above T_g. On a shorter time scale (e.g. ~ 10³ s), at $T_g + 5$ K, single molecule relaxation obeyed a single-exponential law, whereas non-exponentiality increased, wherein β_{KWW} in Equation 1 decreased from 1.0 to 0.4 as the time scale was extended from 10³ to 10⁴ s. In addition, it has been shown that the change from one environment to another occurred in an instantaneous 'jump' manner on the time scale of the experiment. Such a 'jump' was believed to originate from large-scale collective motions. Such heterogeneity of amorphous materials can be visualized from the Adam–Gibbs theory of cooperatively rearranging clusters,[34] with heterogeneity in the size of such clusters. Evidence for clusters and density fluctuations in amorphous materials near T_g has been confirmed by light scattering experiments.[35]

With much evidence of structural heterogeneity in amorphous materials, one might expect that chemical degradation reactions might not follow simple homogeneous kinetics, even in the case of a simple chemical reaction. For example, Shamblin *et al.*[36] suggested that there might be a coupling between structural relaxation (which is usually described by a stretched-exponential function as in

Equation 1) and chemical degradation. Indeed, complex kinetic patterns have been reported for some chemical reactions in amorphous solids which may be affected by such heterogeneities. For example, kinetic data for the hydrolysis of carbapenem antibiotics could not be fitted to a simple reaction model, but rather to a biphasic-based model.[37] Although those authors offered no possible reasons for such behaviour, it should be noted that such biphasic curves can often be described by a stretched-exponential function as in Equation 1. We would suggest that the kinetic curves for the hydrolysis of carbapenem may be associated with the structural heterogeneity observed for most amorphous materials, *i.e.* the distribution of relaxation time properties of structural elements (clusters) that are involved in the chemical reaction. Another example of complex kinetics for amorphous solids is Asp–Pro bond cleavage in the peptide physalaemin which did not follow simple first-order kinetics.[38] This reaction may be another example of the influence of structural heterogeneities on solid sample dynamics. It should be stressed that more detailed kinetic studies are needed to prove that structural and dynamic heterogeneities in amorphous solids result in complex kinetic degradation patterns. Indeed, complex kinetics can sometimes be observed in homogeneous solution reactions that involve, *e.g.* consecutive/parallel/reversible reactions, or if one of the reagents exists in limiting amounts. Hence, in sorting out the role of structural relaxation in controlling chemical reactions, chemical factors must always be kept in mind.

It should also be pointed out that there are many chemical reactions in the solid state for which kinetic curves can be treated by simple homogeneous models. For example, sucrose inversion in glassy sucrose/citric acid mixtures can be fitted to a homogeneous first-order equation with a limited reagent.[11] Another example of simple kinetics in amorphous solids is the hydrolysis of 2-(4-nitrophenoxy)tetrahydrofuran, for which the reaction also follows simple first-order kinetics.[38] Such cases do not necessarily contradict the presence of spatial heterogeneity in such amorphous solids. It just may be that the time scale of structural relaxation is much shorter than that for the chemical reaction. Indeed, it is to be expected that different solid-state properties exist on different scales of relaxation times. For example, it has been shown that volume and enthalpy relaxation can follow different patterns of change on different time scales.[39] Such separation of time scales for different processes, which can be quantified by a 'decoupling index', increases with decreasing temperature.[40]

A special case of heterogeneities in the amorphous state is the possible heterogeneous distribution of water in an amorphous solid. Sugar/water systems represent examples of systems with such heterogeneity of water distribution. Sugar/water systems do not exhibit macroscopic heterogeneity (*e.g.* no liquid/liquid demixing); however, there is evidence of heterogeneity on a microscopic scale. In a pioneering study of sugar and water mobility,[41] the temperature dependence of spin-lattice relaxation times (T_1) for water, D_2O and sucrose was studied as a function of sucrose concentration. It was shown that, at high concentrations of sucrose (> 60 w%), T_1 for water molecules became uncoupled from that for sucrose, when the temperature was near or slightly higher than T_g, whereas at lower concentrations (< 50 w%), the mobilities of sucrose and water were coupled. The T_1 values for water in the concentrated solutions were similar to that for pure D_2O. Based on those observations, it was suggested[41] that sucrose molecules form a gel-like network whereas water molecules retain at least their rotational mobility. It was further suggested that the formation of such 'pools' of water molecules depends on temperature. For example, at a sucrose concentration of 80 w%, decoupling occurred below ~ 295 K, meaning that above 295 K, the T_1 values for sucrose and water indicated a homogeneous solution, which became heterogeneous as the temperature was decreased below 295 K. This conclusion concerning 'pockets' of water at high sucrose concentrations has recently been supported by results from molecular dynamics simulations.[42]

3.1 Significance in the Pharmaceutical Sciences

From a practical perspective, the most important manifestation of heterogeneities in relaxation times, as well as in composition in mixtures, is in their impact on chemical and physical instability, which depend on molecular mobility. Relationships between the distribution of relaxation times and the complex chemical kinetics for some amorphous solids have been suggested, but there is no direct evidence that correlations between relaxation (*e.g.* enthalpy relaxation) and chemical reactivity actually exist. It is also important to recognize, when dealing with mixtures such as drugs and excipients in the amorphous state, that mixing may be homogeneous or heterogeneous on a macroscopic level as well, with the latter case leading to phase separation into individual amorphous components. Such phase separation has serious implications when molecular dispersions are required, as in sugar–protein lyophilized products, or in hydrophobic drug–polymer dispersions designed to improve physical stability and dissolution rates. Thus, more attention needs to be paid to both microscopic and macroscopic tendencies toward heterogeneous structure in amorphous solids.

4 Relationships Between Local Structures of Amorphous and Crystalline States of the Same Material

A majority of studies on the local structure in amorphous solids have been performed with inorganic glasses and polymers. In general, it has been found that the structure of amorphous solids is not random and that there is local order, despite the absence of long-range order as in crystals. For network glasses such as silica, the local structures of glass and crystal are considered to be equivalent (*e.g.* see ref. 43). Experimental methods that have been applied to the study of inorganic glasses include X-ray spectroscopy (EXAFS and XANES),[44] neutron diffraction with analysis of the first sharp diffraction peak[45] and molecular dynamics simulations.[46] Such experiments yield a radial distribution function that gives information on the coordination number and the approximate distribution of bond angles, which appear to be similar for both amorphous and crystalline forms. For synthetic polymers, the prevailing view is that the configurations of macromolecular chains in polymer melts are random with a lack of appreciable order.[47] At temperatures near T_g, however, a polymeric random coil is considered to have a local order that is disrupted at higher temperatures (see also the discussion on T_{ll} in the earlier section on polyamorphism[21,48]).

Low molecular weight organic glasses have been studied to a much lesser extent. There are, however, several structure-related studies that have been performed with spectroscopic methods (IR and Raman). In particular, hydrogen-bonding patterns and molecular conformations have been compared for the glassy and crystalline states in several model systems. For example, FT-IR and FT-Raman measurements have been reported for the drug indomethacin, which contains a carboxylic acid group. There are two major polymorphic forms that appear to be monotropic with respect to each other, with the γ form being the most thermodynamically stable.[49] The amorphous form is easily created by quench-cooling of the melt. Indomethacin was one of the first low molecular weight pharmaceutical glasses shown to exhibit crystallization at temperatures well below its T_g, and, therefore, it has proven to be a useful model compound for studying various properties influenced by amorphous structure.[50,51] It has a relatively simple hydrogen-bonding pattern, which makes it convenient for the study of hydrogen bonding in the amorphous state.

From single-crystal X-ray diffraction and IR and Raman studies, it has been shown that different crystal forms of indomethacin have different hydrogen-bonding patterns.[49] The γ crystals consist of dimers, whereas in the α form (less energetically stable), a weaker dimer is combined with an additional hydrogen bond to a third molecule. Amorphous indomethacin retains the strong dimer structure, with a small portion of molecules also hydrogen bonding in the chain. Hence, the amorphous form appears to retain a local structure very similar to that of the crystal, particularly the more stable γ crystal.

In a recent study on the crystallization of indomethacin,[52] it has been shown that crystallization produces the γ form at temperatures $\leq T_g$, whereas at $T > T_g$, the α crystal form is the dominant species. Combining these results with those of the spectroscopic study[49] suggests that crystallization from the glass is facilitated by structural elements (dimers) that pre-exist in the amorphous state (see Figure 2). It is possible that heating the glass above T_g hinders the formation of the strong hydrogen bonds that characterize the γ form, thus favoring the formation of the less stable α form. Further work on these hydrogen-bonding patterns is required to support these interpretations.

In another set of studies, hydrogen bonding in crystalline and glassy sucrose has been compared.[53] Analysis of the second-derivative IR spectra of crystalline and amorphous sucrose revealed that in crystalline sucrose, there is a hydrogen bond-free OH group (corresponds to an IR band at 1436 nm) that is absent in the amorphous form. Another material with a difference in the hydrogen bonding between crystalline and amorphous forms is indigo, in which all NH groups formed hydrogen bonds in the crystalline state, whereas both free and bound NH groups were observed in the amorphous state.[54]

Another probe for local structure is the assessment of molecular conformation, since different conformations in different crystal structures of the same chemical

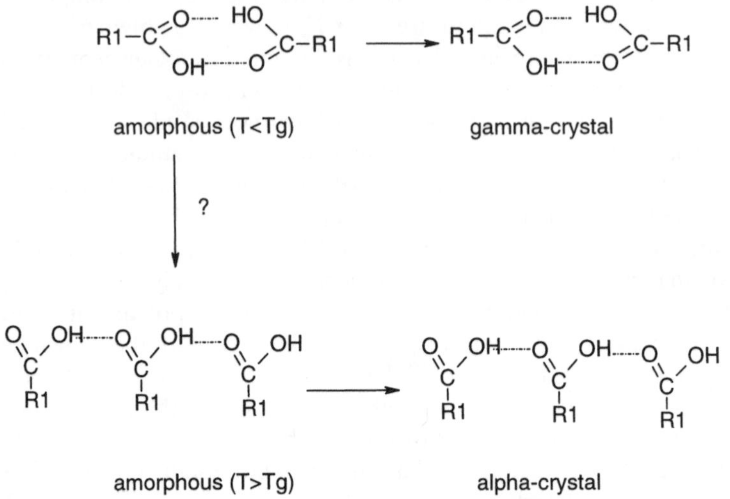

Figure 2 *Proposed structural relationships between amorphous indomethacin, above and below* T_g, *and its two crystal forms*

entity can occur, with one conformation being energetically favored over others.[55] For example, molecular conformations of butanenitrile (C_3H_7CN) in the crystalline, glassy and liquid states have been determined using X-ray diffraction and Raman spectroscopy.[56] In the crystalline state, the molecules adopt a *gauche* conformation, while in the liquid and amorphous solid states a significant distribution of molecular conformations occurs, as judged by a broadening of the 841 cm^{-1} Raman band. The energy differences between *trans* and *gauche* forms appear to be less than 1 kJ mol^{-1}. The fraction of *trans* conformation and the distribution of molecular conformations appear to be similar in the glassy and liquid states. It has been suggested[56] that the distribution of molecular conformations in the glass is static, *i.e.* conformational changes between *trans* and *gauche*, which occur in the liquid state, are 'frozen' (prevented) in the glassy state. When the amorphous sample crystallized, the ratio of Raman bands conforming to the *trans* and *gauche* forms changed significantly. This provided direct evidence that molecular conformations in the crystal and glass are different for butanenitrile. It was further suggested that the difference in conformations between the glassy and crystalline states helps to explain the relative stability of the amorphous form against crystallization.

Another material in which different molecular conformations in the glassy and crystalline states occur is methyl hydroxybutyrate.[57] The more stable conformer (*syn-syn-s-cis*) exists in the crystal, whereas both *gauche-skew-s-cis* and *syn-syn-s-cis* conformers are detected in the glass by Raman spectroscopy. In a glassy material, therefore, two conformers exist, while in the more stable crystal, a single conformer exists.[57] The difference in energy between the two conformers is 6.6 kJ mol^{-1}.

One may speculate that substances that have similar local structures in the glassy and crystalline states should crystallize relatively easily (even at $T < T_g$), whereas crystallization of compounds with different local structures will be hindered, particularly below T_g. Indeed, if a local structure in the glassy material is similar to that in the crystal, a minimal molecular rearrangement would be required for nucleation and crystal growth, whereas if the local structures are different, changes in such structure would require significant molecular mobility. The idea that a high resistance of pure melts and glasses of inorganic materials to crystallization is caused by the differences in short-range order relative to that in the crystal has been discussed.[58] It is natural, therefore, to believe that this should also be true for organic glasses. Unfortunately, there are only a few such studies reported on local structures in amorphous and crystalline organic substances.

4.1 Significance for Pharmaceutical Sciences

A few examples described above indicate that the physical stability of the glassy state against crystallization may be connected to a difference in local structures (*i.e.* hydrogen bonding and molecular conformation) between crystalline and glassy materials. If this is the case, the physical stability of pharmaceutical glasses may be predictable, based on a comparative spectroscopic study of crystalline and amorphous forms of the same substance. In particular, the well-known resistance of a majority of sugars to crystallization may be associated with their rich conformational diversity in the liquid state. For example, a conformational analysis of different alditols indicated that the lowest-energy conformers in solution and in the crystals of the slow-crystallizing alditols (*e.g.* sorbitol) are different.[59] It would be interesting, therefore, to compare the local structures of glasses that are formed by two similar sugar alcohols (*e.g.* sorbitol and mannitol), which are characterized by very different crystallization abilities, in order to verify this hypothesis. In addition, a local structure may have an impact on the chemical stability of amorphous solids, as well. Higher reactivity of amorphous materials, relative to that in the crystalline state, may be connected to a higher conformational variability, in addition to higher molecular mobility and free volume. More direct experimental studies will be needed to support this last suggestion.

5 Crystalline Mesophases

A typical crystalline material is characterized by translational, rotational and conformational order. Between crystalline and amorphous materials, there are three intermediate states of condensed matter which differ in degree of disorder, *i.e.* liquid crystals, orientationally disordered crystals (plastic crystals) and conformationally disordered crystals.[22]

Liquid crystalline materials lack conformational and three-dimensional translational order, but have orientational order. Liquid crystals can be formed by either rod-like or disc-shaped molecules. Depending on the properties of a particular molecule, it can form either a nematic phase that does not have any translational order or a smectic (lamellar) phase with one-dimensional translational order or a hexagonal phase with two-dimensional order.[60] In addition, there are several types of lamellar phases, *e.g.* the simplest smecticA phase, in which molecules are oriented perpendicularly to the lamellar plane, and the tilted smecticB phase.[60] Lamellar liquid crystalline phases of phospholipids may form a more ordered lamellar gel phase that is characterized by *trans/gauche* conformational order.[61]

Plastic crystals (or orientationally disordered crystals) have translational order but rotational and conformational disorder.[62] Plastic crystals are usually formed by almost spherical molecules and often have a simple crystal structure such as face-centered cubic. Molecular plastic crystals are usually easily deformed, whereas ionic orientationally disordered crystals are more rigid.

Conformationally disordered crystals have translational and rotational order but partial or complete conformational disorder, *i.e.* different conformers of a molecule are distributed randomly throughout the crystal lattice. Conformationally disordered crystals can be expected, when two or more conformers have similar overall molecular shapes. Conformational disorder (or conformational synmorphism, as described by Corradini[63]) is one of three cases of conformational diversity in a crystalline state which have been reported. In the other two cases (conformational polymorphism and conformational isomorphism), there is long-range conformational order in the crystal lattice, and these cases are not considered here.

In all three types of crystalline mesophases considered above, the low-temperature-stable phase is an ordered crystal. However, the mesophase may be supercooled and avoid transformation to the more ordered crystalline phase to form a thermodynamically unstable disordered solid in which molecular mobility ceases on the time scale of the experiment. Upon heating, such a 'glassy' crystalline mesophase undergoes a glass-like transition that may be associated with 'unfrozen' translational, rotational or conformational mobility. It should be mentioned that the term 'glassy crystal' has been widely used in the literature in relation to plastic crystalline materials with frozen-in orientational disorder.[64] Depending on the material, the characteristics of the glass transition in a crystalline mesophase and in an amorphous phase may be similar or different from each other. For example, the T_g of the liquid-crystalline glass of polyacryloyloxybenzoic acid is $\sim 50\,°C$ higher than the T_g of the amorphous form of the same material.[65] For ethanol, however, the T_g values for the plastic-crystal glass and amorphous glass are practically identical, but the ΔC_p at T_g is 30–50% higher for the amorphous glass than for the plastic-crystal glass.[66,67]

To recognize if any particular material forms a crystalline mesophase, one can use a combination of X-ray diffraction, thermal analysis (DSC, DMA, *etc.*) and optical microscopy. Crystalline mesophases usually show very few diffraction lines in powder diffraction. In particular, smectic liquid crystalline phases usually show a broad wide-angle diffraction peak and several low-angle diffraction peaks, whereas nematic phases do not show any sharp wide- or low-angle diffraction lines.[60] Plastic crystals[66] and conformationally disordered crystals[68] show a limited number of X-ray and neutron diffraction lines. Under a polarizing microscope, crystalline mesophases (except those exhibiting cubic structures) show birefringence. Liquid crystalline materials and molecular plastic crystals are soft and sometimes flow under their own weight (this is why they have been given their names). Usually, a glass transition can be detected for crystalline mesophases using thermal analysis methods. A summary of the properties of mesophases and a number of examples of crystalline mesophases have been given by Wunderlich.[23]

In pharmaceutical systems, the most well-known examples of crystalline mesophases are liquid crystalline materials that are widely used in cosmetics and drug delivery systems (*i.e.* liposomes, self-emulsifying systems),[69] whereas plastic crystals and conformationally disordered crystals have attracted much less attention. Plastic crystals may be relatively rare in pharmaceutical systems, because they are usually formed by molecules with near-spherical form, which is not necessarily common in pharmaceutical materials. Conformationally disordered crystals, however, may be more common. A lack of reports on conformationally disordered glasses in pharmaceutical systems may exist, because the concept itself is relatively new to the pharmaceutical area. For example, X-ray diffraction and DSC data reported for cefazolin sodium[70] suggest to us that those authors had observed a crystalline mesophase. The X-ray pattern for cefazolin sodium had two weak and relatively broad peaks at approximately 7 and 18° 2Θ. The presence of but a few crystalline lines in X-ray powder diffraction patterns generally may be interpreted as occurring because of the existence of a crystalline mesophase. Another possibility, however, is that the cefazolin sodium sample contained two phases, *i.e.* crystalline and amorphous. Reported DSC data[70] were inconclusive, because a possible T_g event may have been masked by another thermal event caused by water evaporation during the DSC scan. Indirect evidence that crystalline mesophases are quite common for pharmaceutical molecules arises from results of solid-state NMR and thermally stimulated current studies in which significant mobility has been observed in several crystals of pharmaceutical relevance at temperatures lower than their melting temperatures.[71,72] We would also like to mention that protein crystals exhibit some disorder[73] and hence can also be considered as crystalline mesophases.

Crystalline mesophases have a higher molecular mobility, and they are expected to have a poorer stability compared to regular crystals. An example of enhanced chemical reactivity in a crystalline mesophase formed by tetraglycine methyl ester (TGME) has recently been reported.[74] It was found that a disordered state, which was produced by milling and freeze-drying, was likely a crystalline mesophase. Observations of a glass transition on a DSC heating curve and a limited number of X-ray diffraction peaks were considered to be indicative of a crystalline mesophase. The lack of viscous flow, even at 165 °C which is much higher than the T_g (approximately 30 °C), suggested that this material was neither a liquid crystal nor a mixture of crystalline and amorphous phases. Chemical reactivity in the crystal mesophase of TGME was much greater than that in the crystal, although the reaction mechanism remained unchanged when the reaction temperature was 50 to 115 °C.[74,75] At a higher temperature of 165 °C, however, the reaction mechanism changed from a methyl transfer reaction, which was observed in the intact crystal, to a polycondensation reaction in the crystalline mesophase.

Partially or completely disordered states are usually created in typical pharmaceutical processing such as milling, and it is our opinion that, in some cases, a crystalline mesophase may arise as an intermediate state between highly crystalline and the amorphous material produced by milling. It has been suggested[76] that amorphization of molecular crystals occurs by a step mechanism through the formation of disordered states in which ordering of molecules exists only in some crystallographic directions; this description fits the definition of crystalline mesophases. It should be noted also that, if crystalline mesophases are created as a result of milling, this may introduce a further complication in the characterization of the extent of disorder in partially disordered pharmaceutical systems. A common way to characterize disorder in crystalline materials is to use a 'two-state model' in which the disordered crystal is considered to consist of crystalline and amorphous parts, and the extent of disorder is characterized using a calibration curve that is constructed from physical mixtures of crystalline and amorphous phases.[77] If crystalline mesophases contribute to disorder in crystals that are partially disordered by milling, the 'two-state crystalline/amorphous model' would not be applicable, and an already complicated task of characterizing disorder becomes even more challenging.

5.1 Significance for Pharmaceutical Sciences

Crystalline mesophases have attracted attention in different areas of science and technology. In the pharmaceutical area, however, only the liquid crystalline state has been recognized so far. We suggest that conformationally disordered crystals may be more common in pharmaceutical materials than might be anticipated. If this is the case, they should be considered to be undesirable because of a higher risk of chemical and physical instability and batch-to-batch variability which are usually also associated with amorphous materials.

6 Glass Transition in Confined Systems

While it is well-known that the size of a system has a significant effect on first-order phase transitions, *e.g.* melting, the impact of system size on T_g has only been recognized relatively recently. For example, changes in T_g have been reported for free-standing polymer films, polymer films on a solid support and small organic molecules in pores.[78,79] The T_g in a small (confined) system can be either higher or lower than the T_g in the bulk, depending on the balance between the effects of a free surface and of a support (substrate). For example, a significant decrease in T_g for a free-standing film of poly(methylmethacrylate) (PMMA) was observed for film thicknesses in the range of 1–10 nm, whereas for a PMMA film on native silicon oxide, an increase in T_g was observed.[78] It has been suggested that the effect of a free surface is to decrease T_g, whereas the effect of a supporting substrate is to increase T_g.[78,80] The increase in T_g can be explained as the result of an interaction (for example, hydrogen bonding) of the glass-former with the support.[78,80] A recent study provided experimental evidence for the existence of an immobile gel-like layer on a polymer/substrate surface with an attractive polymer/substrate interaction.[81]

Partially amorphous polymers represent a special case of confined systems. In semi-crystalline polymers, two T_g events have been reported to exist: a lower T_g (T_{gL}) observed at the same temperature as T_g in a completely amorphous polymer and a second T_g (T_{gU}) observed at a higher temperature; T_{gU} increased with increasing crystallinity.[82] The increase in T_g was considered to occur because of attractive interactions between polymer segments in the amorphous phase and the surfaces of polymer crystallites[83]

There is no universally accepted theoretical explanation for the effect of domain size on T_g. A possible explanation for the decrease in T_g in confined systems is that the characteristic length scale near T_g (*i.e.* size of a cooperative unit) decreases with the decreasing dimensions of the confining system.[84] Usually, significant changes in T_g are observed for such a system with dimensions of several nanometers, which is a dimension similar to that of a cooperative unit. The dimension of the characteristic length has been reported to be 1–3.5 nm for many polymers and small molecules.[85] However, a noticeable decrease in T_g was reported in much bigger pores of 73 nm radius for *o*-terphenyl and benzyl alcohol,[79] whereas the characteristic length for *o*-terphenyl was reported to be 3.0 nm.[85] Note also that, for polymer systems, a special mechanism for the decrease in T_g was suggested to be due to less chain entanglement in a confined system compared with that in the bulk.[86]

The magnitude of T_g depression depends on the size of the system. For example, the T_g of a free-standing polystyrene film decreased from approximately 360 to 300 K as the film thickness decreased from 60 to 30 nm.[80] Different empirical equations have been suggested to describe relationships between the dimensions of a system and its T_g. T_g was reported to be a linear function of the thickness of a free-standing polymer film.[80] Linearity between T_g of a small organic molecule and the inverse of pore radius,[80,84] and between T_g of a polymer film on a solid support and the inverse of film thickness[87] has been reported as well. In addition, the Michaelis–Menten relationship has been used to describe data on the size-dependence of T_g of polymer films.[88]

In addition to the effect of size, the magnitude of T_g depression in a confined space depends on the material. For example, depression of T_g for *o*-terphenyl was about two times greater than that for benzyl alcohol for the same pore size.[79] A strong dependence of surface T_g on molecular weight of polystyrene has been described.[89] T_g depression of almost 100 °C was observed for a PS polymer of MW 5000, whereas the T_g depression was only 20 °C for a PS polymer of MW $\sim 10^6$.[89]

Other properties of the glass transition may be affected by size as well. For example, the activation energy for a surface α-relaxation was several times smaller than that for the α-relaxation in the bulk.[89] The heat capacity change at

$T_g (\Delta C_{pg})$ in partially crystalline polymers appeared to be affected by the interaction of the surface amorphous polymer layer with the crystalline polymer, when the rigid part of the amorphous portion of the sample (*i.e.* the part with a higher T_g) contributed very little to ΔC_{pg}.[90] If ΔC_{pg} is used as a measure of the proportion of amorphous phase, it would give a significant underestimation of the fraction of amorphous phase (approximately $2 \times$, in the example reported in ref. 90). However, in another study, ΔC_{pg} appeared to be independent of pore size for small molecules.[79] Differences in the effects of confinement on ΔC_{pg} in those two studies may have been related to differences in the interaction between amorphous layer and support. In the first study,[90] the influence of confinement (*i.e.* smaller ΔC_{pg} in the confined system) could be explained by a strong interaction between the amorphous polymer surface layer and the crystal surface. In the second case,[79] when interaction between the sample and support did not appear to be significant, ΔC_{pg} was independent of the size of the system.

In pharmaceutical materials, an amorphous phase may be created (intentionally or by accident) in an initially highly crystalline material by different processing methods, such as milling or dry granulation. If the content of such an amorphous phase is relatively low, the dimensions of that phase would be expected to be small, and the properties of that amorphous phase might be different from those of the bulk amorphous phase. There have been no systematic studies of the properties of such a minor amorphous phase in pharmaceutical crystals. Data on sucrose crystallization[91] provided some indirect support, having shown that molecular mobility in an amorphous portion of a milled crystalline sample may be different from the mobility in the bulk amorphous phase. In the milled material, non-isothermal crystallization of sucrose was detected at 75 °C, whereas in a physical mixture of amorphous and crystalline material, it occurred at 115 °C.[91] The lower crystallization temperature may indicate that molecular mobility in the amorphous portion of the milled sample was higher (and T_g was lower) than that in the bulk amorphous material. However, another possible explanation is that crystallization of the milled sample was facilitated due to the presence of crystalline surfaces acting as seeds.

In addition, it should be mentioned that a common way to measure the fraction of amorphous phase in disordered pharmaceutical crystals is to use a calibration curve, when a physicochemical property (such as density, crystallization enthalpy, *etc.*) is measured for physical amorphous/crystalline mixtures with different fractions of amorphous phase. However, if the properties of the amorphous phase in the disordered material are different from those in the bulk amorphous phase, the use of physical mixtures to model a partially disordered crystal should be treated with caution.

In drug delivery systems, one of the possible ways to solubilize a drug is by entrapping it in liposomes. Drugs entrapped in liposomes may have properties different from those of the same drug in a bulk amorphous phase. For example, they may be more mobile (and less stable) than the amorphous drug in bulk. An example of the depression of T_g (corresponding to increased molecular mobility) in a model sucrose/phospholipid system was recently reported.[92] Sucrose entrapped in the liquid crystalline phase of a phospholipid had a T_g as much as

20 °C lower than that for sucrose in bulk. Hence, a drug in liposome-based systems may have a higher molecular mobility and a lower stability than the bulk drug.

We would like to mention also that T_g in confined systems may be relevant to the still-controversial topic of the thermomechanical properties of frozen aqueous solutions of carbohydrates and other organic compounds. There are two T_g-like thermal events that are usually observed in DSC heating curves for frozen aqueous solutions.[6,93] It is possible that the first, lower temperature transition (which is sometimes referred as T_g'' [93]) corresponds to the glass transition in the bulk phase of freeze-concentrated solution, whereas the second, higher temperature transition (T_g' [6]) results from an overlap of the glass transition on the solution/ice interface and the onset of ice dissolution. Future experimental studies will need to be performed to test this suggestion.

6.1 Significance for Pharmaceutical Science

There are three common types of pharmaceutical systems in which system size may have a significant influence on molecular mobility: the small amorphous portion of initially crystalline materials that had been subjected to common pharmaceutical operations such as milling or dry-granulation; drug entrapped in liposomes; and freeze-concentrated solutions at an ice/solution interface. In all three types of systems, properties of the confined amorphous phase may be different (*e.g.* higher mobility) than those in the bulk amorphous material. However, more direct experimental evidence is needed to evaluate the practical importance of this confinement effect on molecular mobility in real pharmaceutical systems.

Acknowledgements

The authors express their gratitude to Drs F. Franks and T. Auffret for the suggestion to look into the structural aspects of amorphous states, and to Dr M.J. Pikal for making unpublished results available. ES thanks Pfizer colleagues for helpful discussions. GZ acknowledges the financial support of the Purdue/Wisconsin Program on the Physical and Chemical Stability of Pharmaceutical Solids.

References

1. B.C. Hancock and G. Zografi, *J. Pharm. Sci.*, 1997, **86**, 1.
2. L. Yu, *Adv. Drug Delivery Rev.*, 2001, **48**, 27.
3. D.Q.M. Craig, P.G. Royall, V.L. Kett and M.L. Hopton, *Int. J. Pharm.*, 1999, **179**, 179.
4. B.C. Hancock, S.L. Shamblin and G. Zografi, *Pharm. Res.*, 1995, **12**, 799.
5. S.L. Shamblin, X. Tang, L. Chang, B.C. Hancock and M.J. Pikal, *J. Phys. Chem. B*, 1999, **103**, 4113.
6. L. Slade and H. Levine, *Crit. Rev. Food Sci. Nutr.*, 1991, **30**, 115.

7. C.A. Angell, R.D. Bressel, J.L. Green, H. Kanno, M. Oguni and E.J. Sare, *J. Food Eng.*, 1994, **22**, 115.
8. E.Y. Shalaev and G. Zografi, *J. Pharm. Sci.*, 1996, **85**, 1137.
9. Y. Song, R.L. Schowen, R.T. Borchardt and E.M. Topp, *J. Pharm. Sci.*, 2001, **90**, 141.
10. Y. Guo, S.R. Byrn and G. Zografi, *Pharm. Res.*, 2000, **17**, 930.
11. E.Y. Shalaev, Q. Lu, M. Shalaeva and G. Zografi, *Pharm. Res.*, 2000, **17**, 366.
12. M.D. Cohen and G.M.J. Schmidt, *J. Chem. Soc.*, 1966, 164.
13. A.C. Mitus, A.Z. Patashinskii and B.I. Shimilo, *Phys. Lett. A*, 1985, **113**, 41.
14. A.C. Mitus and A.Z. Patashinskii, *Acta Phys. Pol. A*, 1988, **74**, 779.
15. G. Granzese, G. Malescio, A. Skibinsky, S.V. Buldyrev and H.E. Stanley, *Nature*, 2001, **409**, 692.
16. C.A. Angell, *Science*, 1995, **267**, 1924.
17. O. Mishima, L.D. Calvert and E. Walley, *Nature*, 1984, **310**, 393.
18. M. Grimsditch, *Phys. Rev. Lett.*, 1984, **52**, 2379.
19. M. van Thiel and F.H. Ree, *Phys. Rev. B*, 1993, **48**, 3591.
20. S. Aasland and P.F. McMillan, *Nature*, 1994, **369**, 633.
21. R.F. Boyer, *Plastic Engineering*, 1992, **25**, 1.
22. I. Cohen, A. Ha, X. Zhao, M. Lee, T. Fischer, M.J. Stouse, and D. Kivelson, *J. Phys. Chem.*, 1996, **100**, 8518.
23. B. Wunderlich, *Thermochim. Acta*, 1999, **37–52**, 340.
24. E.Y. Shalaev and P.L. Steponkus, *Biochim. Biophys. Acta*, 2001, **1514**, 100.
25. M. Pikal, A.L. Lukes, J.E. Lang and K. Gaines, *J. Pharm. Sci.*, 1978, **67**, 767.
26. I. Tsukushi, O. Yamamuro and H. Suga, *J. Non-Cryst. Solids*, 1994, **175**, 187.
27. A. Saleki-Gerhardt, J.G. Stowell, S.R. Byrn and G. Zografi, *J. Pharm. Sci.*, 1995, **84**, 318.
28. M.J. Pikal, personal communication, 1995.
29. B.C. Hancock, S.L. Shamblin and G. Zografi, *Pharm. Res.*, 1995, **12**, 799.
30. I.M. Hodge, *J. Non-Cryst. Solids*, 1994, **169**, 211.
31. R. Richert, *J. Non-Cryst. Solids*, 1994, **172–174**, 209.
32. M.D. Ediger and J.L. Skinner, *Science*, 2001, **292**, 233.
33. L.A. Deschenes and D.A. Vanden Bout, *Science*, 2001, **292**, 255.
34. G. Adam and J.H. Gibbs, *J. Chem. Phys.*, 1965, **43**, 139.
35. E.W. Fischer, Ch. Becker, J.-U. Haenah and G. Meier, *Progr. Colloid. & Polymer Sci.*, 1989, **80**, 198.
36. S.L. Shamblin, B.C. Hancock, Y. Dupuis and M.J. Pikal, *J. Pharm. Sci.*, 2000, **89**, 417.
37. Ö. Almarsson, R.A. Seburg, D. Godshall, E.W. Tsai and M.J. Kaufman, *Tetrahedron*, 2000, **56**, 6877.
38. L. Streefland, A.D. Auffret and F. Franks, *Pharm. Res.*, 1998, **15**, 843.
39. S. Takahara, M. Ishikawa, O. Yamamuro and T. Matsuo, *J. Phys. Chem. B*, 1999, **103**, 792.
40. C.A. Angell, *J. Non-Cryst. Solids*, 1991, **131–133**, 13.
41. D. Girlich and H.Y.-D. Lüdemann, *Z. Naturforsch*, 1994, **49c**, 250.
42. C.J. Roberts and P.G. Debenedetti, *J. Phys. Chem. B*, 1999, **103**, 7308.
43. E.F. Riebling, 'Phase Diagrams. Materials Science and Technology', A.M. Alper (ed.), Academic Press, 1970, v. III, 253.
44. K.J. Rao and B.G. Rao, *Bull. Mater. Sci.*, 1985, **7**, 353.
45. P.S. Salmon, *Proc. R. Soc. Lond. A*, 1994, **445**, 351.
46. S. Susman, K.J. Volin, D.L. Price, M. Grimsditch, J.P Rino, K.K. Kalia and P. Vashishta, *Phys. Rev. B*, 1991, **43**, 1194.
47. P.L. Flory, *Macromol. Chem.*, 1973, **8**, 1.

48. R.F. Boyer, *J. Macromol. Sci. Phys. B*, 1976, **12**, 253.
49. L. Taylor and G. Zografi, *Pharm. Res.*, 1997, **14**, 1693.
50. V. Andronis and G. Zografi, *Pharm. Res.*, 1997, **14**, 410.
51. V. Andronis and G. Zografi, *Pharm. Res.*, 1998, **15**, 835.
52. V. Andronis and G. Zografi, *J. Non-Cryst. Solids*, 2000, **271**, 236.
53. J.J. Seyer, P.E. Luner and M.S. Kemper, *J. Pharm. Sci.*, 2000, **89**, 1305.
54. A.R. Monahan and J.E. Kuder, *J. Org. Chem.*, 1972, **37**, 4182.
55. J. Bernstein, 'Organic Solid State Chemistry', C.R. Desaraju (ed.), Elsevier, Amsterdam, 1987, 471.
56. K. Ishii, H. Nakayama, K. Koyama, Y. Yokoyama and Y. Ohashi, *Bull. Chem. Soc. Jpn.*, 1997, **70**, 2085.
57. S. Jarmelo, T.M.R. Maria, M.L.P. Leitao and R. Fausto, *Phys. Chem. Chem. Phys.*, 2001, 3, 387.
58. D. Turnbull, *Annals N Y Acad. Sci.*, 1976, **279**, 185.
59. L. Yu, S.M. Reutzel-Edens and C.A. Mitchell, *Organic Proc. Res. Dev.*, 2000, **4**, 396.
60. J.M. Seddon, 'Hanb. Liq. Cryst.', 1998, **1**, 635.
61. J.F. Nagle, *Ann. Rev. Phys. Chem.*, 1980, **31**, 157.
62. R.M. Lynden-Bell and K.H. Michel, *Rev. Modern Phys.*, 1994, **66**, 721; M. Meyer and O.H. Duparc, 'Advances in Solid-State Chemistry', C.R.A. Catlow (ed.), 3, JAI Press, London, 1993, 1.
63. P. Corradini, *Chim. Ind.* (*Milan*), 1973, **55**, 122, cited in J. Bernstein, 'Organic Solid State Chemistry', Elsevier, 1987, 471.
64. H. Suga and S. Seki, *Faraday Discuss.*, 1980, **69**, 221.
65. B. Wunderlich and J. Grebowica, *Adv. Polym. Sci.*, 1984, **1**, 60.
66. F.J. Bernejo, H.F. Fischer, M.A. Ramos, A. de Andrés, J. Dawidowski and R. Rayos, 'Complex Behavior of Glassy Systems', J.J. Rubi and C. Pérez-Vicente (eds.), Springer, Berlin, 1997, 44.
67. G.P. Johari, *Phase Transitions*, 1985, **5**, 277.
68. B. Wunderlich, M. Moeller, J. Grebowicz and H. Baur, 'Conformational Motion and Disorder in Low and High Molecular Mass Crystals', Springer-Verlag, Berlin, 1988, 137.
69. J.C. Shah, Y. Sadhale and D.M. Chilukuri, *Adv. Drug Del. Rev.*, 2001, **47**, 229.
70. S. Furlanetto, P. Mura, P. Gratteri and S. Pinzati, *Drug Dev. Ind. Pharm.*, 1994, **20**, 2299.
71. G.M. Venkatesh, M.E. Barnett, C. Owusu-Fordjour and M. Galop, *Pharm. Res.*, 2001, **18**, 98.
72. D.C. Chatfield and S.E. Wong, *J. Phys. Chem. B*, 2000, **104**, 11342.
73. D.L.D. Caspar and J. Badger, *Curr. Opinion Struct. Biol.*, 1991, **1**, 877.
74. E.Y. Shalaev, M. Shalaeva and G. Zografi, *J. Pharm. Sci.*, 2001, submitted.
75. E.Y. Shalaev, M. Shalaeva, S.R. Byrn and G. Zografi, *Intern. J. Pharm.*, 1997, **152**, 75.
76. T.P. Shakhtshneider and V.V. Boldyrev, 'Reactivity of Molecular Solids', E. Boldyreva and V.V. Boldyrev (eds.), John Wiley & Sons, 1999, 271.
77. R. Suryanarayanan and A.G. Mitchell, *Int. J. Pharm.*, 1985, **24**, 1.
78. J.K. Keddie, R.A.L. Jones and R.A. Cory, *Faraday Discuss.*, 1994, **98**, 219.
79. C.L. Jackson and G.B. McKenna, *J. Non-Cryst. Solids*, 1991, **131–133**, 221.
80. J.A. Forrest, K. Dalnoki-Veress, J.R. Stevens and J.R. Dutcher, *Phys. Rev. Lett.*, 1996, **77**, 2002.
81. Y. Zhang, S. Ge, B. Tang, M.H. Rafailovich, J.C. Sokolov, D.G. Peiffer, Z. Li, A.J. Dias, K.O. McElrath, S.K. Satija, M.Y. Lin and D. Nguyen, *Langmuir*, 2001, **17**, 4437.
82. R.F. Boyer, *J. Macromol. Sci. – Phys.*, 1973, **B8**, 503.

83. S.M. Aharoni, *Polym. Adv. Technol.*, 1998, **9**, 169.
84. E. Hempel, S. Vieweg, A. Huwe, K. Otto, C. Schick, and E. Donth, *J. Phys. IV France*, 2000, **10**, 7.
85. E. Hempel, G. Hempel, A. Hensel, C. Schick and E. Donth, *J. Phys. Chem.*, 2000, **104**, 2460.
86. G. Reiter, *Europhys. Lett.*, 1993, **23**, 579.
87. J.K. Keddie, R.A.L. Jones and R.A. Cory, *Europhys. Lett.*, 1994, **27**, 59.
88. J.H. Kim, J. Jang and W.-C. Zin, *Langmuir*, 2000, **16**, 4064.
89. K. Tanaka, A. Takahara and T. Kajiama, *Macromolecules*, 2000, **33**, 7588.
90. H. Suzuki, J. Grebowicz and B. Wunderlich, *Brit. Polym. J.*, 1985, **17**, 1.
91. A. Saleki-Gerhardt, C. Ahlneck and G. Zografi, *Int. J. Pharm.*, 1994, **101**, 237.
92. E. Shalaev and P.L. Steponkus, *Langmuir*, 2001, **17**, 5137.
93. E.Y. Shalaev and F. Franks, *J. Chem. Soc., Faraday Trans.*, 1995, **91**, 1511.

Analytical Model for the Prediction of Glass Transition Temperature of Food Systems

Vinh Truong*, Bhesh R. Bhandari*, Tony Howes**
and Benu Adhikari**

* FOOD SCIENCE AND TECHNOLOGY, SCHOOL OF LAND AND
FOOD SCIENCES, UNIVERSITY OF QUEENSLAND, GATTON,
QLD 4345, AUSTRALIA
** DEPARTMENT OF CHEMICAL ENGINEERING, UNIVERSITY OF
QUEENSLAND, ST. LUCIA, QLD 4072, AUSTRALIA

1 Introduction

Glass transition temperature (T_g) is well known as an important parameter in food preservation and processing.[1] One of the interesting problems in this area is how to predict the T_g of a mixture from the known T_gs and fractions of individual components. There are some equations that have been used widely in research on foods as well as synthetic polymers.

1.1 Gordon–Taylor Equation (1952)

The Gordon–Taylor equation[2] was originally developed for polymer blends and was based on expansion coefficients (β) and the assumption of ideal volume mixing. The Gordon–Taylor equation has been rewritten in the following form:[3–5]

$$T_g = \frac{x_1 T_{g1} + K x_2 T_{g2}}{x_1 + K x_2} \tag{1}$$

For an *anhydrous system*, subscript 1 is component 1 and subscript 2 is component 2. For an *aqueous system*, subscript 1 is the solid component 1 (or a dry mixture of various solid components), subscript 2 is water, x is the mole fraction (or weight fraction) and K is an arithmetic average of K values that are obtained by solving Equation 1 for each binary (component 1:component 2) system.[5,6] This equation has been applied to polymer blends such as suc-

31

rose–maltodextrin,[7] to glucose–fructose[6] and to solute–diluent systems such as sucrose–water and fructose–water.[5,7–10]

1.2 Mandelkern, Martin and Quinn Equation (1957)[11]

$$\frac{1}{T_g} = \frac{1}{x_1 + Rx_2}\left(\frac{x_1}{T_{g1}} + \frac{Rx_2}{T_{g2}}\right) \tag{2}$$

If $R = K(T_{g2}/T_{g1})$, this equation is identical to the Gordon–Taylor equation (1). When $R = 1$, this relationship becomes Equation 3, proposed by Fox (1956).[12]

$$\frac{1}{T_g} = \frac{x_1}{T_{g1}} + \frac{x_2}{T_{g2}} \tag{3}$$

1.3 Free Volume Theory Equation

Kelley and Bueche[13] gave an equation for estimating K in the Gordon–Taylor equation, as follows:

$$K = \Delta\beta_{g2}/\Delta\beta_{g1} \tag{4}$$

where β is the free volume expansion coefficient: $\beta = (1/V)(\partial V/\partial T)_p$, and V and T are specific volume and temperature of the component, respectively. The β used here is different from that in the original Gordon–Taylor equation by an amount $(1/V)$.

1.4 Couchman–Karasz Equation (1978)[14]

$$T_g = \frac{x_1 \Delta C_{p1} T_{g1} + x_2 \Delta C_{p2} T_{g2}}{x_1 \Delta C_{p1} + x_2 \Delta C_{p2}} \tag{5}$$

In the Couchman–Karasz equation,[14] ΔC_{pi} is the change in heat capacity of component i between its liquid-like and glassy states. This equation was developed based on classical thermodynamic theory, with an assumption that the system is purely conformational. It is identical to the Gordon–Taylor equation when $K = \Delta C_{p2}/\Delta C_{p1}$.

1.5 Couchman–Karasz Equation for Ternary Systems

$$T_g = \frac{x_1 \Delta C_{p1} T_{g1} + x_2 \Delta C_{p2} T_{g2} + x_3 \Delta C_{p3} T_{g3}}{x_1 \Delta C_{p1} + x_2 \Delta C_{p2} + x_3 \Delta C_{p3}} \tag{6}$$

This is an expansion of Equation 5, where in an aqueous system, the subscripts

1, 2 and 3 refer to components 1, 2 and water, respectively. For polymer blends of three components, the subscript 3 refers to the component of lowest T_g. Equation 6 can be rewritten in the following form, by division of numerator and denominator by ΔC_{p1}:

$$T_g = \frac{x_1 T_{g1} + x_2 K_{12} T_{g2} + x_3 K_{13} T_{g3}}{x_1 + x_2 K_{12} + x_3 K_{13}} \tag{7}$$

The constants, K_{12} and K_{13}, can be estimated by the Simha–Boyer rule,[15] where ρ is the density of each component:

$$K_{12} = \rho_1 T_{g1} / \rho_2 T_{g2} \text{ and } K_{13} = \rho_1 T_{g1} / \rho_3 T_{g3} \tag{8}$$

The Gordon–Taylor equation is popular and reliable for many binary food systems. However, application of this equation (using either binary form (1) or expansion from (7)) to ternary or higher mixtures requires a large number of experiments. For instance, using Equation 1 for prediction of a ternary mixture, one component should be selected as a plasticizer and the other components are considered as solids. For each ratio of solid components, a set of experiments should be done to estimate a constant, K. Through variation of the ratio of solid components, a set of constants, K, will be obtained empirically. Eventually, this procedure requires a large number of experiments. Alternatively, one can use Equation 7 for prediction of T_g for a ternary mixture. In turn, many experiments are required to obtain good estimates of K_{12} and K_{13}, using a complicated non-linear regression procedure. Other models are identical to the Gordon–Taylor equation. Expansion of the Couchman–Karasz equation appears to be a good model in terms of requiring a smaller number of experiments. It requires only the data for individual components (T_g and ΔC_p). Unfortunately, it works well only for ideal mixing systems. In this study, we propose a model for prediction of T_g for ternary mixtures, based on known relationships for the constituent binary mixtures. Experiments have been conducted to compare the predicted values from the proposed model with those from others such as Gordon–Taylor, Couchman–Karasz and an empirical polynomial equation. Further expansion of the proposed model to quaternary and higher mixtures is discussed.

2 Interaction Factor for Binary Mixture

Before introduction of the interaction factors for non-ideal mixing systems, we derive the equations for ideal mixing systems based on the macroscopic properties of the vitreous system, in order to show the equivalence between the original Gordon–Taylor and Couchman–Karasz equations. Figure 1 is an enthalpy diagram illustrating the mixing of two components with mass fractions X_1 and X_2. Based on the assumption that the mixing is ideal, *i.e.* there is no interaction between the two components, the enthalpy of the mixture in the glassy state (H_G) can be calculated from the enthalpy of the components (H_{G1} and H_{G2}):

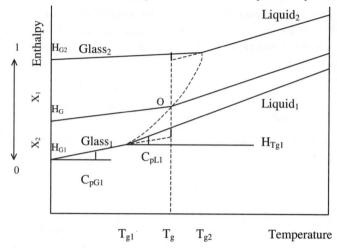

Figure 1 *Enthalpy diagram showing the mixing properties of the two components at the glass transition*

$$H_G = X_1 H_{G1} + X_2 H_{G2} \tag{9}$$

or

$$H_G - H_{G1} = X_1 H_{G1} + X_2 H_{G2} - (X_1 + X_2) H_{G1} \tag{10}$$

and then

$$X_2 = (H_G - H_{G1})/(H_{G2} - H_{G1}) \tag{11}$$

Similarly,

$$X_1 = (H_{G2} - H_G)/(H_{G2} - H_{G1}) \tag{12}$$

Finally,

$$X_2/X_1 = (H_G - H_{G1})/(H_{G2} - H_G) \tag{13}$$

With the assumption of an ideal mixing condition in the liquid state, and similarly in the glassy state, we have:

$$X_2/X_1 = (H_L - H_{L1})/(H_{L2} - H_L) \tag{14}$$

At T_g of the mixture, $H_G = H_L$; therefore, by subtracting the numerator and denominator of the right-hand side of Equation 13 from 14, we get:

$$X_2/X_1 = (H_{L1} - H_{G1})/(H_{G2} - H_{L2}) \tag{15}$$

Conversion of the enthalpy in Equation 15 to specific heat ($C_p = dH/dT$) in the transition region yields:

$$\frac{X_2}{X_1} = \frac{\displaystyle\int_{T_{g1}}^{T_g} (C_{PL1} - C_{PG1}) dT}{\displaystyle\int_{T_{g2}}^{T_g} (C_{PG2} - C_{PL2}) dT} \tag{16}$$

where T_{g1}, T_{g2}, and T_g are the glass transition temperatures of component 1, component 2 and the mixture, respectively. With the assumption that C_{pG} and C_{pL} are independent of temperature in the transition region, Equation 16 becomes:

$$\frac{X_2}{X_1} = \frac{\Delta C_{P1}}{\Delta C_{P2}} \frac{T_g - T_{g1}}{T_{g2} - T_g} = \frac{\Delta C_{P1}}{\Delta C_{P2}} \alpha \tag{17}$$

From Equation 17, the Couchman–Karasz equation (5) can easily be obtained. By replacing specific heat, C_p, with the expansion coefficient, β ($\beta = dV/dT$), we obtain the original Gordon–Taylor equation (the enthalpy, H, is replaced by specific volume, V). This approach can be applied to other macroscopic properties such as refractive index or electrical conductivity for ideal systems. However, there are many binary systems that show interactions during mixing; for instance, the existence of hydrogen-bond interactions in sucrose–dextran blends,[16] or ionic interactions in polymer–diluent–salt systems.[17] For non-ideal mixing systems in which the interactions are not too strong, we introduce an interaction factor, ε, into Equation 17 to obtain the following expression:

$$\frac{X_2}{X_1} = \varepsilon \frac{\Delta C_{P1}}{\Delta C_{P2}} \alpha = K_{21} \alpha \tag{18}$$

In this paper, predictions using ΔC_p are referred to as derived from the Couchman–Karasz equation, and predictions that use K are referred to as derived from the Gordon–Taylor equation. It should be noted that $K_{21} = 1/K_{12}$. The empirical constant, K, can be obtained by a least-squares method, *i.e.* solving Equation 19:

$$\frac{d}{dK} \sum_{i=1}^{n} \left(\frac{x_{1i} T_{g1} + K x_{2i} T_{g2}}{x_{1i} + K x_{2i}} - T_{gmi} \right)^2 = 0 \tag{19}$$

where n is the number of data points, T_{gmi} is the measured T_g of the model mixture i, x_{1i} and x_{2i} are the fractions of components 1 and 2, respectively, of the model mixture i, and T_{g1} and T_{g2} are the T_gs of components 1 and 2, respectively. The interaction factor, ε, is then calculated from Equation 20:

$$\varepsilon = K_{21}/(\Delta C_{p1}/\Delta C_{p2}) \tag{20}$$

$$= K_{12}/(\Delta C_{p2}/\Delta C_{p1})$$

If the interaction factor ε is close to 1, the system is considered to be ideal mixing. It is non-ideal when ε is far from 1. Thus, ε is an indicator of the non-ideal mixing behavior of binary mixtures.

3 Derivation of the Equation for a Ternary Mixture

A mixture of sucrose, glucose and fructose is considered as an example in this section. Figure 2 shows a T_g diagram for a ternary sucrose–glucose–fructose (S–G–F) system. The T_gs for sucrose–fructose (T_{gSF}), glucose–fructose (T_{gGF}) and the ternary mixture (T_{gSGF}) are plotted against fructose fraction. At any point X_F (mass fraction of fructose), $Y_1 = T_{gSGF} - T_{gGF}$, $Y_2 = T_{gSG} - T_{gSGF}$ and $\alpha = Y_1/Y_2$.

$$T_{gSGF} = \frac{T_{gGF} + \alpha T_{gSF}}{1 + \alpha} \tag{21}$$

For the case where binary mixtures G–F and S–F follow the Gordon–Taylor equation, if the third-order interactions between the three components are negligible, then sucrose and glucose are considered to be plasticized separately by fructose,[18] and Equation 21 becomes:

$$T_{gSGF} = \frac{1}{1+\alpha}\left[\frac{(1-X_F)T_{gG} + K_{GF}X_F T_{gF}}{(1-X_F) + K_{GF}X_F} + \alpha\frac{(1-X_F)T_{gS} + K_{SF}X_F T_{gF}}{(1-X_F) + K_{SF}X_F}\right] \tag{22}$$

This equation is assumed to be applicable for any fraction of fructose. When $X_F = 0$, Equation 22 reduces to that for the binary mixture of sucrose and glucose:

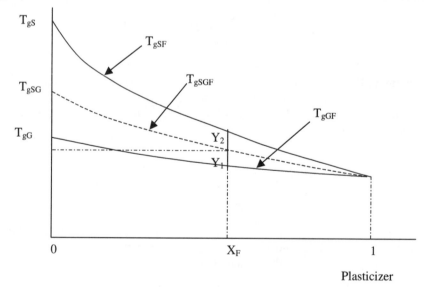

Figure 2 *Diagram for derivation of the equation to determine the T_g of a ternary mixture of sucrose–glucose–fructose*

$$T_{gSG} = \frac{T_{gG} + \alpha T_{gS}}{1 + \alpha}$$ (23)

or

$$\alpha = \frac{T_{gSG} - T_{gG}}{T_{gS} - T_{gSG}}$$ (24)

Again, if third-order interactions can be ignored, *i.e.* sucrose is plasticized separately by glucose, then α in Equation 24 is equivalent to α in Equation 17, and from Equation 18, we obtain the following expression for α:

$$\alpha = \frac{1}{K_{SG}} \frac{X_S}{X_G}$$ (25)

Therefore, when all three binary mixtures, sucrose–fructose, glucose–fructose and sucrose–glucose, follow the Gordon–Taylor equation, by substitution of α in Equation 25 into Equation 22, a final form of the equation for prediction of the T_g of a ternary mixture is:

$$T_{gSGF} = \frac{K_{SG}X_G}{K_{SG}X_G + X_S} \frac{(1 - X_F)T_{gG} + K_{GF}X_F T_{gF}}{1 - X_F + K_{GF}X_F}$$

$$+ \frac{X_S}{K_{SG}X_G + X_S} \frac{(1 - X_F)T_{gS} + K_{SF}X_F T_{gF}}{1 - X_F + K_{SF}X_F}$$ (26)

where X_S, X_G and X_F are the mass fractions of sucrose, glucose and fructose, respectively, and K_{SG}, K_{SF} and K_{GF} are the K values in the Gordon–Taylor equation for binary mixtures of sucrose–glucose, sucrose–fructose and glucose–fructose, respectively. These constants are determined empirically, using a least-squares method, from data for binary mixtures, as described previously. The interaction factors in this equation are inferred from each binary mixture, as shown in Equation 20.

4 Experimental Validation of the Ternary Equation

4.1 Preparation of Amorphous Materials

Food-grade, solid, crystalline sucrose (Bundaberg Sugar LTD, Australia), fructose (ADM Corn Processing, USA), glucose (Boots Healthcare, Australia) and citric acid (ADM Australia) were used in the experiments. For anhydrous amorphous systems, solutions of food components at 50% solids concentration were frozen at $-18\,°C$ for 20 hrs and then freeze-dried at 30 Pa for four days. Heat was applied at a temperature of $60\,°C$ during the last two days of drying. For aqueous systems with moisture contents less than 10% by weight on a dry basis (db), the freeze-dried samples were placed in desiccators and equilibrated over different saturated salt solutions. For aqueous systems with moisture contents greater than 15% db, the sample was heated in an oven until a clear solution was obtained.

4.2 Experimental Design

Anhydrous binary systems of sucrose–glucose, sucrose–fructose and glu-cose–fructose (7/1, 6/2, 5/3, 4/4, 3/5, 2/6 and 1/7 (w/w fractions)) and glu-cose–citric acid (4/1, 3/2, 2/3 and 1/4 (w/w)) and ternary mixtures of suc-rose–glucose–fructose (5/3, 4/4 and 3/5 (w/w) of sucrose:glucose, and for each sucrose:glucose ratio, the fructose/(sucrose–glucose) was 1/5, 2/4, 3/3, 4/2 or 5/1) were studied. For aqueous systems, glucose–citric acid–water (4/1, 3/2, 2/3 and 1/4 (w/w) of glucose:citric acid) and sucrose–fructose–water (7/1, 5/3, 2/6 and 1/7 (w/w) of sucrose:fructose) mixtures were studied. For glucose–citric acid–water mixtures, six levels of moisture were chosen. The moisture ranges were 2.7 to 29% db, 2.7 to 26% db, 3.1 to 22.5% db and 1 to 20% db for glucose:citric acid ratios of 4/1, 3/2, 2/3 and 1/4, respectively. The moisture range was changed due to the limited capability of our instrumental cooling system. Three to five levels of moisture below 9% db were selected for sucrose–fructose–water mixtures.

4.3 Differential Scanning Calorimetry (DSC)

Differential scanning calorimetry (Perkin–Elmer Pyris 1 DSC equipped with an Intracooler II) was used to determine T_g. The purge gas used was dry nitrogen. Indium and zinc (Perkin–Elmer standards) were used for temperature and heat-flow calibration. Hermetically sealed, 50 μL aluminium sample pans (Perkin–Elmer) were used in all experiments. Sample weights of 10–30 mg were used. A typical DSC scanning rate of 10 °C min^{-1} [8,18] was selected for T_g measurements for all samples. For anhydrous systems, all samples were scanned twice. The first scan was carried out from -10 to 130 °C, in order to completely remove the effect of thermal history. The second scan was carried out from -10 °C to $T_g + 20$ °C. The midpoint of the transition observed in the second scan was reported as T_g. For aqueous systems, samples were first cooled from 25 °C to -65 °C and then reheated immediately to 25 °C for determination of T_g.

5 Results and Discussion

5.1 Data Collection

The anhydrous samples obtained from the freeze-drying procedure in this study were assumed to contain no moisture, as a drying temperature of 60 °C was applied for the final two days of drying. Thus, the effect of moisture on T_g could be neglected. Partial crystallization in some samples was observed for binary mixtures of glucose–fructose (5/3, 6/2, 7/1), sucrose–fructose (5/3, 6/2, 7/1), sucrose–glucose (7/1, 1/7) and glucose–citric acid (4/1, 1/4). These samples were first melted completely and then cooled to -10 °C, prior to a second scan. Interestingly, all the samples of ternary sugar mixtures were completely amorphous. The completely amorphous samples of binary and ternary mixtures were scanned twice, as described in the DSC method section. The maximum temperature of 130 °C in the first scan was selected based on our experience with fructose and glucose. Heating of a fructose glass to 130 °C (or a glucose glass to 80 °C) in a DSC can completely remove the effect of thermal history on T_g, due to physical aging, without causing degradation. Thus, a maximum temperature of 130 °C was used for the first scan of mixtures containing fructose or glucose. Even though partial crystallization occurred in some binary mixtures, the constants, K, derived from data for those binary mixtures worked well for ternary mixtures, as shown below.

5.2 Interaction Factors for Binary Mixtures

The interaction factors calculated from Equation 20 are shown in Table 1 for different binary mixtures. The K values in the Gordon–Taylor equation for binary mixtures were calculated using Equation 19. The predicted T_g values, obtained using the Gordon–Taylor equation, agreed well with the experimental data ($R^2 = 0.96$–0.99), as shown in Table 2. However, the prediction was not so good when the glucose fraction was high. For the cases of sucrose–glucose and sucrose–fructose, the difference in calculated T_g values between the ideal mixing model ($\Delta C_{p1}/\Delta C_{p2}$) and the non-ideal mixing model (K) was only 3% (Table 1). Therefore, these two model mixtures were considered as ideal mixing systems. The other model mixtures (glucose–fructose, glucose–citric acid, glucose–water and citric acid–water) were non-ideal mixing systems, for which the interaction factors ε varied from 1.20 to 2.64. The differences between the experimental data and predicted values are shown in Figure 3.

5.3 Anhydrous Ternary Mixtures

The experimental T_g data for anhydrous ternary mixtures of sugars were compared with the predicted values (Table 3) from the Couchman–Karasz equation (6), the Gordon–Taylor equation using the Simha–Boyer rule (7), an analytical equation (26) and an empirical second-order polynomial equation (27) obtained from regression analysis (of the data in Tables 2 and 3):

Table 1 *Interaction factors (ε) for different binary mixtures*

| | Mixture | | | | | |
Parameter	S–G	S–F	G–F	G–C	G–W	C–W
$\Delta C_{p1}/\Delta C_{p2}$	1.35	1.44	0.94	1.21	3.35	2.77
K	1.31	1.49	2.48	1.59	4.07	3.31
E	0.97	1.04	2.64	1.32	1.22	1.20
Comment	ideal mixing systems		non-ideal mixing systems			——

Note: ΔC_p for sucrose (S), glucose (G), fructose (F), citric acid (C) and water (W) are 0.43, 0.58, 0.62, 0.70 and 1.94 J g^{-1} °C, respectively. $\Delta C_{p1}/\Delta C_{p2} = \Delta C_p(\text{water})/\Delta C_p(\text{solid})$. The interaction factors (ε) for the S–G and S–F systems are close to 1; therefore, these systems are considered to be ideal mixing systems.

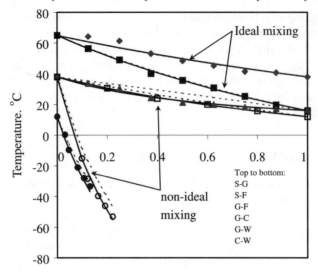

Fraction of plasticizer

Figure 3 *Predicted* T_g *values for binary mixtures, obtained from the Gordon–Taylor (solid line) and Couchman (dotted line) equations. S, G, F, C, W denote sucrose, glucose, fructose, citric acid and water, respectively. S–G and S–F are considered as ideal mixing systems, because the interaction factors (ε) are close to 1*

Table 2 *Experimental and predicted values for* T_g *of binary mixtures of sugars, the latter obtained using the Gordon–Taylor equation*

Mass fraction		DSC T_g (°C)			Predicted T_g (°C)		
Solid	Plasticizer	Sucrose–Fructose	Glucose–Fructose	Sucrose–Glucose	Sucrose–Fructose	Glucose–Fructose	Sucrose–Glucose
0	1	15.8±0.6	15.8±0.6	38.0±0.5	15.8	15.8	38.0
0.125	0.875	19.8±0.5	17.2±0.6	40.7±1.1	20.1	17.0	40.7
0.250	0.750	25.3±0.6	18.2±0.6	41.5±0.9	24.8	18.4	43.5
0.375	0.625	30.8±0.6	20.0±1.7	45.3±0.9	29.9	20.1	46.5
0.500	0.500	35.7±0.5	21.3±1.2	48.5±2.8	35.6	22.2	49.7
0.625	0.375	40.1±0.8	24.3±0.2	53.1±0.3	41.8	24.7	53.2
0.750	0.250	49.0±0.8	30.4±1.6	61.3±1.8	48.7	28.0	56.8
0.875	0.125	56.4±1.3	33.7±1.1	64.3±0.9	56.4	32.2	60.8
1	0	65.0±0.5	38.0±0.5	65.0±0.5	65.0	38.0	65.0
R^2					0.998	0.981	0.957

Note: Plasticizer is the lower T_g component in the binary mixture.

$$T_g = A_S X_S + A_G X_G + A_F X_F + A_{SF} X_S X_F + A_{SG} X_S X_G + A_{GF} X_G X_F \qquad (27)$$

where $A_S = 67.2$, $A_G = 38.2$, $A_F = 16.6$, $A_{SF} = -27.8$, $A_{SG} = -14.9$ and $A_{GF} = -23.1$. The parameters used in Equation 26 were obtained from experiments on binary mixtures ($K_{SG} = 1.30$, $K_{GF} = 2.48$ and $K_{SF} = 1.49$, $T_{gS} = 65\,^\circ\text{C}$, $T_{gG} = 38\,^\circ\text{C}$ and $T_{gF} = 15.8\,^\circ\text{C}$). The coefficients of determination, R^2, and the differences between experimental data and predicted values among the models are shown in Table 4. R^2 varied in a wide range for the Gordon–Taylor (0.867 to 0.992) and Couchman–Karasz equations (0.810 to 0.965), but in a narrow range for the analytical (0.967 to 0.993) and polynomial equations (0.983 to 0.998). The correlation for a sucrose–glucose ratio of 3/5 was low ($R^2 = 0.967$) due to the high amount of glucose. This result agreed well with the result for the binary mixture at high glucose content, as discussed previously. The maximum difference from the analytical equation (1.3–2.6 °C) was comparable to that from the polynomial equation (0.8–2.4 °C), whereas it was much larger from the Gordon–Taylor (1.6–6.2 °C) and Couchman–Karasz equations (3.6–5.1 °C). The comparison among the Gordon–Taylor, Couchman–Karasz and analytical equations is also shown in Figure 4. The experimental data were lower than the predicted values from the Couchman–Karasz equation, but higher than those from the Gordon–Taylor equation using the Simha–Boyer rule. The predicted values from the analytical equation were close to the DSC values and were located in between the above models.

5.4 Aqueous Ternary Components

Interaction factors, ε, and constants, K, for glucose–citric acid–water systems in which water was taken as the plasticizer are shown in Table 5. The interaction values varied from 1.16 to 1.19. The prediction from the Gordon–Taylor equation, using the binary form (solid–water), was compared to that from the Couchman–Karasz equation (6) and the analytical equation (26), as shown in Table 6. The values used for water were $-135\,^\circ\text{C}$ and 1.94 J g^{-1} °C for T_g and ΔC_p, respectively.[5,7-10] The experimental data fitted well the analytical equation ($K_{GC} = 1.59$, $K_{GW} = 4.07$ and $K_{CW} = 3.31$, as shown in Table 1, $T_{gG} = 38\,^\circ\text{C}$, $T_{gC} = 12\,^\circ\text{C}$ and $T_{gW} = -135\,^\circ\text{C}$) and the Gordon–Taylor equation with empirical constants ($K_{\text{solid–water}} = 3.39$ to 3.72, as shown in Table 5). The average difference between the predicted values and DSC data was less than 1.2 °C for the analytical equation and less than 1.3 °C for the Gordon–Taylor binary form with empirical K. These two models were comparable, but the Gordon–Taylor equation requires more experimental data and was applicable only for each solids ratio. However, use of the Couchman–Karasz equation (binary form) with $\Delta C_{p1}/\Delta C_{p2}$ (Table 5) and the Couchman–Karasz equation (6) gave poor predicted values, with an average difference greater than 4.3 °C (Table 6). This comparison was plotted in Figure 5 for the Couchman–Karasz equation using $\Delta C_{p1}/\Delta C_{p2}$, the Couchman–Karasz equation (6) and the analytical equation.

There was no significant difference between the Gordon–Taylor equation using empirical K values ($K_{\text{solid–water}} = 4.33$, 4.10, 3.60 and 3.52 for sucrose/fruc-

Table 3 *Experimental and predicted* T_g *values for ternary mixtures of sugars*

$X_S/$	Mass fraction			DSC T_g	Predicted T_g (°C)			
X_G	Sucrose	Glucose	Fructose	(°C)	(a)	(b)	(c)	(d)
5/3	0.625	0.375	0	53.1±0.3	53.2	51.8	52.9	52.8
	0.5208	0.3125	0.1667	43.8±2.5	43.7	38.2	45.4	43.7
	0.4167	0.25	0.3333	34.7±2.7	36.1	30.1	38.5	35.7
	0.3125	0.1875	0.5	31.0±1.5	29.8	24.8	32.2	29.1
	0.2083	0.1275	0.6667	24.6±1.6	24.5	21.0	26.4	23.6
	0.1042	0.0625	0.8333	18.6±0.7	20.0	18.2	21.0	19.5
4/4	0.5	0.5	0	48.5±2.8	49.7	48.4	49.5	49.0
	0.4167	0.4167	0.1667	39.8±1.1	40.7	36.1	42.9	40.6
	0.3333	0.3333	0.3333	33.2±1.0	33.8	29.3	36.8	33.3
	0.25	0.25	0.5	27.5±1.4	28.1	24.3	31.0	27.3
	0.1667	0.1667	0.6667	22.1±1.4	23.4	20.8	25.7	22.5
	0.0833	0.0833	0.8333	18.5±1.5	19.5	18.1	20.7	18.9
3/5	0.375	0.625	0	45.3±0.9	46.5	45.4	46.3	45.6
	0.3125	0.5208	0.1667	35.4±0.3	38.0	35.0	40.5	37.8
	0.25	0.4167	0.3333	30.0±2.1	31.6	28.4	35.0	31.2
	0.1875	0.3125	0.5	25.3±1.3	26.5	23.9	29.9	25.7
	0.125	0.2083	0.6667	21.1±0.6	22.4	20.5	25.0	21.5
	0.0625	0.1042	0.8333	17.2±1.1	19.0	18.0	20.4	18.4

Note: X_S/X_G is the ratio of mass fraction of sucrose to glucose in the mixture (5/3, 4/4 and 3/5).
(a) Analytical equation, where $m = 1/K_{SG} = 1/1.3 = 0.769$ for all cases, $K_{SF} = 1.49$ and $K_{GF} = 2.48$, as shown in Table 1.
(b) Gordon–Taylor equation using Simha–Boyer rule.
(c) Expansion of Couchman–Karasz equation ($\Delta C_p = 0.43, 0.58$ and 0.62 (J g^{-1} °C) for sucrose, glucose and fructose, respectively).
(d) Polynomial equation (27).

Table 4 *Coefficient of determination* (R^2) *and the differences between experimental data and predicted values among the equations for ternary sugar mixtures*

Model	Maximum difference (°C)			Average difference (°C)			R^2		
X_S/X_G	5/3	4/4	3/5	5/3	4/4	3/5	5/3	4/4	3/5
(a)	1.41	1.34	2.60	0.70	0.96	1.57	0.993	0.991	0.967
(b)	6.24	3.90	1.56	3.61	2.02	0.79	0.867	0.940	0.989
(c)	3.89	3.61	5.12	1.85	2.84	3.78	0.965	0.915	0.810
(d)	1.98	0.75	2.40	0.87	0.41	0.98	0.992	0.998	0.983

Note: X_S/X_G is the ratio of the mass fraction of sucrose to glucose. (a), (b), (c) and (d) are as given in Table 3.

tose of 7/1, 5/3, 2/6 and 1/7, respectively), the Couchman–Karasz equation (6) and the analytical equation ($K_{SW} = 4.42$, $K_{FW} = 3.18$ and $K_{SF} = 1.49$) for prediction of sucrose–fructose–water systems, because of the narrow range of experimental T_g values (freeze-dried sucrose–fructose samples placed over salt sol-

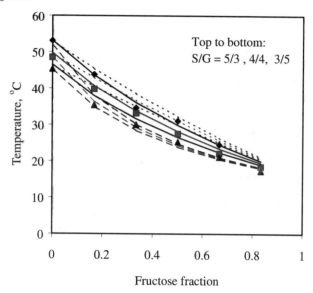

Figure 4 *Comparison between experimental T_g data and predicted values for ternary sucrose–glucose–fructose mixtures. S/G is the mass ratio of sucrose to glucose. Dotted line is the Couchman equation, dashed line is the Gordon–Taylor equation using the Simha–Boyer rule and solid line is the analytical equation*

Table 5 *Interaction factors (ε) for glucose–citric acid–water systems in which water is taken as the plasticizer*

Parameter	Glucose/Citric acid			
	4/1	3/2	2/3	1/4
ΔC_p	0.604	0.628	0.652	0.676
$\Delta C_{p1}/\Delta C_{p2}$	3.21	3.09	2.98	2.87
K	3.73	3.61	3.54	3.39
ε	1.16	1.17	1.19	1.18

Note: Constant K was obtained from Equation 16 for binary mixtures.

utions gave a moisture range below 9% db). The prediction based on the analytical equation is shown in Figure. 6.

5.5 Comparison Between Second- and Third-order Interactions

As we discussed previously, the analytical equation was developed based on disregarding third-order interactions. In order to estimate the difference between second- and third-order interactions, a third-order polynomial equation (28), obtained from regression analysis of data for anhydrous systems (binary and

Table 6 *Coefficient of determination (R^2) and the differences between experimental data and predicted values among the equations for glucose–citric acid–water systems*

Model X_G/X_C	Average difference (°C)				R^2			
	4/1	3/2	2/3	1/4	4/1	3/2	2/3	1/4
(1)	1.17	0.88	0.83	0.55	0.997	0.998	0.998	0.999
(2)	1.30	0.84	0.69	0.79	0.997	0.990	0.998	0.998
(3)	4.59	4.27	4.31	4.66	0.967	0.779	0.950	0.946
(4)	5.57	5.24	5.24	5.20	0.957	0.623	0.940	0.938

Note: X_G/X_C is the ratio of the mass fraction of glucose to citric acid.
(1) Analytical equation using constant K from Table 1.
(2) Gordon–Taylor equation in binary form, using empirical constant K from Table 5.
(3) Couchman–Karasz equation in binary form, using $\Delta C_{p1}/\Delta C_{p2}$ from Table 5.
(4) Couchman–Karasz equation in ternary form, (6), using ΔC_p data noted in Table 1.

ternary mixtures in Tables 2 and 3), using Statgraphic 7.0 software, was used for comparison:

$$T_g = A_S X_S + A_G X_G + A_F X_F + A_{SF} X_S X_F + A_{SG} X_S X_G + A_{GF} X_G X_F + A_{SGF} X_S X_G X_F \tag{28}$$

where $A_S = 66.6$, $A_G = 37.7$, $A_F = 16.2$, $A_{SF} = -23.6$, $A_{SG} = -10.6$, $A_{GF} = -18.9$ and $A_{SGF} = -41.4$. The maximum difference in predicted values from Equations 28 and 27 was less than 0.6 °C. This difference is equivalent to the magnitude of the error of instrumentation. Since the analytical equation was comparable to the second-order polynomial equation (27), we can say that the maximum error from the analytical equation (26), due to ignoring third-order interactions, is approximately 0.6 °C.

In general, prediction using the analytical equation, for both anhydrous and aqueous systems, is an economical method, as it requires only experiments on binary systems plus a few experiments on ternary mixtures for validation.

6 Proposed Model for Quaternary or Higher-order Systems

For quaternary or higher-order systems, a similar approach can be used. For example, T_g of the four-component sucrose–glucose–fructose–citric acid system (SGFC) can be described by the following equations (where $T_{gS} > T_{gG} > T_{gF} > T_{gC}$):

$$T_{gSGFC} = \frac{T_{gGFC} + \alpha_1 T_{gSFC}}{1 + \alpha_1} \tag{29}$$

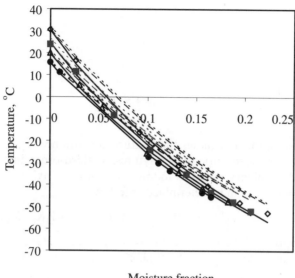

Moisture fraction

Figure 5 *Comparison between experimental* T_g *data and predicted values for ternary glucose–citric–water systems. From top to bottom, the ratio of glucose to citric is* (◇) *4/1,* (■) *3/2,* (Δ) *2/3 and* (●) *1/4, respectively. Dotted line is the Couchman–Karasz equation, dashed line is the Couchman–Karasz equation in binary form with water as the plasticizer and solid line is the analytical equation*

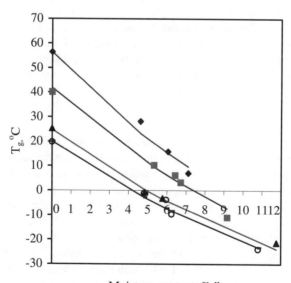

Moisture content, %db

Figure 6 *Prediction using an analytical equation (lines) and* T_g *data (dots) for ternary sucrose–fructose–water systems. From top to bottom, the ratio of sucrose/fructose is 7/1, 5/3, 2/6 and 1/7*

where

$$T_{gGFC} = \frac{T_{gFC} + \alpha_2 T_{gGC}}{1 + \alpha_2} \tag{30}$$

and

$$T_{gSFC} = \frac{T_{gFC} + \alpha_3 T_{gSC}}{1 + \alpha_3} \tag{31}$$

The subscript of the T_g denotes the mixture; for instance, T_{gGFC} is the T_g of ternary glucose–fructose–citric acid mixtures. If third-order interactions are negligible, and all binary mixtures follow the Gordon–Taylor equation, then constants ε_i ($i = 1$ to 3) can be determined as follows:

$$\alpha_1 = 1/K_{SG}(X_S/X_G), \ \alpha_2 = 1/K_{GF}(X_G/X_F) \text{ and } \alpha_3 = 1/K_{SF}(X_S/X_F) \tag{32}$$

A similar approach is then carried out to obtain the final form of the equation for predictions for quaternary mixtures (33):

$$T_{gSGFC} = \frac{1}{1+\alpha_1} \left\{ \frac{1}{1+\alpha_2} \left[\frac{(1-X_C)T_{gF} + K_{FC}X_C T_{gC}}{1 - X_C + K_{FC}X_C} + \alpha_2 \frac{(1-X_C)T_{gC} + K_{GC}X_C T_{gC}}{1 - X_C + K_{GC}X_C} \right] \right.$$
$$\left. + \frac{1}{1+\alpha_3} \left[\frac{1-X_C)T_{gF} + K_{FC}X_C T_{gC}}{1 - X_C + K_{FC}X_C} + \alpha_3 \frac{(1-X_C)T_{gS} + K_{SC}X_C T_{gC}}{1 - X_C + K_{SC}X_C} \right] \right\} \tag{33}$$

where X_i ($i = S, G, F, C$) is the mass fraction of component i in the mixture. This equation requires six values of K for six binary mixtures created from four components. A similar approach can be used for n-component mixtures, where $n(n-1)/2$ values of K are required for prediction of the T_g of n-component mixtures. The equation is long but simple, because it is based on the knowledge of constants K, fractions X, and T_gs of the individual components.

7 Conclusion

The binary mixtures of sucrose–fructose and sucrose–glucose are close to ideal-mixing systems, for which the Couchman–Karasz equation can be applied. The binary mixtures of glucose–fructose, glucose–citric acid and solid–water are non-ideal mixing systems, for which an empirical constant, K, should be used in the Gordon–Taylor equation. Using this knowledge for prediction of T_g for ternary systems, by means of an analytical equation in which third-order interactions are negligible, gave a better result, compared to expansion of the Couchman–Karasz equation and to the Gordon–Taylor equation using the Simha–Boyer rule. The Gordon–Taylor equation in binary form, including an empirical constant K, is equivalent to the analytical equation for predicting T_g for ternary mixtures, but it requires more experimental data and is applicable only for each solids ratio. In contrast, our analytical equation requires only the K values for binary mixtures in order to predict T_g over the whole range of ternary mixtures. A large reduction in the number of experiments is considered to be an advantage of this method. In addition, it can be expanded to quaternary or higher-order mixtures, using a similar approach.

Acknowledgment

We are grateful to the staff of Food Science and Technology, University of Queensland, for making experimental facilities available to us.

References

1. L. Slade and H. Levine, *Crit. Revs. Food Sci. Nutr.*, 1991, **30**, 115.
2. M. Gordon and J.S. Taylor, *J. Appl. Chem.*, 1952, **2**, 493.
3. M. d. P. Buera, G. Levi and M. Karel, *Biotechnol. Prof.*, 1992, **8**, 144.
4. K. Jouppila and Y.H. Roos, *J. Dairy Sci.*, 1994, **77**, 2907.
5. Y. Roos and M. Karel, *Int. J. Food Sci. Technol.*, 1991, **26**, 553.
6. I. Arvanitoyannis, J.M.V. Blanshard, S. Ablett, M.J. Izzard and P.J. Lillford, *J. Sci. Food Agric.*, 1993, **63**, 177.
7. Y. Roos and M. Karel, *J. Food Sci.*, 1991, **56**, 266.
8. Y. Roos and M. Karel, *J. Food Sci.*, 1991, **56**, 38.
9. Y. Roos, *J. Food Sci.*, 1987, **52**, 146.
10. Y. Roos, *Carbohydr. Res.*, 1993, **238**, 39.
11. L. Mandelkern, G.M. Martin and F.A. Quinn, *J. Res. Nat. Bur. Stand.*, 1957, **58**, 137.
12. T.G. Fox, *Bull. Am. Phys. Soc.*, 1956, **1**, 123.
13. F.N. Kelley and F. Bueche, *J. Polym. Sci.*, 1961, **1**, 549.
14. P.R. Couchman and F.E. Karasz, *Macromolecules*, 1978, **11**, 117.
15. R. Simha and R.F. Boyer, *J. Chem. Phys.*, 1962, **37**, 1003.
16. S.L. Shamblin, L.S. Taylor and G. Zografi, *J. Pharm. Sci.*, 1998, **87**, 694.
17. J.Y. Kim, S.U. Hong and Y.S. Kang, *Macromolecules*, 2000, **33**, 3161.
18. L. FineGold, F. Franks and R.H.M. Hatley, *J. Chem. Soc., Faraday Trans. 1*, 1989, **85**, 2945.

Microstructural Domains in Foods: Effect of Constituents on the Dynamics of Water in Dough, as Studied by Magnetic Resonance Spectroscopy

Yang Kou,[a] Edward W. Ross[b] and Irwin A. Taub[b]

[a]DEPARTMENT OF FOOD SCIENCE, UNIVERSITY OF MASSACHUSETTS, AMHERST, MA 01003, USA
[b]OTD, U.S. ARMY NATICK SOLDIER SYSTEMS CENTER, NATICK, MA 01760, USA

1 Introduction

Unlike many synthetic polymers, most food materials are complex in chemical composition, heterogeneous in structure and reactive. The stability of a food matrix depends strongly on its microstructure, local viscosity and associated molecular mobility.[1,2] Moisture content, concentration of constituents and temperature are key factors that determine the structure and distribution of microstructural domains in an amorphous food matrix, and control local viscosity and molecular mobility. Microstructural domains are defined as regions of differing local viscosity and molecular mobility, which are randomly distributed.

The concept of a distribution of microstructural domains within an amorphous food matrix provides a basis for explaining any observed dispersion in its volume-averaged properties. Consequently, the volume-averaging of these localized properties determines, in turn, the global values of such properties as glass transition temperature (T_g), water mobility as reflected in proton spin–lattice (T_1) and spin–spin (T_2) relaxation time constants, and reaction rate constants (k). These properties are best described in terms of an average and a dispersion. Since physical, chemical, and microbiological changes can take place in regions of low viscosity or high reactivity, the distribution of microstructural domains should be considered in assessing food safety, quality, and stability.

In the present work, in order to understand the nature of microstructural domains, their distributions were either discerned or inferred from measurements made on dough matrices of different moisture contents and constituent concentrations and at different temperatures. The changes in T_1, T_2, and k (for a reduction reaction), as moisture content, constituent concentration and temperature were changed, were measured, primarily using time-domain nuclear magnetic resonance (NMR) and electron spin resonance (ESR) spectroscopy. These measured values were then correlated where appropriate, and distribution functions for the microstructural domains were obtained from the T_1, T_2, and k values.

2 Materials and Methods

2.1 Materials and Sample Preparation

Wheat flour (12% protein, 11% moisture, 0.49% ash; Manildra Milling, Shawnee Mission, KS, USA) was used to make (flour/water) dough samples of different moisture contents (28, 30, 33 and 35%, total weight basis). Wheat starch and vital wheat gluten (Manildra Milling, Shawnee Mission, KS, USA) were used to make mixtures of starch/water, gluten/water and starch/gluten/water of different concentrations.

In the NMR experiment, samples were prepared prior to use, and approximately 10 grams of sample was weighed and placed into a glass test tube. In the kinetic study using ESR spectroscopy, erythorbic acid (Aldrich Chemical, Milwaukee, WI, USA) and TEMPO (tetramethyl piperidine nitroxide, Aldrich Chemical, Milwaukee, WI, USA) were added to separate samples that were then mixed together by kneading equal volumes just prior to use.

2.2 NMR Experiment

A 20 MHz PCT 20/20 NMR Analyser (Process Control Technology, Ft. Collins, CO, USA) was used to perform all NMR experiments. A saturation recovery pulse sequence[3] was used to acquire free induction decay (FID) data for T_1. A 90° pulse sequence and a Carr–Purcell–Meiboom–Gill (CPMG) pulse sequence[4,5] were combined and used for acquisition of FID data for T_2. Experimental parameters were set appropriately to maximize signal-to-noise ratio and to cover the entire relaxation range as completely as possible.

2.3 ESR Experiment

An ESR spectrometer equipped with a VT-2000 temperature controller (Model EMX, Bruker Instruments, Billerica, MA, USA) was used to study the spectral pattern and reduction of TEMPO. All samples were placed in a special plastic holder that fits into a resonator, the temperature of which was controlled by cold nitrogen gas. Temperature accuracy was ± 0.1 K. Microwave power was set to 0.63 mW. Scan range, scan rate, time constant and field modulation amplitude were adjusted so that distortion of spectra was avoided.

2.4 Mathematical Model

Our model describes a mathematical procedure for analysing data obtained from experiments on the time-dependent behavior of dough samples subjected to various moisture and temperature conditions. The data consisted of either proton NMR or ESR signals arising from the reaction of tetramethyl piperidine nitroxide (TEMPO) with erythorbic acid in the dough. The dough can be viewed as an assemblage of many domains with random properties. In each domain, a first-order reaction is assumed to take place, and the NMR or ESR signal is taken as the sum of all these reactions. The reactions are assumed to occur at different, random rates that may be affected by moisture and temperature, the levels of which may be constant or variable over time.

We assume that $N+1$ domains, numbered from 0 to N, are present in the dough. The first-order process in the j-th domain leads to a signal intensity, $I_j(t)$, that depends on time t as follows:

$$I_j(t) = I_j(0) \, e^{-\Phi j(t)} \tag{1}$$

where $\Phi j(t)$ governs the time dependence of I_j and may embody the effects of temperature and moisture variation. For consistency, we assume $\Phi j(0) = 0$. For the moment, we shall assume that only temperature is varying, and the Arrhenius Law models its effect. Then, it is convenient to define T as the absolute temperature, T_r as a reference temperature, k_j as the rate constant at the reference temperature in the j-th domain, and $x(t) = [T_r/T(t)] - 1 = $ dimensionless, inverse temperature. The Arrhenius Law can be written as:

$$\Phi j(t) = k_j \, F_j(t) \tag{2}$$

and

$$F_j(t) = \int_0^t e^{-Ej.x(u)} du \tag{3}$$

where $E_j = E_{aj}/(RT_r)$, E_{aj} is activation energy in the j-th domain and R is the universal gas constant.

The measured signal, $I(t)$, is taken to be a weighted sum of the contributions of all the individual domains, *i.e.*:

$$I(t) = \sum_0^N Wj \cdot Ij(t)/Ij(0) = \sum_0^N Wj \cdot e^{-kj.Fj(t)} \tag{4}$$

Wj is the weight function, which specifies the contribution of the j-th first-order process to the overall $I(t)$. Alternatively, we can think of Wj as the probability of selecting the j-th process when randomly choosing among the various processes.

We assume that many domains are present, each with its own rate constant, k_j. Then it is appropriate to let $N \rightarrow \infty$, and the differences in rate constants approach 0, and approximate the sums by integrals over all positive values of k:

$$I(t) = \int_0^\infty W(k) \cdot e^{-k.F(t,k)} dk \tag{5}$$

where

$$F(t,k) = \int_0^t e^{-E(k).x(u)} du \qquad (6)$$

In the important, special case where temperature is constant and equal to T_r, $x(t) = 0$ and so $F(t,k) = t$. Then we have:

$$I(t) = \int_0^\infty W(k) \cdot e^{-k \cdot t} dk \qquad (7)$$

We recognize $I(t)$ as the Laplace transform of $W(k)$ in this case. For example, if $W(k) = [a^b/\Gamma(b)] k^{(b-1)} e^{-ak}$, where $k > 0$, $a > 0$ and $b > 0$. Then:

$$I(t) = [1/(1 + t/a)]^b \qquad (8)$$

This $W(k)$ is a true probability density function (the gamma or chi-square density) and has $k_{ave} = b/a$, and $\sigma(k) = b^{1/2}/a$. Data from experiments on dough samples using NMR and ESR spectroscopy appear to be well-fitted by the above form of $I(t)$, provided that a and b are estimated accurately.

The above procedure can be generalized for the case where more than one reaction of this sort is going on simultaneously. Similar extensions can be made when two or more reactions are thought to take place. However, the parameter estimates become less reliable as the number of parameters increases, so there is a limit (essentially governed by the noise-level of the data), beyond which it is unproductive to apply this model.

3 Results and Discussion

3.1 Probing Molecular Mobility Using ¹H NMR Spectroscopy

In previous NMR studies, different pulse sequences have been used to measure spin–lattice (T_1) and spin–spin (T_2) relaxation time constants, values of which have been used as indicators of water mobility (or the state of water).[6-11] In general, a higher T_1 or T_2 value indicates a higher water mobility (however, in certain cases, such as for the glassy state, a higher T_1 value does not necessarily mean higher mobility). It is widely believed that water molecules in foods are in some way associated with different sites on constituents or are in exchange with such associated water (so-called 'bound water') and experience faster relaxation than that in bulk water and thus have much lower T_2 than does bulk water.[12,13] Therefore, use of a single relaxation time constant to describe complex and heterogeneous food systems may be an oversimplification. In a heterogeneous system, spins exist in a large variety of different environments, giving rise to a wide range of relaxation time constants. Thus, it is reasonable to assume that a continuum of relaxation time constants would arise from a continuum of different environments and exchange rates.[14-16]

In the present experiment, a 90° or Carr–Purcell–Meiboom–Gill (CPMG)[4,5] pulse sequence set the proton magnetization signals to their maximum values, which began to decay immediately after the pulse was removed. The resulting decay curve is usually termed free induction decay (FID), representing both the fast (rigid proton signals) and slow (mobile proton signals) relaxation processes in samples, and can be described by a single- or multi-exponential equation and can give a single T_2 or several discrete T_2 values. Rigid proton signals (*i.e.* in a μsec range) originated from solids (*i.e.* gluten or starch in this case) or from water molecules tightly associated with solids, while mobile proton signals (*i.e.* in a msec range) originated from water molecules with relatively high mobility.[17]

The experimental data were fitted to our mathematical model in order to obtain predicted values and the probability density function for T_2. There was good agreement between the predicted and experimental values, in both the μsec and msec ranges. Figure 1 shows the distribution of values of probability density function for T_2, in the μsec range (*i.e.* rigid proton signals), for a 30% moisture dough and for mixtures of starch/water, gluten/water and starch/gluten/water in the same ratio as in the 30% moisture dough. Model-fitting resulted in one T_2 peak for both starch/water and gluten/water samples, with average T_2 values of 19.2 and 20.5 μsec, respectively. For dough and starch/gluten/water samples, model-fitting resulted in two T_2 peaks with different dispersions. Average T_2 values for both samples were in the range of 18.5–29.1 μsec. The appearance of a second peak in the rigid proton signal suggested that new physical or chemical environments were formed within the system as a result of the addition of a new constituent.

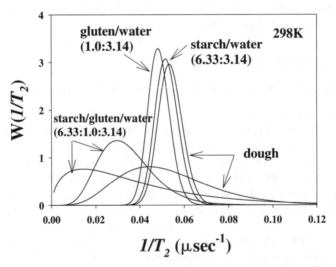

Figure 1 *The distribution of values of probability density function for* T_2, *in the μsec range (i.e. rigid proton signals), for a 30% moisture dough and for mixtures of starch/water, gluten/water and starch/gluten/water in the same ratio as in the 30% moisture dough*

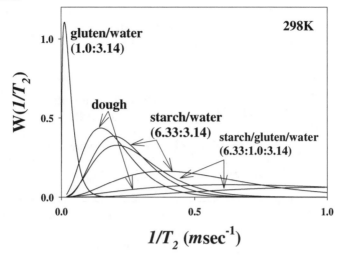

Figure 2 *The distribution of values of probability density function for* T$_2$, *in the msec range (i.e. mobile proton signals), for a 30% moisture dough and for mixtures of starch/ water, gluten/water and starch/gluten/water in the same ratio as in the 30% moisture dough*

Figure 2 shows the distribution of values of probability density function for T_2, in the msec range (*i.e.* mobile proton signals), for a 30% moisture dough and for mixtures of starch/water, gluten/water and starch/gluten/water in the same ratio as in the 30% moisture dough. Model-fitting resulted in two T_2 peaks with different dispersions for each sample type. Average T_2 values ranged from 0.7 to 5.6 msec, which suggested that the detected signals were from water molecules with relatively high mobility.

Figure 3 shows the distribution of values of probability density function for T_1 for a 30% moisture dough and for mixtures of starch/water, gluten/water and starch/gluten/water in the same ratio as in the 30% moisture dough. Average T_1 values for the samples were 0.088 sec for dough, 0.115 sec for starch/gluten/water 0.133 sec for starch/water and 0.155 sec for gluten/water.

A comparison of these averaged T_1 and T_2 values with values obtained from an exponential model showed good agreement between the two different approaches. However, the 'continuum' approach used here is more consistent with the 'continuum' nature of water mobility in food systems.[2] Furthermore, additional information may be obtained from a continuum model. For example, a spectrum with a large number of peaks or broader peaks would be expected for heterogeneous samples, rather than for relatively homogeneous samples. In other words, the number of peaks and their breadth could be used as a measure of the homogeneity of a sample under analysis.

In order to study the effect of constituent concentration on molecular mobility, an NMR experiment was conducted on 35% moisture dough with varying gluten content (9.1 to 65%) at 298 K. Another NMR experiment on 35% moisture dough at 258 to 298 K was conducted to investigate the effect of temperature on molecular mobility. In the case of the concentration effect, results

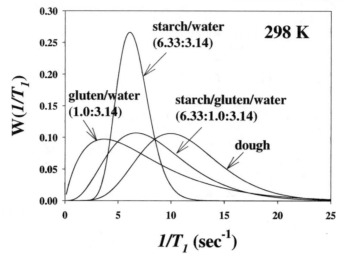

Figure 3 *The distribution of values of probability density function for* T_1 *for a 30% moisture dough and for mixtures of starch/water, gluten/water and starch/gluten/water in the same ratio as in the 30% moisture dough*

indicated that average T_1 values decreased (*i.e.* lower mobility) and their dispersion increased (*i.e.* less homogeneity) with increasing solids (gluten) concentration. In the case of the temperature effect, average T_1 values decreased and their dispersion increased with decreasing temperature. In general, T_2 analysis discerned different spin environments in dough samples. T_2 values varied with moisture and solids contents, previously interpreted as reflecting four dominant fractions of associated water molecules.[10]

3.2 Kinetic Study Using ESR Spectroscopy

In order to investigate further the microscopic distribution of local viscosity and the inhomogeneous reaction kinetics in a dough matrix, the spin probe, TEMPO, was dissolved in the dough matrix and then monitored with ESR spectroscopy.[18-20] In this kinetic study, erythorbic acid was used as the reductant, in excess amount. Dough samples were prepared by mixing two equal-weight portions, one containing TEMPO and the other the reductant, right before an experiment.

Both the shapes of the ESR spectral patterns for TEMPO and the changes in signal intensity for TEMPO, upon reduction to its spin-inactive form through reaction with erythorbic acid, were monitored at several combinations of temperature and moisture content. Although the ESR spectral pattern for TEMPO was, in all cases, affected by temperature and moisture content, under appropriate conditions, there was only a slight change in the spectral pattern, as the ESR signal for TEMPO was reduced by reaction with erythorbic acid. This finding suggested that the decrease in fast-line intensity could be used to monitor the reaction.

Since localized differences in matrix viscosity can affect the rates of reaction between two diffusing molecules, the reduction of spin-active TEMPO by erythorbic acid was monitored in dough matrices at different moisture contents and temperatures. Such reduction converts TEMPO to a spin-inactive state, so the spectral intensity decreased as the reaction proceeded (Figure 4). To simplify the kinetics, excess erythorbic acid was used, so a pseudo-first-order process should have been observed. However, as shown in Figure 4 for 30% moisture dough samples, a semi-logarithmic plot of normalized signal intensity, I_t/I_0, against time was curved rather than straight. This result indicated that the reduction of TEMPO by erythorbic acid did not conform to homogeneous kinetics. The ever-decreasing rate of reaction with increasing time was characteristic of dispersive kinetics, and it could be modelled by taking into account the distribution of microstructural domains. In other words, the data could be analysed on the basis of a microscopic distribution of local domains, each with

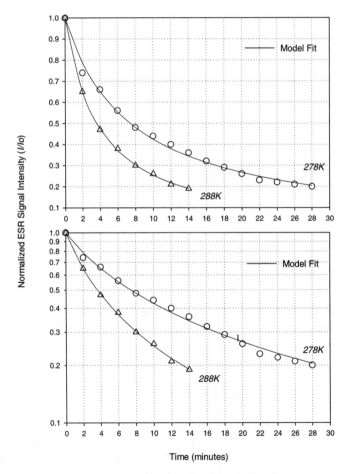

Figure 4 *A comparison of predicted reaction rate constants with experimental data for a 30% moisture dough at 278 and 288 K*

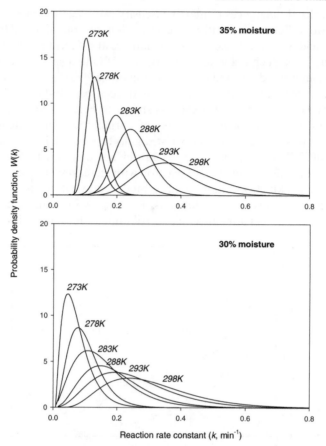

Figure 5 *The distribution of values of probability density function for reaction rate constants as a function of temperature, for 35% and 30% moisture dough samples*

characteristic local viscosity and associated rate constant. The fit of this dispersive kinetic model to the experimental data is illustrated by the smooth curves in
Figure 4. Similarly good fits were obtained for other dough samples.

Our mathematical model assumes that a localized viscosity and a corresponding *pseudo*-first-order rate constant, k_j, characterize each domain for the reacting
partners in the *j*-th domain. It further assumes that, for a matrix with a particular
composition and history at a constant temperature, the distribution of *k* values
can be described by a probability density function, $W(k)$. The concentration of
TEMPO at any time, *t*, would be reflected in the signal intensity, $I(t)$, defined by
Equation 7, with integration from $k = 0$ to ∞. As for the distribution function, an
incomplete gamma function was found to best fit the experimental data. Substitution of this function for $W(k)$ in the integration, followed by a Laplace transformation, gave $I_t/I_0 = [1/(1 + t/a)]^b$, where *a* and *b* characterize the distribution
of *k* values, and b/a defines the average *k* value, k_{ave}.

Experimental data for ESR signal intensity of TEMPO as a function of reaction time were obtained at combinations of six temperatures (*i.e.* 273 to 298 K) and four moisture contents (28 to 35%). As expected, the higher the temperature, the faster the reduction of TEMPO, because of lower overall average viscosity. The same trend was observed for the effect of moisture content (*i.e.* higher moisture contents resulted in faster degradation).

The distribution of values of probability density function for reaction rate constants as a function of temperature, for 35% and 30% moisture dough samples, is shown in Figure 5. The mean reaction rate constant, k_{ave}, increased with increasing temperature due to lower overall average viscosity, while peak dispersion also increased with increasing temperature, which indicated that the degree of sample homogeneity decreased. In another experiment on the effect of moisture content, the results indicated that the mean reaction rate constant, k_{ave}, increased with increasing moisture content due to lower overall average viscosity. It was also observed that peak dispersion decreased with increasing moisture content, which indicated that the degree of sample homogeneity increased with increasing moisture content.

From the model analyses, the derived values of *a* and *b* could be used to plot the number-density of *k* values against *k*, which showed how the distribution shifted to larger *k* values with increasing moisture content and temperature. Moreover, the derived k_{ave} values could be plotted semi-logarithmically against reciprocal temperature. It appeared to be clear that the observed reduction reaction followed an Arrhenius relationship, and activation energies for both 35% and 28% moisture dough samples were essentially the same. This finding suggested that temperature was here a barrier to both diffusion and reaction (following encounter and complex formation). Overall, the kinetic results were consistent with other rheological and relaxation phenomena, all of which reflect dispersion in the data associated with spatial inhomogeneities in microstructure.

4 Conclusions

The concept of a distribution of microstructural domains has been used here to understand the dispersive, volume-averaged properties of an amorphous food matrix. Measurements of proton spin–lattice and spin–spin relaxation time constants and reduction rate constants were made on dough matrices at different moisture contents and temperatures. A consistency in the results emerged, which reflected interrelationships among structure, local viscosity, molecular mobility and reactivity. The data could be explicitly defined in terms of an average and a dispersion. The main implication of this concept of microstructural domains is that such localized differences in structure and properties need to be quantitatively considered, in order to ensure that the physically, chemically, and microbiologically least-stable part of a matrix does not compromise the quality or safety of the entire matrix.

Acknowledgements

This work was financially supported by the US Department of Defense –

Combat Feeding Program. Access to the NMR and ESR facilities at the US Army Natick Soldier Systems Center is greatly appreciated.

References

1. O. Fennema, 'Food Chemistry', Marcel Dekker, NY, 1996, Chapter 2, 17.
2. L. Slade and H. Levine, *Crit. Rev. Food Sci. Nutr.*, 1991, **30**, 115.
3. R. Vold, J. Waugh, M. Klein and D. Phelps, *J. Chem. Phys.*, 1968, **48**, 3831.
4. H. Carr and E. Purcell, *Phys. Rev.*, 1954, **94**, 630.
5. S. Meiboom and D. Gill, *Rev. Sci. Instrum.*, 1958, **29**, 688.
6. M. Kalichevsky, E. Jaroszkiewicz, S. Ablett, J. Blanshard and P. Lillford, *Carbohydr. Polym.*, 1992, **18**, 77.
7. M. Kalichevsky, E. Jaroszkiewicz and J. Blanshard, *Polymer*, 1993, **34**, 346.
8. B. Hills and K. Pardoe, *J. Mol. Liquids*, 1995, **63**, 229.
9. R. Lloyd, X. Chen and J. Hargreaves, *Int. J. Food Sci. and Technol.*, 1996, **31**, 305.
10. R. Ruan and P. Chen, 'Water in Foods and Biological Materials: A Nuclear Magnetic Resonance Approach', Technomic Publishing, Lancaster, PA, 1998, 269.
11. Y. Kou, P. Molitor and S. Schmidt, *J. Food Sci.*, 1999, **64**, 950.
12. S. Tanner, B. Hills and R. Parker, *J. Chem. Soc. Faraday Trans.*, 1991, **87**, 2613.
13. S. Ablett, A. Darke and P. Lillford. 'Water Relationships in Foods', H. Levine and L. Slade (eds.), Plenum Press, New York, 1991, 453.
14. P. Lillford, A. Clark and D. Johns, *ACS Symposium Series*, 1980, **127**, 177.
15. B. Hills, S. Takacs and P. Belton, *Food Chem.*, 1990, **37**, 95.
16. R. Menon and P. Allen, *J. Magn. Reson.*, 1991, **86**, 214.
17. Y. Kou, L. Dickinson and P. Chinachoti, *J. Agric. Food Chem.*, 2000, **48**, 5489.
18. M. Hemminga and I. van den Dries, 'Spin Labeling: The Next Millennium', Plenum Press, New York, 1998, 339.
19. I. van den Dries, P. de Jager and M. Hemminga, *J. Magn. Reson.*, 1998, **131**, 241.
20. P. Belton, A. Grant, L. Sutcliffe, D. Gillies and X. Wu, *J. Agric. Food Chem.*, 1999, **47**, 4520.

Supplemented State Diagram for Sucrose from Dynamic Mechanical Thermal Analysis

I. Braga da Cruz,[1] William M. MacInnes,[2]* Jorge C. Oliveira[3] and F. Xavier Malcata[1]

[1]ESCOLA SUPERIOR DE BIOTECNOLOGIA, UNIVERSIDADE CATÓLICA PORTUGUESA, PORTO, PORTUGAL
[2]NESTLÉ RESEARCH CENTER, LAUSANNE, SWITZERLAND
[3]UNIVERSITY COLLEGE CORK, CORK, IRELAND
* Corresponding author

1 Introduction

The supplemented state diagram is often used to characterise the physical state of food materials as a function of temperature and water content. Franks et al.[1] and Levine and Slade[2] first showed that, for amorphous food materials, in addition to the equilibrium lines bordering solid, liquid and gaseous phases, a glass transition line separates metastable states that differ greatly in mechanical rigidity and viscosity, and thus in molecular mobility and stability. Such glass transition lines provide a relationship between product composition and physical behaviour, and may underlie the design of process and formulation of food products tailored to impart desired changes either on the level of processing or of preservation of qualities during processing and subsequent storage.

Supplemented state diagrams are commonly constructed using DSC data exclusively. However, these diagrams are of little help, e.g. for industrial freeze-drying, because that is generally carried out at temperatures above the glass transition temperature (T_g) but below the temperature at which 'freeze-drying collapse' occurs. This latter temperature is well-defined when DSC is applied to solutions containing ice, but poorly defined when no ice is present. In this chapter, we shall show how the frequency dependence of DMTA scans, coupled with solution viscosity measurements, can be used advantageously in the definition of a global Williams, Landel, Ferry (WLF)[3] equation able to estimate the 'average' molecular relaxation time, and viscosity, over the whole state diagram. These 'supplementary' data can then be shown on the state diagram and can help

59

define optimum processing pathways, as well as the time scale variations applicable for collapse and crystallisation phenomena.

2 Materials and Methods

Sucrose aqueous solutions in the range 2–87.5% (w/w) were prepared by weighing and adding exact amounts of sucrose (Merck) to known weights of de-ionised water; after sealing in bottles, microwave-heating with stirring was continued until clear solutions were obtained. Samples were then injected between two thin plastic discs in a DMTA disc-bending sample holder, as illustrated in Figure 1. For accurate temperature measurements, two thermocouples were placed directly inside the sample in the holder, through each of two side holes in the centring ring.

For highly concentrated solutions, *i.e.* above 60% (w/w), samples were injected hot, and the apparatus was immediately submerged in liquid nitrogen so as to have the sample achieve the glassy state and hence avoid ice crystal formation. Amorphous samples of pure sucrose were obtained by pressing crystalline sucrose powder into the 2 mm-thick disc-bending sample holder, followed by mounting in the DMTA device and performance of *in situ* melting, and then by rapid cooling.

Bending tests were performed by applying an oscillatory strain of 0.5% at the centre of a sample, at frequencies of 0.1, 1 and 10 Hz. The temperature was scanned at 0.3 °C min^{-1}, during both cooling and heating scans. In order to equilibrate samples, six thermal cycles were performed (including up and down scans), thus promoting formation of the maximum amount of ice; only data from the last heating scans were used to produce the supplemented state diagram and for correlation with DSC data.

The viscoelastic moduli, *viz.* the storage modulus (E'), the loss modulus (E'') and the δ-phase angle, were measured as a function of temperature and frequency and were thus used to obtain thermomechanical profiles.

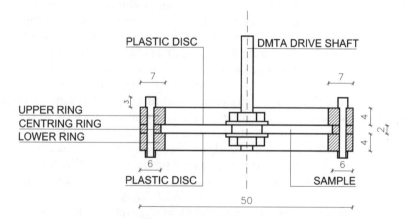

Figure 1 *DMTA disc-bending sample holder (dimensions in mm)*

3 Results

The thermomechanical profiles for sucrose aqueous solutions, ranging in concentration from 2 to 100% (w/w), were determined by DMTA; Figures 2 and 3 show the typical mechanical results obtained for 30 and 80% (w/w) sucrose solutions, in the presence or absence of ice, respectively.

For sucrose concentrations above 60% (w/w), samples were quenched in liquid nitrogen, so as to form the glassy state immediately, by avoiding ice crystal

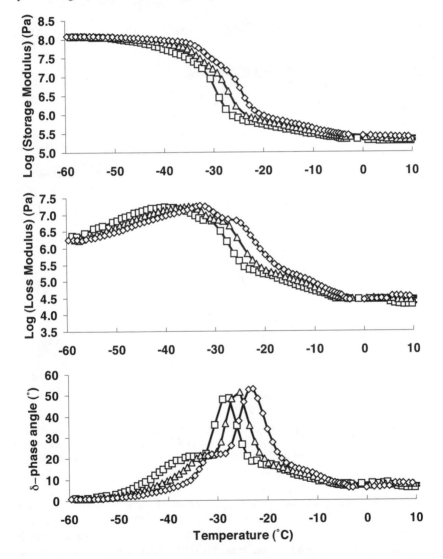

Figure 2 Thermomechanical profile for an aqueous *30% (w/w) sucrose solution; heating scan at 0.3 °C min*[−1]: *the storage modulus* (E′), *loss modulus* (E″) *and δ-phase angle were obtained at 0.1* (□), *1* (△), *and 10* (◇) *Hz*

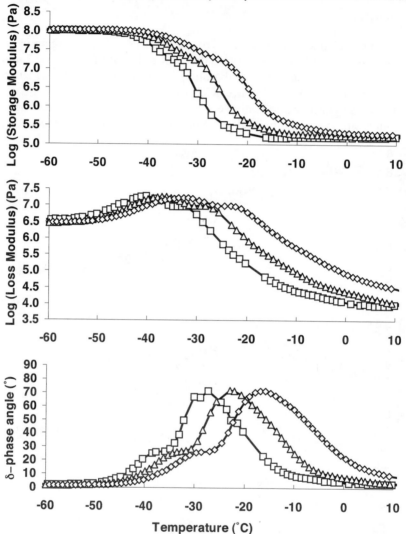

Figure 3 Thermomechanical profile for an aqueous *80% (w/w) sucrose solution; heating scan at 0.3 °C min⁻¹, after quenching in liquid nitrogen: the storage modulus* (E′), *loss modulus* (E″), *and δ-phase angle were obtained at 0.1* (□), *1* (△), *and 10* (◇) *Hz*

formation. On subsequent slow heating, the onset of the glass transition corresponds to the start of the fall in the real component of the modulus (*i.e.* E′); the peak temperature of the imaginary component of the modulus (*i.e.* E″) can be used to define the T_g at each frequency, as this peak corresponds to the maximum energy absorption at that frequency. Similar results were obtained for sucrose concentrations below 60% (w/w), for which cooling was not sufficiently rapid to avoid ice crystal formation. The cooling and heating scans exhibited very little hysteresis, even after six thermal cycles were carried out in order to approach the maximum amount of ice formation. The T_g at each frequency was taken as the peak temperature of E''. They were identified with the glass transition of the maximally freeze-concentrated solution, T_g',[1,2] with concentration C_g'.[2]

While the thin plastic discs in the disc-bending sample holder enabled successful measurements on the solutions, the measured decrease in E' above the glass transition was, in fact, limited by the rigidity of those discs. As a result, upon further heating, the δ-phase angle exhibited a clear peak that correlated, rather fortuitously, with the usual freeze-drying 'collapse' temperatures of frozen solutions. In ice-containing solutions, these peak temperatures were identified with the transition known as 'onset of ice melting', T_m',[4] or the 'freeze-drying collapse', T_c'.[2,4] In solutions without ice, these peaks were associated with the 'softening' or 'collapse' temperature, T_c.[5]

4 Discussion

4.1 Lines and Transitions in the Supplemented State Diagram

The supplemented state diagram is currently used to characterise the physical state of amorphous materials as a function of temperature and water content. It represents not only true equilibrium transitions, but rather a combination of both equilibrium and metastable (or kinetically stable) transitions.

When a dilute sucrose solution is cooled, the formation of ice progresses as a function of temperature, following the ice-liquid equilibrium curve (T_m), until the viscosity of the unfrozen liquid becomes so high that the freeze-concentrated material passes into the glassy state. The melting point depression can be adequately described by Equation 1, also known as a 'stretched exponential' function (A. Besson, private communication), *viz.*:

$$T_m(x) = -\left[-A\ln\left(1 - \frac{x}{C_0}\right)\right]^{1/B} \tag{1}$$

where A, B and C_0 are parameters, and x denotes solute concentration. The 'stretched exponent' of the melting point depression line is the parameter, B; C_0 is also adjusted in the fitting procedure, and can be seen as a limiting solids concentration, with the calculated melting point diverging at C_0. The above equation has been fitted to melting point data given by Blond *et al.*,[6] Luyet and Rasmussen[7] and Shalaev;[8] the best estimates obtained for the three parameters were $A = 4.69$, $B = 0.68$ and $C_0 = 0.87$.

The concentration dependence of T_g is described by a Gordon–Taylor equation,[4] *viz.*:

$$T_g(x) = \frac{xT_{gs} + k(1-x)T_{gw}}{x + k(1-x)} \tag{2}$$

where x is the fraction of solute, k is a constant, T_{gs} is the T_g of the pure solute and T_{gw} is the T_g of water. In this study, the values obtained for sucrose by Roos[4] using DSC, *i.e.* $T_g$sucrose $=62\,°C$ and $k=5.42$, and using $T_g$water $=-135\,°C$, were assumed as valid.

The intersection of the glass transition and melting curves gives $T_g' = -45.8\,°C$ and $C_g' = 81.8\%$ (w/w) (note that C_g' is not necessarily the same as C_0). These values are essentially identical to those given by Roos[4] from DSC ($T_g' = -46\,°C$, $C_g' = 81.7\%$ (w/w)), but T_g' is lower than the values reported by Simatos and Blond[9] ($T_g' = -37\,°C$, $C_g' = 81.0\%$ (w/w)) from the UNIQUAC model, Ablett *et al.*[10] ($T_g' = -40.0\,°C$, $C_g' = 81.2\%$ (w/w)) from DSC, and recently by Liesebach and Lim[11] ($T_g' = -40.7\,°C$, $C_g' = 81.0\%$ (w/w)).

Both melting and glass transition curves are plotted in Figure 4. For solutions without ice, DMTA E'' loss modulus peak temperatures and loss tangent peak temperatures, for the three frequencies of measurement, are also included. The mechanical data are located above the DSC T_g line, although they clearly show a similar curvature. In order to establish a correlation between DSC and DMTA measurements, an Arrhenius model was used to describe the dependency of the peak temperatures on frequency; the activation energy, E_{act}, was calculated from the slope of this dependence, and used to extrapolate to lower frequencies. A good correlation between the DSC T_g onset and the extrapolated DMTA E'' peak temperature was found to lie in the vicinity of 0.001 Hz. At that frequency, the average molecular relaxation time, τ, is 159 s (calculated as $\tau = 1/2\pi f$), which is very close to the commonly assumed value for the molecular relaxation time at the onset of the DSC glass transition (200 s).[12] As measurements at 0.001 Hz are not feasible, extrapolation of mechanical measurements must always be done to obtain an estimate analogous to the 'DSC T_g', for the construction of any

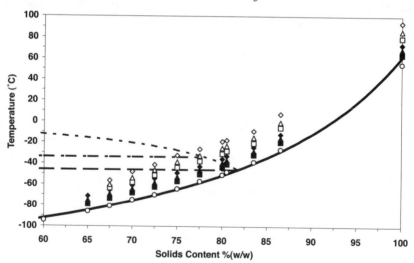

Figure 4 *Supplemented state diagram for sucrose, showing: 'stretched exponential' melting line (- -); Gordon–Taylor T_g line, $T_g(x)$ (—); Roos' $T_c' = -34\,°C$ (·—) and $T_g' = -46\,°C$ (——) lines;[4] E'' peak temperatures measured at: 0.1 (■), 1 (▲) and 10 (◆) Hz and extrapolated to 0.001 (○) Hz (T_g); and δ-phase angle peak temperatures at 0.1(□), 1 (△) and 10 (◇) Hz (associated with Tc)*

supplemented state diagram. Loss modulus peak temperatures at usual frequencies (as well as 'crossover' temperatures and, obviously, loss tangent peak temperatures) are in the unstable rubbery state, and cannot be used for the construction of supplemented state diagrams. Inspection of Figure 4 indicates that for the E'' peak temperatures, the extrapolated values fall on the Gordon–Taylor line that describes Roos' DSC data.[4]

The supplemented state diagram, covering sub-zero temperatures, for aqueous sucrose solutions containing ice, is shown in Figure 5. The E'' peak temperatures were extrapolated to 0.001 Hz for definition of T_g', associated with the maximally freeze-concentrated solution, and the δ-phase angle peak temperatures were extrapolated to 0.001 Hz to define the collapse temperature (T_c'); the extrapolated values are explicitly depicted in this plot. All peaks corresponded again to the final slow heating scan, after six cycles, intended to produce maximum ice formation.

The concentration dependencies of T_g' and T_c' were found to be:

$$T_g'(0.001\ \text{Hz}) = -38 - 0.25 * x \tag{3}$$

$$T_c'(0.001\ \text{Hz}) = -32.1 - 0.113 * x \tag{4}$$

respectively. The concentration dependence of T_c' was rather weak, and T_c' compared well with that based on DSC data,[4] $T_c' = -34\,°C$; indeed, good agreement was found between our experimental data and theoretical predictions, for concentrations from 2 to 55% (w/w), but this agreement broke down at higher concentrations. For $T_g'(0.001\text{Hz})$, a stronger concentration dependence

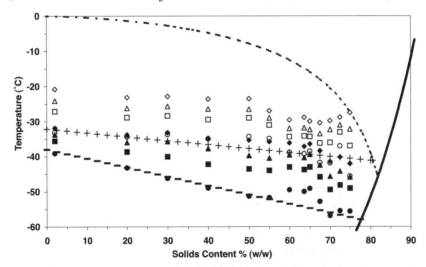

Figure 5 *Supplemented state diagram for aqueous sucrose solutions in the presence of ice: 'stretched exponential' melting line (- -); Gordon–Taylor T_g line, $T_g(x)$ (—); E'' peak temperatures at 0.1 ($\square$), 1 ($\triangle$) and 10 ($\diamond$) Hz and extrapolated to 0.001 ($\bigcirc$) Hz (T_g'); δ-phase angle peak temperatures at 0.1 ($\blacksquare$), 1 ($\blacktriangle$) and 10 ($\blacklozenge$) Hz and extrapolated to 0.001 ($\bullet$) Hz (T_c'); linear fit for $T_g'(0.001\ Hz)$ (+ +); linear fit for $T_c'(0.001\ Hz)$ (— —)*

was found, which could not be eliminated by our cycling/annealing treatments (Roos' DSC data yielded $T_g' = -46\,°C$).[4] A possible explanation could be that DMTA transitions are associated with viscous relaxation, whereas the DSC transition is associated with enthalpic relaxation. In solutions without ice, it was observed that the transition temperatures were essentially the same when the viscous and enthalpic relaxation times were equal.

4.2 Global WLF Model

Although the viscosity of sucrose aqueous solutions has been extensively studied at temperatures above $0\,°C$, little information is available to date for sub-zero temperatures and high sucrose concentrations. Parker and Ring[13] used published data for sucrose solution viscosities to model the temperature- and concentration-dependence of viscosity across the supplemented state diagram. In our study, DMTA data were combined with solution viscosity data[14] to give a complete description of the temperature- and concentration-dependence of the 'average' viscous molecular relaxation time (and corresponding viscosity). The WLF equation was used to model such dependencies, and it was inverted for the calculation of iso-viscosity lines:

$$\log\tau = \log\tau_g - \frac{C_1(x)(T-T_g(x))}{C_1(x)+T-T_g(x)} \tag{5}$$

The sucrose viscosity data were converted to viscous molecular relaxation times using the Maxwell relationship:

$$\tau(T) = \frac{\eta(T)}{G_{inf}} \tag{6}$$

where η is shear viscosity, and G_{inf} is the limiting modulus. These data were then combined with discrete mechanical data, *viz*.:

$$\tau(T_{peak}) = \frac{1}{2\pi f_{peak}} \tag{7}$$

and plotted *versus* $T_g(0.001\,Hz)/T$. In order to improve the fit to the non-linear dependence predicted by the WLF model, δ-phase angle values were also included in the fit, using a shift factor for each set of DMTA data.

At low sucrose concentrations, no mechanical data could be obtained; at high concentrations, in turn, no viscosity data were available, due to sucrose crystallisation. However, the adjustable parameters in the WLF equation were fitted, so as to cover the whole concentration range.

Initially, the WLF equation was fitted to the data by imposing the known concentration dependence of $T_g(x)$, and an individual fit was carried out for the data pertaining to an aqueous 70% (w/w) sucrose solution, as shown in Figure 6. As both DMTA and viscosity measurements were available for this concentration, a good first estimate was obtained for the other individual fits. The value for $\log\tau$ at T_g was initially set to 2.2, as this corresponded to a molecular relaxation

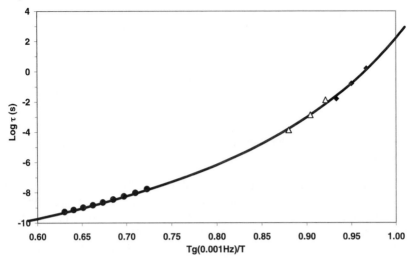

Figure 6 *Relaxation time, τ, vs. (T$_g$/T) for an aqueous 70% (w/w) sucrose solution: WLF fit (—); τ from E″ peak temperatures (◆); τ from δ-phase angle peak temperatures (△); and τ from sucrose viscosity data (●)*

time of 159 sec at T_g; C_1 and C_2 were initially made equal to the universal WLF constants, 17.4 and 51.6,[3] respectively, and allowed to float for the subsequent fit. Then, log $G_{inf} = 8.34$ was determined, and this was used as a fixed parameter for the individual fits at other concentrations. For these individual fits, a strong non-linear dependence of C_2 on concentration was found. Therefore, in the global fit, Gordon–Taylor concentration dependencies were imposed on C_1 and C_2, as:

$$C_i = \frac{C_{i1}x + k_i C_{i0}(1-x)}{x + k_i(1-x)} \tag{8}$$

where C_{i1} is the C parameter (1 or 2) for pure sucrose, C_{i0} is the C parameter (1 or 2) for water, x is solids content, and k_i is the constant describing the curvature.

For the global fit of the WLF equation, the mechanical and viscosity data for the various sucrose concentrations were fitted by a single expression, with all parameters allowed to float, and the Gordon–Taylor parameters for $C_i(x)$, as well as $\log(G_{inf})$ and $\log(\tau_g)$, were adjusted. The latter two parameters were considered to be independent of concentration. The final results from the WLF global fit are shown in Table 1.

A weak concentration dependence was found for C_1, increasing linearly from 17.5 for water to 19.0 for sucrose. For C_2, a strong concentration dependence was noted, with $C_2 = C_{20} = 17.4$ for water and $C_2 = C_{21} = 56.8$ for pure sucrose, corresponding to a dramatic decrease in 'fragility' with concentration,[15] as the WLF temperature dependence approached Arrhenius behaviour ($C_2 = 335$).

Table 1 *WLF global fit parameters*

	C_1		C_2		
C_{10}	17.53	C_{20}	17.39	$\log(G_{inf})$ Pa	9.54
C_{11}	18.96	C_{21}	56.78	$\log(\tau_g)$ s	3.22
K_1	1.00	K_2	1.82	$\log(\eta_g)$ Pa s	12.76

4.3 Iso-viscosity Lines Overlaid on the Supplemented State Diagram

Iso-viscosity lines were determined by calculation of $T(x)$ from the WLF global fit, and for fixed values of τ and viscosity. $T_g(x)$ was imposed as an iso-viscosity line by the WLF equation, and the viscosity at T_g was found to be $10^{12.8}$ Pa s, compared with $10^{10.3}$ Pa s reported by Parker and Ring,[13] who used only 'high temperature' sucrose viscosity data. The combining of our DMTA data with sucrose viscosity data available from the literature enabled a considerable improvement in the determination of the viscosity at T_g. Our value agreed very well with that of Soesanto and Williams,[16] who found $10^{13.1}$ Pa s for 100% sucrose. However, we did not confirm their finding of a concentration-dependent viscosity at T_g, or concentration-independent 'universal values' for C_1 and C_2. Additionally, good agreement was found between Shalaev and Franks'[5] softening temperature data at high concentrations and our estimate for $T_c(0.001\ Hz)$, which corresponded to the iso-viscosity line at $10^{9.2}$ Pa s, obtained from the WLF global fit, as shown in Figure 7.

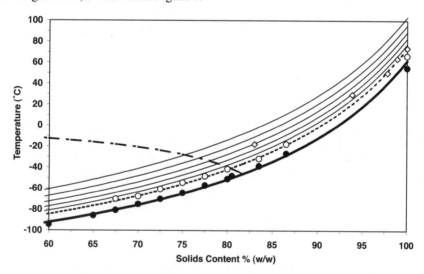

Figure 7 *Supplemented state diagram for sucrose, showing: T_m line (- —); $T_g(x)$ line (—); $T_g(0.001\ Hz)$ line (●); $T_c(0.001Hz)$ line (○); iso-viscosity line at $10^{9.2}$ Pa s (- -); $T_{softening}$ line from Shalaev and Franks[5] (◇); as well as other iso-viscosity lines (—) separated by a decade in viscosity*

5 Conclusions

With respect to the establishment of supplemented state diagrams, it is clear that determination of E'' and δ-phase angle peak temperatures *via* DMTA is far more definitive and accurate than *via* DSC onset (endothermic) glass transition temperatures; in fact, the former technique results in well-defined peaks. The disc-bending sample holder allows measurements to be performed on samples of very low solids contents, as well as repeated cycling on concentrated solutions with and without ice. In the case of powders, the amorphous structure can be prepared directly in the disc-bending sample holder by pressing the powder into a disc and melting it *in situ*.

The Arrhenius extrapolation of the E'' peak temperatures to 0.001 Hz agreed well with the DSC onset T_g values, thus corresponding to an 'average' molecular relaxation time (τ) of 159 s at T_g. The sharply defined peaks of the δ-phase angle, when extrapolated to T_c' (0.001 Hz) and $T_c(0.001$ Hz), exhibited good correlations with the freeze-drying collapse and DSC softening temperatures, respectively.

When a global WLF equation was fitted to DMTA mechanical data combined with viscosity data, a strong concentration dependence was found for C_2, and a weak concentration dependence for C_1 improved the fit. Both G_{inf} and $\log \tau_g$ were taken to be concentration-independent, and thus the viscosity at T_g, $10^{12.8}$ Pa s was also concentration-independent. Note that because of the strong WLF temperature dependence (or 'fragility'), the actual fitted value for the 'average' relaxation time at T_g was an order of magnitude greater than that found by Arrhenius extrapolation, $\tau_g = 10^{3.22} = 1660$ seconds. Iso-viscosity lines could be generated from the WLF equation and added to the supplemented state diagram. T_g was an iso-viscosity line ($\eta = 10^{12.7}$ Pa s), and $T_c'(x)$ and $T_c(x)$ ($\eta = 10^{9.2}$ Pa s) transition lines were then identified as the maximum temperatures (or viscosities) for successful freeze-drying without collapse of porous powdered structures, as was also concluded by Shalaev and Franks.[5] Other iso-viscosity lines could eventually help to define regions of accessible viscosity for processing, for example in extrusion, crystallisation and concentration.

References

1. F. Franks, M.H. Asquith, C.C. Hammond, H.B. Skaer and P. Echlin, *J. Microsc.*, 1977, **110**, 223.
2. H. Levine and L. Slade, *Carbohydr. Polym.*, 1986, **6**, 213.
3. M.L. Williams, R.F. Landel and J.D. Ferry, *J. Am. Chem. Soc.*, 1955, **77**, 3701.
4. Y. Roos, 'Phase Transitions in Foods', Academic Press, San Diego, 1995, Chapter 5, 109.
5. E.Y. Shalaev and F. Franks, *J. Chem. Soc. Faraday Trans.*, 1995, **91**, 1511.
6. G. Blond, D. Simatos, M. Catté, C.G. Dussap and J.B. Gros, *Carbohydr. Res.*, 1997, **298**, 139.
7. B. Luyet and D. Rasmussen, *Biodynamica*, 1968, **10**, 167.

8. E.Y. Shalaev, PhD Thesis, Institute of Molecular Biology, Kolsovo, 1991 and personal communication.

9. D. Simatos and G. Blond, 'The Glassy State in Foods', J.M.V. Blanshard and P.J. Lillford (eds.), Nottingham University Press, Nottingham, 1993, Chapter 19, 395.

10. S. Ablett, M.J. Izzard, P.J. Lillford, *J. Chem. Soc. Faraday Trans.*, 1992, **88**, 789.

11. J. Liesebach and M. Lim, poster presented at 'The Amorphous State – A Critical Review', Churchill College, Cambridge, May, 2001.

12. C.T. Moynihan, N. Balitactac, L. Boone and T.A. Litovitz, *J. Chem. Phys.*, 1971, **55**, 3013.

13. R. Parker and S.G. Ring, *Cryo-Lett.*, 1995, **16**, 197.

14. R.S. Norrish, 'Selected Tables of Physical Properties of Sugar Solutions', BFMIRA, Scientific and Technical Surveys, Leatherhead, 1967, No. 51, 127.

15. C.A. Angell, *Polymer*, 1997, **38**, 6261.

16. T. Soesanto and M.C. Williams, *J. Phys. Chem.*, 1981, **85**, 3338.

Glassy State Dynamics and its Significance for Stabilization of Labile Bioproducts

Glassy State Dynamics, its Significance for Biostabilisation and the Role of Carbohydrates

Roger Parker, Yvonne M. Gunning, Bénédicte Lalloué, Timothy R. Noel and Steve G. Ring

FOOD MATERIALS SCIENCE DIVISION, INSTITUTE OF FOOD RESEARCH, NORWICH RESEARCH PARK, COLNEY LANE, NORWICH NR4 7UA, UK

1 Introduction

The aim of this chapter is to review recent work on the physical chemistry linking amorphous carbohydrates, glassy state dynamics and biostabilisation. While the stabilisation of labile biomolecules in amorphous carbohydrate glasses is a well developed technology,[1,2] aspects of its physico-chemical basis remain poorly understood. In this chapter, we first review the relaxation dynamics of glassy and near-glassy carbohydrates, using recent experimental results.[3-8] Secondly, focusing on chemical stability, we describe some theoretical approaches connecting chemical reaction kinetics and glassy state dynamics.[9] For simplicity, that section is limited to reactions in a single homogeneous phase; reactions in heterogeneous systems,[10,11] e.g. vitrified emulsions, are beyond the scope of the current review, though, as is described later, immiscibility and phase separation are common in carbohydrate-rich mixtures. Seeking a link between biostabilisation and glassy state dynamics is one way in which the 'vitrification hypothesis'[1,2] can be elaborated; another complementary hypothesis, first applied to anhydrobiotic organisms, is the 'water replacement hypothesis'.[12,13] Conversion of an aqueous flavour emulsion into a carbohydrate-encapsulated flavour is a further example of a stabilisation system in which water is replaced by a glassy carbohydrate. Studies of the effect of water content on release from carbohydrate-encapsulated flavours indicate extensive flavour–carbohydrate immiscibility.[14,15] Characterisation of the phase behaviour of amorphous carbohydrates with water and aliphatic alcohols shows to what extent the solvent properties of amorphous carbohydrates are similar to those of water. One role of carbohydrates in biostabilisation is as a non-volatile solvent and diluent, a role that is clearly

73

related to the 'water replacement hypothesis'. Finally, some future research directions and potential developments of glassy state technology are indicated.

2 Glassy State Dynamics of Amorphous Carbohydrates

Collective molecular motions in amorphous solids give rise to relaxation processes and time-dependent material properties.[16] The rates of these collective motions are characterised by (mean) relaxation times. For example, a shear-stress relaxation time might be measured in a step-response experiment in which a small instantaneous shear strain is applied to a material, resulting in a jump in shear stress, which subsequently decays (relaxes) with time. For the primary relaxation of amorphous solids, the shear-stress relaxation curve is commonly non-exponential and can be fitted with a stretched exponential response function, $\exp(-(t/\tau)^{\beta})$, to determine the relaxation time, τ.[17] Viscosity measurements and mechanical and dielectric spectroscopy can be used to characterise relaxation times.[3,4] The relaxation times are temperature-dependent and conventionally plotted in an Arrhenius plot known as a 'relaxation map'. A particularly comprehensive study of glucitol is shown in Figure 1.

As liquid glucitol is cooled, the mean relaxation time for the primary or α-relaxation process initially increases in a non-Arrhenian manner, with an increasingly large activation energy, before reaching a discontinuity at about 10^4 s. This discontinuity indicates the glass transition, and in this example, the glass transition temperature, T_g, is defined as the temperature at which the mean relaxation time is 10^4 s. At temperatures above T_g, the temperature dependence of the relaxation time can be described by the Vogel–Tammann–Fulcher (VTF)

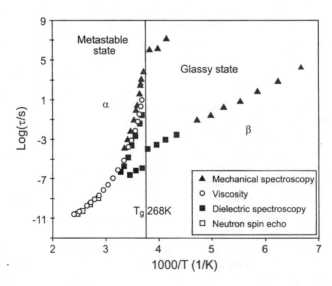

Figure 1 *Relaxation map of glucitol (sorbitol) above and below its glass transition. Based upon A. Faivre,[3] with permission*

equation, $\tau = \tau_0 \exp(B/(T-T_0))$, or equivalently by the Williams–Landel–Ferry (WLF) equation.[18] A second, faster relaxation process, termed the secondary or β-relaxation, is observed at lower temperatures; it shows Arrhenius temperature dependence. While this relaxation is observed predominantly in the glassy state, it can be observed at temperatures above T_g, where it ultimately merges with the α-relaxation process. A detailed assignment of the molecular motions giving rise to these relaxations has not yet been achieved. Roughly speaking, the β-relaxation has some of the characteristics of a localised, short-time, cage-rattling motion (see Angell[19] 'question 4'), whereas the α-relaxation, the main structural relaxation, occurs over longer timescales and corresponds to the cages formed by nearest neighbours relaxing and allowing diffusive motions. During structural relaxation, the system is exploring its energy landscape, and for this reason, when the structural (primary) relaxation time exceeds experimental timescales, the system becomes non-equilibrium, *i.e.* in the glassy state the system no longer fully explores its energy landscape during an experiment.

In complementary studies on amorphous glucose[5] and glucose–water mixtures,[6] specialised NMR techniques have probed the slow rotational motions associated with the primary relaxation. Moran and Jeffrey[6] found that, at temperatures close to T_g, conventional NMR relaxation time measurements probed processes that were faster than the primary relaxation, and at temperatures below T_g, these faster relaxations had lower activation energies than those of secondary relaxations observed by dielectric techniques. Even single-component glass-formers have complicated dynamics that require specialised techniques to probe them.[5] Dielectric spectroscopic studies of secondary relaxations in carbohydrates and their water mixtures show variation with molecular structure.[7]

Figure 2 shows temperature scans of the dielectric $\tan \delta$ (= dielectric loss, ε''/dielectric constant, ε') at 1 kHz, for a range of dry amorphous carbohydrates. The main peak in $\tan \delta$ is due to the primary relaxation, and as these measurements are at 1 kHz, it peaks at a temperature above the calorimetrically determined T_g. In these scans, T_g corresponds to the position of the low-temperature shoulder of the primary relaxation peak. Peaks in $\tan \delta$ at temperatures below the primary relaxation peak are due to secondary relaxation processes. In glucitol, an acyclic carbohydrate, the secondary relaxation is strong and occurs at temperatures not far below the primary relaxation, so that the two relaxations overlap, even at 1 kHz. A comparison of the secondary relaxations in the monosaccharide D-mannose, which has a hydroxymethyl group at C6, and its deoxy-sugar L-rhamnose, which has a methyl group at C6, indicates that the presence of a polar exocyclic group has a strong influence on the secondary relaxation. In the disaccharide maltose, the primary relaxation is relatively weak, and the secondary relaxation peak is deep in the glassy state ($-50\,°C$). The full significance of secondary relaxation processes to functionality is not yet understood, though polymer scientists associate strong secondary relaxations with enhanced transport properties, such as high gas permeability.[20]

Glassy state dynamics has an extensive and fast-growing literature. For access to this, the reader is directed to other recent reviews.[17,21–23] One topical issue is

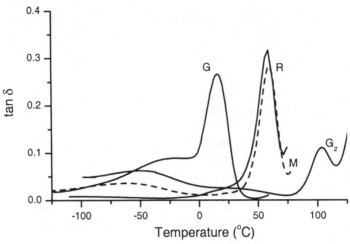

Figure 2 *Primary and secondary dielectric relaxations for a range of dry amorphous carbohydrates at 1 kHz. Key: G, glucitol; R, L-rhamnose; M, D-mannose; G_2, maltose*

the non-exponential nature of the primary relaxation, which has been associated with dynamic heterogeneity.[22,23] The dynamical structure is one in which groups of molecules rearrange cooperatively. A recent DSC-based technique gives a length scale of 3–5 nm for the size of these regions in glucitol at T_g, which corresponds to about 100 molecules rearranging cooperatively.[8] Progress has also been made in applying Tool–Narayanaswamy–Moynihan (TNM) models of physical aging to glassy carbohydrates,[24] which should allow strategies for achieving optimal aging to be developed.

3 Chemical Kinetics and the Glassy State

Preservation can be characterised in terms of the timescale of deteriorative reactions. Typical desirable timescales are in excess of 10^6 s (11.6 days) to 10^8 s (3.2 years). In a glass, the timescale of the main structural relaxation (see Figure 1) is comparable to that of preservation, but would the chemical reactions of species dispersed in a glass be expected to occur over a similar timescale? Distinct approaches to unimolecular and bimolecular reactions must be taken.[9] Whereas a unimolecular reaction can be modelled as a process of passing over an energy barrier, a bimolecular reaction also involves a diffusive step, in which the reactants diffuse together prior to reaction. Thus, the very high viscosity of a glass may affect a reaction by decreasing the frequency of diffusive encounters of molecules and by slowing the rate at which reactive molecules pass over energy barriers.

For diffusion-controlled bimolecular reactions, Smoluchowski's theory[25,26] can be used to estimate reaction rate. This predicts that the second-order rate constant $k_2 = 8RT/3\eta$, where R is the gas constant, T the absolute temperature and η the viscosity. At T_g, η is about 10^{12} Pa s.[21] The half-life for the reaction is $t_{1/2} = 1/k_2[A]_0$,[27] where $[A]_0$ is the initial reactant concentration. The half-lives for some typical reactant concentrations are shown in Table 1. The half-life varies between systems, simply because $[A]_0$ is varying. This has a dramatic effect, although the emulsion and protein are predicted to be stable (with respect to aggregation) for timescales greater than years; the molecular reactant is predicted to react over a timescale of 17.5 days, an unacceptably short half-life for many applications. There are, however, a number of assumptions implicit in these predictions, which require further examination.

One assumption in Smoluchowski theory is that of angular-independent reactivity. While this is acceptable for applications to emulsions and, possibly, to proteins, it is inappropriate for most chemical reactions that are subject to steric constraints. Solc and Stockmayer[28,29] developed a theory in which the reactivity was angular-dependent; it depended upon the size of circular reactive patches. Reactive patch sizes can plausibly be estimated using simple geometric arguments. This effect is estimated to reduce the diffusion-controlled rate by a factor of 10^2 to 10^3 for small molecular reactants.[30]

Another potential shortcoming of Smoluchowski theory is the assumption of diffusion control. In aqueous solution, diffusion-controlled reactions appear to be unusual, the few examples being well-documented,[25] and most reactions are reaction- or activation-controlled. Collins and Kimball[31] modified the Smoluchowski theory to include the effects of finite chemical reactivity, *i.e.* in contrast to the Smoluchowski approach, on molecular encounter, there is a possibility that molecules may simply diffuse apart again without reaction. At steady state, the overall bimolecular rate constant, k, for the reaction is predicted to be

$$k = \frac{4\pi r_c D k_{act}}{4\pi r_c D + k_{act}} \tag{1}$$

where $4\pi r_c D$ is the diffusion-controlled rate, r_c the collision diameter, D the relative diffusivity and k_{act} the reaction-controlled rate constant. This theory predicts that, as the relative diffusive mobility of reacting species is reduced, there is a crossover from reaction- to diffusion-control, as shown in Figure 3.

The crossover occurs when $k_{act} = \pi r_c D$. An outcome of this theory is that slow

Table 1 *Half-lives for diffusion-controlled reaction at* T_g, *predicted using Smoluchowski theory*

Material	Concentration	$t_{1/2}$
Emulsion	20% v/v 1 μm diameter	7.5×10^6 years
Protein	10% w/w 0.75 cm^{3}g^{-1} 5 nm diameter	2.3 years
Molecule	100 mM	17.5 days

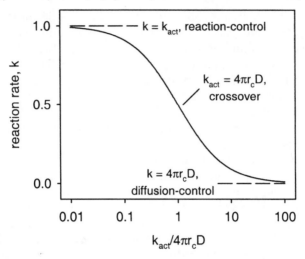

Figure 3 *Crossover between reaction-control and diffusion-control for a bimolecular reaction with a finite rate*

reactions will remain reaction-controlled, until the diffusive mobility is sufficiently small for the crossover to occur. It may be that, even at the glass transition, there is sufficient mobility for a reaction to remain reaction-controlled, and under these conditions, it would be expected to exhibit classical Arrhenius temperature-dependence.[27,30]

For reactions of small molecules, the most serious shortcoming of the Smoluchowski prediction of reaction rate is the use of the Stokes–Einstein (SE) relationship,[32] $D = kT/6\pi\eta r_h$, where k is Boltzmann'sconstant and r_h a hydrodynamic radius (modelling the molecule as a spherical particle). The SE relationship underestimates the diffusivity, both in the case of small molecules and when the viscosity is high.[32] Figure 4 shows a test of the SE relationship for fluorescein diffusing in sucrose-water mixtures, measured using fluorescence recovery after photo-bleaching (FRAP).[33]

The ratio $T/D\eta$ ($\propto$ hydrodynamic radius) is constant at temperatures well above the glass transition ($T \geq T_g/0.86$), as predicted by the SE relationship; however, at temperatures close to the glass transition ($T < T_g/0.86$), the relationship breaks down. At T_g, the diffusivity is about 10^7 faster than that predicted by the SE relationship. For small molecules, the factor by which the SE relationship underestimates diffusion can be larger,[34] *e.g.* for ethanol, $\sim 10^7$ and water, $\sim 10^9$. Conductivity measurements can also be applied to measure translational mobility in near-glassy amorphous carbohydrates. The molar conductivity, Λ_m, of a symmetrical electrolyte (ion charge ze) is related to the self-diffusion coefficients of the ions, D_+ and D_-, through the Nernst–Einstein relationship, $\Lambda_m = z^2 F^2 (D_+ + D_-)/RT$, where F is the Faraday constant.[35] Figure 5a shows a T_g-scaled Arrhenius plot of the molar conductivity of KCl in a series of amorphous carbohydrate-10% w/w water mixtures. At T_g, the molar conductivity varies in the order monosaccharides < disaccharide < trisaccharides. Some care

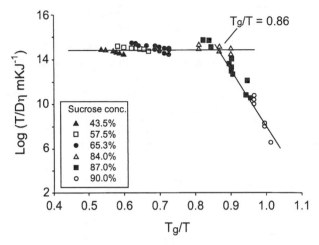

Figure 4 *Testing the Stokes–Einstein relationship for fluorescein diffusion in sucrose–water mixtures. Based upon Champion* et al.,[33] *with permission*

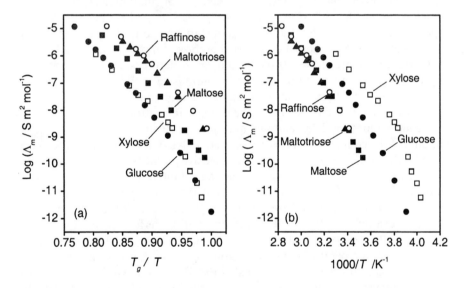

Figure 5 *Molar conductivity of KCl in amorphous carbohydrate–10% w/w water mixtures at temperatures above* T_g*: (a)* T_g*-scaled Arrhenius plot and (b) Arrhenius plot*

needs to be taken in interpreting the change in molar conductivity, as, in addition to the changing ion mobility, there may potentially be a change in ion pairing. Calculations based upon Bjerrum theory[36] suggest that the ion-pairing effect is small, but this is worthy of further investigation. Figure 5b shows the same data as in Figure 5a, but plotted as a conventional Arrhenius plot.

In Figure 5b, the order is reversed; at constant temperature, molar conductivity varies in the order trisaccharides and disaccharide < monosaccharides. Thus, using conductivity as an index of mobility, it would be predicted that, at T_g, preservation in a glassy monosaccharide would minimise mobility, whereas at constant temperature, preservation in a di- or trisaccharide would minimise mobility. However, the magnitude of the conductivities indicates considerable uncoupling of the ionic motions from the main structural relaxation and diffusion coefficients (D_+ and D_-), which would indicate that there are many ion–ion encounters over preservation timescales.

Hagen et al.[9] argued that the effects of glassy state dynamics on the rates of barrier crossing can be predicted using Kramers theory:

$$k = \frac{A}{\eta}\exp(-E_0/RT) \tag{2}$$

where A is a constant and E_0 an energy barrier. This equation (slightly modified) was successfully applied to the rearrangements of electronically excited myoglobin in a trehalose glass, a unimolecular reaction.[37] It could also be applied to the reaction step of bimolecular reactions (k_{act} in Equation 1). We are not aware whether the theory has been tested for reactions of small molecules. Kramers theory is a model for diffusion over an energy barrier, and the inverse viscosity dependence originates from the SE relationship, and so is likely to suffer from similar problems of SE breakdown as the above Smoluchowski predictions. Furthermore, the breakdown is likely to be more severe, since the SE relationship is being applied to more localised sub-diffusive motions, as the reacting system passes through the transition state, motions that occur over a shorter length scale than that characterising the primary relaxations.

In conclusion, the SE breakdown and limitations in our theoretical understanding of the connection between glassy state dynamics and chemical kinetics mean that we are some way from a comprehensive physico-chemical basis for the effect of vitrification on deteriorative reactions of interest to food and pharmaceutical scientists. It should be stressed that diffusion-controlled bimolecular reactions and 'WLF kinetics' are well-founded in the physical chemistry literature. For example, in the 1970s, Dainton et al.[38] performed photochemical studies of diffusion-controlled energy transfer kinetics in glass-forming solvents at near-glassy temperatures. The diffusion-controlled kinetics showed 'WLF' dependence, and further analysis revealed the SE breakdown described above. The experimentally measured second-order rate constants indicated that, depending upon reactant concentration, diffusion-controlled reactions can have half-lives of the order of hours in near-glassy solvents.

On the basis of this discussion, what advice can be given for controlling reaction rate through vitrification? The answer to this question can be no more than our best guess. Firstly, minimise mobility by putting the system into a glass.[1,2] This can be achieved by mixing the material with a carbohydrate in aqueous solution and then drying to a low water content amorphous glass.[1,2] Since aging reduces the diffusivity of gases through glassy polymers, annealing the glass to achieve maximal aging/density should be performed.[39] This can be optimised using TNM models.[24] Minimise mobility by optimising storage temperature and the molecular structure of the carbohydrate (see Figure 5). If storage at ambient temperatures is a constraint, then use higher molecular weight carbohydrates, *e.g.* at least di- or trisaccharides.[40] In order to minimise mobility, it is simpler if only a single phase is present, *i.e.* partially miscible mixtures lead to the complication of minimising mobility in multiple phases. There is now a body of literature showing that chemical reactions can occur in the glassy state,[30,41,42] from which we conclude that vitrification technology does not replace the need to understand the chemistry of the deteriorative reactions in a system. Acid–base catalysis is common, and so pH should be optimised to minimise reaction rate. If the reaction is bimolecular, dilution of the reactants with amorphous carbohydrate increases reaction half-life. If chemical deterioration persists, then analyse the chemistry and look for chemical preservation strategies (there are recent examples of faster deterioration upon encapsulation[43]).

4 Encapsulation and Carbohydrates as Solvents

4.1 Flavour Encapsulation in Glassy Carbohydrates

Liquid flavours are encapsulated in glassy carbohydrates to prevent their evaporation, to protect them from adverse chemical reactions and to convert them into easily handled free-flowing powders.[44] Carbohydrate-encapsulated flavours are prepared by processes such as spray-drying and extrusion.[45–47] To act as encapsulation matrices, glassy carbohydrates must be, to a large extent, impermeable to the components of a flavour. Impermeability could result from flavour components having a low solubility in the matrix, a limited diffusive mobility in the matrix or a combination of both. Furthermore, the flavour must not plasticise the matrix, making it sticky and liable to caking.[40] Thus, miscibility and component partitioning in flavour–carbohydrate–water mixtures are important to the understanding of the function of encapsulation systems.

Flavours contain components of varying hydrophobicity, a property that can be characterised using octanol–water partition coefficients, *P*. Table 2 shows values of log *P* for components present in a cherry flavour; they vary from hydrophobic components that are poorly miscible with water, *e.g.* benzaldehyde and benzyl alcohol, to relatively hydrophilic components, *e.g.* diacetyl, which is fully miscible with water.[15,48,49]

Table 2 *Octanol-water partition coefficients,* P, *for components of a cherry flavour*

Component	Log P
Diacetyl	-1.3
Acetaldehyde	-0.22
Ethyl acetate	1.04
Benzyl alcohol	1.10
Benzaldehyde	1.48

Comparison of partition coefficients in benzyl alcohol–glucitol (a model of the flavour–carbohydrate matrix system) with those in octanol–water (Figure 6) shows that the effect of substituting octanol for benzyl alcohol, and water for glucitol, is to increase the partitioning of species into the hydrophobic phase by at least an order of magnitude.

Overall, only the most hydrophilic components show significant partitioning into the amorphous carbohydrate matrix, and, from the solubility viewpoint, it is only these components that will permeate the matrix. As these components are at low concentrations, and there are small amounts of unencapsulated flavour oil, leakage rates through matrices are difficult to detect. However, it is straightforward to measure release of encapsulated flavours in response to changes in water content (and temperature).[14] This can be achieved by conditioning samples in different humid atmospheres, followed by sealing samples in headspace vials and analysing the headspace concentration after 24 hours. Figure 7 shows that release increases, in response to both increases and decreases in the 'as prepared' water content (3.5% w/w).

After an increase in water content, microscopy showed that the sucrose in the matrix had crystallised, causing droplets of flavour oil to break the surface of the matrix, thus allowing release into the headspace. Release upon reduction of matrix water content was thought to be due to the glassy matrix cracking under drying stresses. Even the 'as prepared' material showed small levels of release that was attributed to the evaporation of small amounts of surface oil and sub-surface oil in cracks. Overall, there was no unambiguous evidence for the matrix being permeable to any of the flavour components. The release of all components followed essentially the same pattern, and could be explained in terms of processes by which the flavour phase gained direct access to the headspace. Experiments using single-phase mixtures would aid unambiguous identification of matrix permeability.

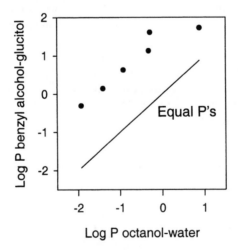

Figure 6 *Correlation between benzyl alcohol–glucitol partition coefficients (70 °C) and octanol–water partition coefficients (20 °C) for a series of alcohols*

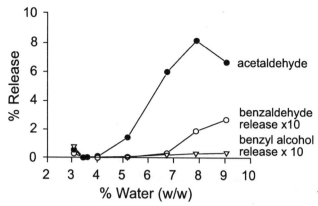

Figure 7 *Effect of water content on headspace release of a model cherry flavour encapsulated in a initially amorphous sucrose–maltodextrin matrix*[14]

4.2 Solvent Properties of Amorphous Carbohydrates

The above partitioning studies show that, although amorphous carbohydrates, like water, are extensively hydrogen-bonded, their solvent properties are not identical. Maltotriose is a convenient material to use to study the general solvent behaviour of a pure amorphous carbohydrate, since, to our knowledge, it has never been observed to crystallise (unless derivatised[50]), and so only amorphous states appear in its state diagram. 50:50 w/w mixtures of dry glassy maltotriose ($T_g = 134\,°C$) with water, dry methanol, ethanol, *n*-propanol, and *n*-butanol, at ambient temperature, show an interesting range of phenomena. Whereas maltotriose and water are completely miscible and form a single solution phase, the alcohol mixtures are only partially miscible and form two separate phases. Methanol is sufficiently miscible to swell the maltotriose particles, plasticising the mixture through its glass transition and causing the viscous, maltotriose-rich, liquid phase to collapse. In contrast, the ethanol, *n*-propanol and *n*-butanol mixtures are insufficiently miscible to plasticise the maltotriose-rich phase through the glass transition. On stirring these mixtures, the maltotriose-rich particles remain discrete, with no apparent stickiness. Analysis of the alcohol-rich phase does reveal some solubility of the glassy maltotriose-rich particles in the alcohols; it is not yet known whether this non-equilibrium solubility is sensitive to aging. The physics of the situation is similar to water sorption isotherms of carbohydrate glasses, wherein one phase is in equilibrium, while the other is not. By variations in composition in ternary maltotriose–alcohol–water mixtures, behaviour can be varied continuously. Figure 8 shows a ternary state diagram for maltotriose–ethanol–water mixtures.

Phase separation only occurs below about 25% w/w water, and below this limit, the water content has a strong influence on the miscibility of the two phases and on the viscosity of the maltotriose-rich phase. Water and ethanol plasticise maltotriose similarly, so that when the combined water and ethanol content is

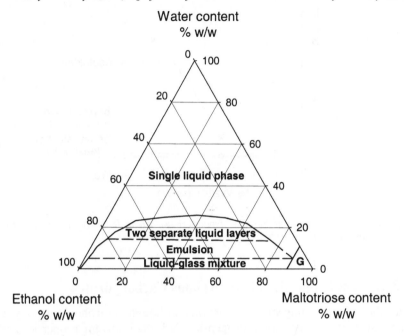

Figure 8 *Maltotriose–ethanol–water state diagram at 20 °C*

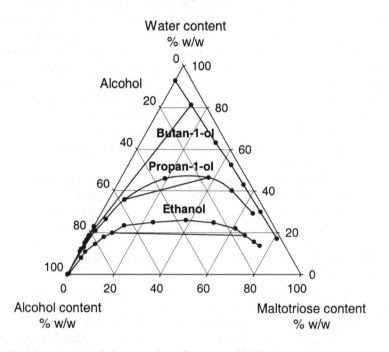

Figure 9 *Maltotriose–alcohol–water phase diagram at 20 °C*

less than 7% w/w, the mixture is glassy. The equilibrium portions of the ethanol, *n*-propanol and *n*-butanol state diagrams are shown in Figure 9.

The tie-lines show that, whereas water partitions fairly equally between the ethanol- and maltotriose-rich phases, as the hydrophobicity of the alcohol increases, the water progressively partitions predominantly into the maltotriose-rich phase. Although these systems are models, they are not far removed from some practical systems, and they exhibit a rich variety of phenomena involving phase separation and glass transitions (*e.g.* critical points, spinodals, sintering, glassy precipitation). Studies of isothermal drying ('selective' diffusion[51]) and mixing and temperature (including freezing) behaviour would all be revealing.

5 Future Directions

5.1 Fundamentals

This chapter has sought to show that we are some way from a fundamental understanding of the relationship between amorphous carbohydrates, glassy state dynamics and biostabilisation. Despite a growing number of studies on molecular mobility, there remain relatively few on diffusion, arguably the simplest form of mobility. For glassy synthetic polymers, Hopfenberg and Berens[52] showed that diffusion coefficients decrease strongly with increasing molecular size, a relationship that has not been extensively examined for small molecule glass-formers such as carbohydrates.[34,51] An improved molecular understanding of diffusion and the SE breakdown would be a worthwhile goal. A recent comparative study of the predictive capabilities of free volume and multi-dimensional transition state theories, with regard to diffusion in polymers, found some shortcomings in each approach.[53]

A more comprehensive understanding of the relationship between glassy state dynamics and the whole range of chemical reaction kinetics is much needed. In isolated areas such as diffusion-controlled energy transfer[38] and the relaxation of the electronically excited states of globular proteins,[37] there is relatively good understanding, but this leaves many areas poorly studied, including those of relevance to practical problems of biostabilisation. For example, there is a striking correlation between spin-probe mobility (a partially uncoupled motion) and plant seed viability, which demands a more mechanistic understanding.[54]

Low-water-content and dry amorphous carbohydrates have their own distinct solvent properties. Drying into the glassy state causes solute repartitioning (Figures 8 and 9 and Golovina *et al.*[55]) and results in a lower dielectric constant[56] and increased ion-pairing,[36] which will affect pH. In studies of chemical reaction kinetics, these particular properties might usefully be studied as 'kinetic medium effects'.[57] Research in this general area should relate to the 'water replacement hypothesis'.[13]

5.2 Applications

Some current ideas that could be developed include: (1) Calorimetric measurement and TNM modelling of physical aging[24] could lead to optimised annealing and storage processes, with potentially beneficial results. (2) Non-crystallising, higher malto-oligomers such as maltohexaose may find uses, as they combine high glass transition temperatures with reduced hygroscopicity. They are currently considerably more expensive than the closely related cyclodextrins. The analogous maltodextrin may suffice, but these tend to contain higher molecular weight oligomers that can potentially aggregate and crystallise. (3) The maltotriose–ethanol–water state diagram (Figure 8) suggests that it may be possible to precipitate glassy maltotriose from solution, by either mixing or quenching a solution into the liquid–glass coexistence region, a process of 'glassy precipitation'. This process should be arrested by the glass transition and would result in particularly small particles. (4) In addition to providing stabilisation, encapsulation matrices need to release encapsulated species as required.[46,47,58] With increased knowledge of individual nutritional and health requirements, this may create a demand for foods with site-directed release functionalities. Flavour encapsulation systems that trigger release in response to more diverse changes in physico-chemical environment can also be envisaged.

References

1. F. Franks, *Biotechnology*, 1994, **12**, 253.
2. H. Levine and L. Slade, *BioPharm*, 1992, **5(4)**, 36.
3. A. Faivre, PhD Thesis, INSA de Lyon, France, 1997.
4. A. Faivre, G. Niquet, M. Maglione, J. Fornazero, J.F. Jal and L. David, *Eur. Polym. J. B*, 1999, **10**, 277.
5. D. van Dusschoten, U. Tracht, A. Heuer and H.W. Speiss, *J. Phys. Chem. A*, 1999, **103**, 8359.
6. G.R. Moran and K.R. Jeffrey, *J. Chem. Phys.*, 1999, **110**, 3472.
7. T.R. Noel, R. Parker and S.G. Ring, *Carbohydr. Res.*, 2000, **329**, 839.
8. E. Hempel, G. Hempel, A. Hensel, C. Schick and E. Donth, *J. Phys. Chem. B*, 2000, **104**, 2460.
9. S.J. Hagen, H.J. Hofrichter, H.F. Bunn and W.A. Eaton, *Transfusion Clinique et Biologique*, 1995, **2**, 423.
10. S. Anandaraman and G. A. Reineccius, *Food Technology*, 1986, **40(11)**, 88.
11. J. Imagi, K. Muraya, D. Yamashita, S. Adachi and R. Matsuno, *Biosci. Biotech. Biochem.*, 1992, **56**, 1236.
12. J.H. Crowe, J.F. Carpenter and L.M. Crowe, *Ann. Rev. Physiol.*, 1998, **60**, 73.
13. J.H. Crowe, J.S. Clegg and L.M. Crowe, in 'The Properties of Water in Foods – ISOPOW6', D.S. Reid (ed.), Blackie, London, 1998, 440.
14. Y.M. Gunning, P.A. Gunning, E.K. Kemsley, R. Parker, S.G. Ring, R.H. Wilson and A. Blake, *J. Agric. Food Chem.*, 1999, **47**, 5198.
15. Y.M. Gunning, R. Parker, S.G. Ring, N.M. Rigby, B. Wegg and A. Blake, *J. Agric. Food Chem.*, 2000, **48**, 395.
16. J.D. Ferry, 'Viscoelastic Properties of Polymers', 3rd ed., Wiley, New York, 1980.
17. P.G. Debenedetti and F.H. Stillinger, *Nature*, 2001, **410**, 259.
18. M.L. Williams, R.F. Landel and J.D. Ferry, *J. Am. Chem. Soc*, 1955, **77**, 3701.
19. C.A. Angell, *J. Phys.: Cond. Mat.*, 2000, **12**, 6463.
20. B.D. Freeman and A.J. Hill, *ACS Symposium Series*, 1998, **710**, 306.

21. P.G. Debenedetti, 'Metastable Liquids – Concepts and Principles', Princeton University Press, Princeton, 1996.
22. H. Sillescu, *J. Non-Cryst. Solids*, 1999, **243**, 81.
23. M.D. Ediger, *Ann. Rev. Phys. Chem.*, 2000, **51**, 99.
24. T.R. Noel, R. Parker, S.M. Ring and S.G. Ring, *Carbohydr. Res.*, 1999, **319**, 166.
25. S.A. Rice. 'Diffusion-limited Reactions', Elsevier, Amsterdam, 1985.
26. W.B. Russel, D.A. Saville and W.R. Schowalter, 'Colloidal Dispersions', Cambridge University Press, Cambridge, 1989, 267.
27. P.W. Atkins, 'Physical Chemistry', Oxford University Press, Oxford, 1978.
28. K. Solc and W.H. Stockmayer, *J. Chem. Phys.*, 1971, **54**, 2981.
29. H.X. Zhou, *Biophys. J.*, 1993, **64**, 1711.
30. I.D. Craig, R. Parker, N.M. Rigby, P. Cairns and S.G. Ring, *J. Agric. Food Chem.*, 2001, **49**, 4706.
31. F.C. Collins and G.E. Kimball, *J. Colloid Sci.*, 1949, **4**, 425.
32. H.J.V. Tyrrell and K.R. Harris, 'Diffusion in Liquids', Butterworths, London, 1984.
33. D. Champion, H. Hervet, G. Blond, M. LeMeste and D. Simatos, *J. Phys. Chem. B*, 1997, **101**, 10674.
34. Y.M. Gunning, R. Parker and S.G. Ring, *Carbohydr. Res.*, 2000, **329**, 377.
35. T.R. Noel, R. Parker and S.G. Ring, *J. Chem. Soc., Faraday Trans.*, 1996, **92**, 1921.
36. R.A. Robinson and R.H. Stokes, 'Electrolyte Solutions', 2nd ed., Butterworths, London, 1959, 392.
37. S.J. Hagen, J. Hofrichter and W.A. Eaton, *Science*, 1995, **269**, 959.
38. F.S. Dainton, M.S. Henry, M.J. Pilling and P.C. Spencer, *J. Chem. Soc., Faraday Trans.*, 1977, **73**, 243.
39. M.R. Tant and A.J. Hill (eds.), 'Structure and Properties of Glassy Polymers', ACS Symposium Ser. 710, 1999.
40. H. Levine and L. Slade, *Carbohydr. Polym.*, 1986, **6**, 213.
41. L. Streefland, A.D. Auffret and F. Franks, *Pharm. Res.*, 1998, **15**, 843.
42. C. Schebor, M.D. Buera, M. Karel and J. Chirife, *Food Chem.*, 1999, **65**, 427.
43. M.C. Lai, R.L. Schowen, R.T. Borchardt and E.M. Topp, *J. Peptide Res.*, 2000, **55**, 93.
44. C. Wharton, ACS Symposium Ser., 1995, **590**, 134.
45. S.J. Risch, ACS Symposium Ser., 1988, **370**, 103.
46. H. Levine, L. Slade, B. Van Lengerich and J.G. Pickup, US Patent 5009900, 1991.
47. H. Levine, L. Slade, B. Van Lengerich and J.G. Pickup, US Patent 5087461, 1992.
48. A.J. Leo, C. Hansch and D. Elkins, *Chem. Rev.*, 1971, **71**, 525.
49. S.C. Valvani, S.H. Yalkowsky and T.J. Roseman, *J. Pharm. Sci.*, 1981, **70**, 502.
50. W. Pangborn, D. Langs and S. Perez, *Int. J. Biol. Macromol.*, 1985, **7**, 363.
51. L.C. Menting, B. Hoogstad and H.A.C. Thijssen, *J. Food Tech.*, 1970, **5**, 111.
52. A.R. Berens and H.B. Hopfenberg, *J. Membrane Sci.*, 1982, **10**, 283.
53. M.P. Tonge and R.G. Gilbert, *Polymer*, 2001, **42**, 501.
54. J. Buitink, O. Leprince, M.A. Hemminga and F.A. Hoekstra, *Proc. Nat. Acad. Sci.*, 2000, **97**, 2385.
55. E.A. Golovina, F.A. Hoekstra and M.A. Hemminga, *Plant Physiol.*, 1998, **118**, 975.
56. R.K. Chan, K. Pathmanathan and G.P. Johari, *J. Phys. Chem.*, 1986, **90**, 6356.
57. R.W. Alder, R. Baker and J.M. Brown, 'Mechanisms in Organic Chemistry', Wiley, London, 1971, 40.
58. S.G. Ring, D.B. Archer, M.C. Allwood and J.M. Newton, US Patent 5294448, 1994.

Influence of Physical Ageing on Physical Properties of Starchy Materials

Denis Lourdin,[a] Paul Colonna,[a] Geoff Brownsey,[b] and Steve Ring[b]

[a]INSTITUT NATIONAL DE LA RECHERCHE AGRONOMIQUE, BP 71627, 44316 NANTES CEDEX, FRANCE
[b]INSTITUTE OF FOOD RESEARCH, COLNEY, NORWICH NR4 7UA, UNITED KINGDOM

1 Introduction

Starch and its hydrolysis products have widespread industrial uses, often in low-moisture systems in which the materials are glassy or partially crystalline. Biscuits, breakfast cereals and snacks are typical low-moisture, starch-based products, in which textural properties are of fundamental importance to the consumer. In this context, the primary objective for industry is to maintain a constant product quality during storage. Often prepared under drastic conditions, including cooking or extrusion cooking, drying, freeze-drying, etc., the glassy state of such products under ambient conditions is particularly unstable, with the potential for time-dependent changes in material properties during storage. In addition to the effects of changes in water content, which influence the properties of glassy starch and have been much studied,[1–4] processes such as recrystallisation, chemical degradation or physical ageing can occur.

The physical ageing of a glassy material is a spontaneous and progressive evolution to a thermodynamically more favorable energetic state. The slow processes leading to structural relaxation of a glass are linked to the unstable thermodynamic state of the glass, but also to molecular mobility. Physical ageing is associated with a change in material properties of products:[5,6] e.g. density, mechanical properties[7] and diffusional behaviour.[8] Because there is not necessarily a simple relationship between the observed time-dependent behaviour and its associated effect on mechanical behaviour,[9,10] a complete study of the ageing phenomenon requires investigation of these different properties. This phenomenon has been extensively studied for synthetic polymers and reviewed,[11–14] but rarely examined, for biopolymers.[4] The purpose of the present paper is to show the effects of physical ageing on a starch-based material, assessed by different instrumental methods. The system – starch/sorbitol/water – chosen for this study was taken to be representative of low-moisture, starch-based food products.

1.1 Principle of Structural Relaxation

Figure 1 shows a schematic of the dependence of volume and enthalpy on temperature, as an undercooled liquid is cooled below T_g. As a polymer in a rubbery state is cooled to a temperature below the calorimetric glass transition temperature, T_g, the local viscosity of the polymer increases progressively, leading to reduced molecular mobility and vitrification. At T_g, there is a sharp change in behaviour, as the liquid structure becomes arrested over the timescale of cooling. The structure of the glass will slowly evolve with time to that of the 'equilibrium' liquid. This relaxation will be dependent, in part, on temperature and on liquid structure, characterised on a temperature scale through the concept of fictive temperature, T_f, the temperature at which a particular structure would be fully relaxed. Structural relaxation can be probed by measurements of enthalpy or volume change.

2 Experimental Methods

2.1 Materials

Sorbitol-plasticised amorphous starch was prepared by extrusion of native potato starch containing 25% (w/w) water and the appropriate sorbitol content. For the dilatometric study, extruded starch rods were granulated. Samples were conditioned in an atmosphere controlled by a saturated NaBr solution (57% relative humidity (RH) at 25 °C), and water content was determined gravimetrically, after drying in a vacuum oven over P_2O_5 at 80 °C for 12 h. For thermom-

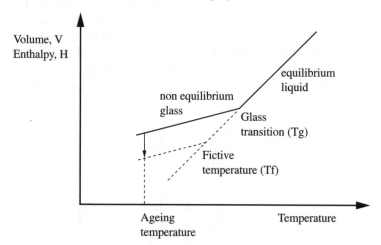

Figure 1 *Schematic of the dependence of volume and enthalpy on temperature, as an undercooled liquid is cooled below* T$_g$

echanical measurements, thick films (approximately 0.5 mm) were produced by thermo-moulding (140 °C, 10 min) of a native potato starch–20% water mixture, without additional plasticiser. Amorphous starch films were conditioned at an RH of 84% obtained with a saturated KCl solution.

2.2 Calorimetry

Samples ($\sim$10 mg) were sealed in pre-weighed aluminium pans in an argon atmosphere. DSC experiments were performed using a Perkin–Elmer DSC 7 fitted with a robotic auto-sampler. The DSC was calibrated for temperature, at each heating rate used, from the melting points of indium (156.6 °C) and dodecane (-9.65 °C). Heat flow was calibrated from the heat of melting of indium (28.45 J g^{-1}), and the calibration for heat capacity was checked against a sapphire standard. Heat capacity was measured in the temperature range 0–100 °C at a scanning rate of 10 °C min^{-1}. Data analysis was performed, using Perkin–Elmer thermal analysis software, to obtain heat capacity data.

2.3 Dilatometry

A dilatometer of borosilicate glass was used. The design of the apparatus, illustrated in Figure 2, consisted of a capillary (1) with a precision bore (0.2 mm diameter), located between two side-arms that served as the sample chamber (2). The arms were sealed with Teflon taps. The entire dilatometer was filled with degassed decane, and the sample was introduced into the decane. It had been determined that decane had no effect on T_g of the sample. The dilatometer was initially equilibrated to a temperature 15 °C above T_g of the sample for 10 min. Then the dilatometer was quickly immersed in the water bath (3), which was maintained at the ageing temperature and controlled to $\pm$0.001 °C. The measured time for complete temperature equilibration was $\sim$0.28 h. Time zero for the physical ageing experiment was taken as the time at which the dilatometer was placed in the bath. After thermal equilibration, one tap was closed, and the other limb was pressurised to introduce decane into the capillary, and then the other tap was closed. The contraction or expansion of the sample was followed

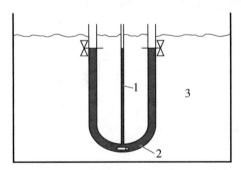

Figure 2 *Schematic of the dilatometer used for volume change measurements*

through the movement of decane in the capillary, using a cathetometer; this device allowed measurement of a volume change of 3×10^{-5} mm^3.

2.4 Mechanical Measurements

Amorphous starch/sorbitol mixtures, extruded into rods of 6 mm diameter and 30 mm length, were used. The mechanical properties of these rods were examined with an Instron 1122, using a 3-point-bend test over a total sample length of 26 mm, at a temperature of 27.7 °C.

2.4.1 Large Deformation Behaviour. The mechanical behaviour of samples was examined, using a 3-point-bend test at a crosshead speed of 2 mm min^{-1}, until break occurred.

2.4.2 Small Deformation Behaviour. An initial load was applied, and the load required to maintain the resulting deformation was recorded as a function of time. In order to avoid water diffusion from and to the ambient atmosphere, mechanical measurements were made on samples immersed in a temperature-controlled bath of silicone oil during the entire period of the ageing experiments. In preliminary short-term experiments, the oil was shown to have no effect on material properties.

2.5 DMTA

Thermomechanical measurements were performed with a DMTA MKIV apparatus (Rheometric Scientific, USA). The 3-point-bending mode was used. Vibration frequency in bending mode was set at 1 Hz, and the heating rate was 3 °C min^{-1}. Films were coated with a silicone-based hydrophobic grease to limit dehydration during experiments above room temperature. Previous experiments had shown that this device allowed negligible dehydration below 70 °C and at the scanning rate applied. Above 70 °C, water loss could affect the ageing phenomenon investigated in this study. It was determined that a thin coating of grease had no effect on thermomechanical properties.

3 Results and Discussion

3.1 DSC Experiments

DSC enables easy, routine measurements to be made with good control of hydration and thermal history in food and pharmaceutical products. DSC is frequently used to study enthalpy relaxation, not by direct observation of the relaxation, but by measurement of the recovery of the enthalpy lost during ageing, which results in a peak in heat capacity prior to T_g or in an overshoot at T_g.

A plot of normalised heat capacity, $C_{p,n}$, as a function of temperature is shown

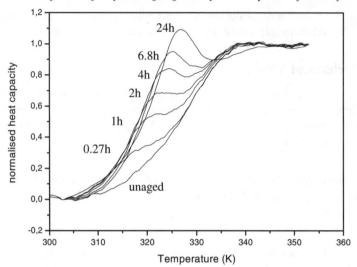

Figure 3 *Plot of experimental normalised heat capacity, for amorphous starch containing 15% sorbitol and 5% water, as a function of temperature for various ageing times (indicated on the figure) at 27 °C*

in Figure 3 for a starch/sorbitol/water (80:15:5) mixture. There was a sharp change in heat capacity in the region of 50 °C, indicative of a glass transition. The transition was broad, and the thermograms appeared to result from a combination of different effects. This behaviour could be due to the complexity of the system, which contained two polymers (amylose and amylopectin), water and a low molecular weight component, and which could exhibit phase separation phenomena.[15–17] The samples were aged for different times at 27 °C. As is usually observed, with increasing ageing time, there was an increasing overshoot in heat capacity at the glass transition. The peak temperature increased with ageing duration, in accord with typical observations on the physical ageing of synthetic systems.

3.2 Volume Relaxation

The change in volume of a starch/sorbitol/water (80:15:5) mixture, $\Delta V/V$, with ageing time is shown in Figure 4. The quantity $\Delta V/V$ was calculated from:

$$\Delta V/V = |V_{(t)} - V_{(0)}|/V_{(0)}$$

where $V_{(t)}$ is the specific volume at time t, and $V_{(0)}$ is the initial specific volume measured by pycnometry after quenching and equilibration of the sample to the ageing temperature. The measured value of $V_{(0)}$ is a slight underestimate, as time is required for thermal equilibration. Volume relaxation measurements were made at different ageing temperatures in the range $T_g - 28\,°C$ to $T_g - 20\,°C$ (the calorimetric T_g of the mixture was 50 °C for the sample considered). Volume decreased with increasing ageing time. The initial behaviour, up to ~ 0.3 h, was

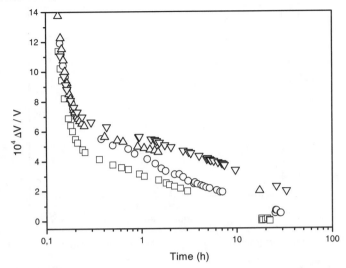

Figure 4 *Relative volume change, as a function of time, for starch/sorbitol/water (80:15:5) samples at 21.5 °C (∇), 25 °C (Δ), 27.5 °C (○) and 30 °C (□). The 25, 27.5 and 30 °C data are displaced upwards by 0.5, 1.0 and 1.5 units, respectively*

dominated by the effects of thermal equilibration. After that time, the volume change was essentially linear with log *t*. The equilibration of liquid structure did not occur within the timescale observed. The slope of $\Delta V/V$ *vs* log *t* did not show a significant dependence on temperature within the range examined. As was expected, these results showed that physical ageing has a densification effect on materials. Such a contraction of structure could lead to fractures appearing with time, as sometimes observed for glassy polymers.

3.3 Mechanical Relaxation

Samples were initially equilibrated, at a temperature 15 °C above T_g, by immersion in a silicone oil bath for 2 min. Samples were then rapidly cooled, by the same method of immersion, at 27.7 °C.

3.3.1 Large Deformation Behaviour. The force at break and the deformation were recorded for samples aged to different extents at 27.7 °C (Figure 5). Reproducibility was determined from five experiments for each ageing duration. With increased ageing, the deformation at break decreased, from about 6 mm for a fresh sample to about 2 mm for samples aged for 60 days. Simultaneously, the strength at break increased from 112 to 150 N. This observation indicated a progressive embrittlement on ageing.

3.3.2 Small Deformation Behaviour. Figure 6 shows a schematic of the loading sequence. A load was applied, and the load required to maintain the deformation

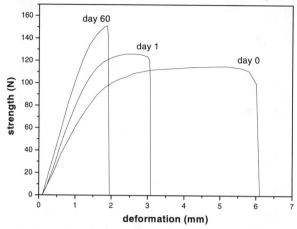

Figure 5 *Results of strength/deformation experiments for samples aged to different extents*

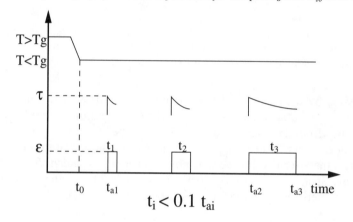

Figure 6 *Schematic of the loading sequence used to analyse physical ageing after a temperature jump. τ is the initial load, and ε is the corresponding deformation. t_{a1} is the ageing time and t_i is the duration of the test*

was recorded as a function of time. An initial load of 30 N was determined to be within the linear region of stress–strain behaviour. As recommended by Struik,[12] the duration, t_i, of each mechanical test was less than 10% of the total ageing time, t_{ai}, in order to ensure that the time-dependence of ageing could be determined.

Figure 7 shows the mechanical relaxation behaviour for the starch/sorbitol/water mixture as a function of ageing time. The different curves show the mechanical relaxation, in terms of Young's modulus (E), on ageing at 27.7 °C. Mechanical relaxation time increased with ageing, which indicated that the material became stiffer with time. This behaviour was generally similar to that observed for amorphous synthetic polymers.

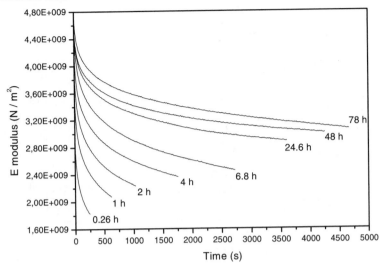

Figure 7 *Plot of mechanical relaxation at 27.7 °C for amorphous starch containing 15% sorbitol and 5% water. Different curves correspond to various ageing times (indicated in hours) for the same sample*

3.4 DMTA

The sample studied was composed of amorphous potato starch containing 15% water (as measured by thermogravimetry). The T_g of the material, as measured by DSC, was 70 °C. As with the mechanical measurements, samples were initially equilibrated at 70 °C by immersion in a silicone oil bath for 2 min. Samples were then rapidly cooled at 20 °C, before loading them in the apparatus. The behaviour of fresh sample, sample aged for one day, and sample aged for 30 days was compared. Ageing observations were made on different pieces of film stored at 20 °C. Figure 8 shows the variations in E' and E'' moduli (left y axis) and tan δ (right y axis) as a function of temperature. The temperature region investigated comprised the tail of the α or main relaxation, the peak maximum of which appeared at ~ 70 °C. The α relaxation is related to the glass transition and involves long-range molecular movements. Because of uncertainties related to the extent of sample dehydration during heating, the portion of the curves above 70 °C could not be analysed. However, it was clear that the tail of the relaxation, from ambient temperature to 70 °C, was affected by physical ageing, as also observed for synthetic polymers.[18] The longer the ageing time, the lower the tan δ values, which indicated that ageing decreased molecular mobility at the shortest relaxation times. The values of storage modulus (E') at 25 °C increased from 10^{10} Pa for a fresh sample to 1.6×10^{10} Pa for a sample aged for 30 days. The mechanical modulus remained higher for aged samples over a wide temperature range, until the glass transition was approached. From 60 to 80 °C, a drop in E' associated with a peak in tan δ was typical of a vitreous-to-rubbery state change.

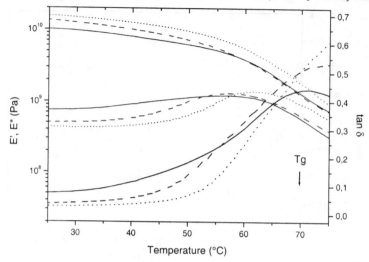

Figure 8 *Plot of thermomechanical properties, measured for samples of amorphous starch (containing 15% water) aged to different extents: fresh, ·····; 1 day, ----; 30 days, ——*

4 Conclusions

A combination of several independent techniques has been used to obtain a broad view of the physical ageing phenomenon. In the case of starchy materials, important effects on functional properties have been demonstrated. As was expected, structural relaxation led to decreases in enthalpy and volume. In the temperature domain examined ($T_g - 20\,^\circ\text{C}$ to $T_g - 28\,^\circ\text{C}$), no equilibration was achieved in the timescale studied. Measurements of mechanical properties showed that starchy materials became stiffer and more brittle during physical ageing. Thermomechanical measurements indicated that changes in mechanical properties could be attributed to a loss in molecular mobility, which was a precursor of the main relaxation associated with the glass transition. Thus, physical ageing has a direct effect on the quality of glassy biopolymer products.

References

1. H. Levine and L. Slade, *Water Sci. Rev.*, 1988, **3**, 79.
2. L. Slade and H. Levine, *J. Food Eng.*, 1995, **24**, 431.
3. H. Bizot, P. LeBail, B. Leroux, J. Davy, P. Roger and A. Buleon, *Carbohydr. Polym.*, 1997, **32**, 33.
4. Y.H. Roos. 'Phase Transitions in Foods', Academic Press, London, 1995.
5. A.J. Kovacs, *Fortschr. Hochpolym.-Forsch.*, 1963, **3**, 394.
6. G.B. McKenna, Y. Leterrier and C.R. Schultheisz, *Polym. Eng. Sci.*, 1995, **35**, 403.
7. D.J. Plazek, J.H. Magill, I. Echeverria and I.-C. Chay, *J. Chem. Phys.*, 1999, **110**, 10445.

8. P. Pekarski, J. Hampe, H.G. Böhm and R. Kirchheim, *Macromolecules*, 2000, **33**, 2192.

9. J.M.G. Cowie, S. Harris and I.J. McEwen, *Macromolecules*, 1998, **31**, 2611.

10. S.L. Simon, D.J. Plazek, J.W. Sobieski and E.T. McGregor, *J. Polym. Sci., Part B: Polym. Phys.*, 1997, **35**, 929.

11. G.B. McKenna, *J. Res. Nat. Inst. Stand. Technol.*, 1994, **99**, 169.

12. L.C.E. Struik. 'Physical Ageing in Amorphous Polymers and Other Materials', Elsevier, Amsterdam, 1978.

13. J.M. Hutchinson, *Prog. Polym. Sci.*, 1995, **20**, 703.

14. I.M. Hodge, *J. Non-Cryst. Solids*, 1994, **169**, 211.

15. G.K. Moates, T.R. Noel, R. Parker and S.G. Ring, *Carbohydr. Polym.*, 2001, **44**, 247.

16. D. Lourdin, L. Coignard, H. Bizot and P. Colonna, *Polymer*, 1997, **38**, 5401.

17. S. Gaudin, D. Lourdin, P. Forssell and P. Colonna, *Carbohydr. Polym.*, 2000, **43**, 33.

18. E. Muzeau, J.Y. Cavaille, R. Vassoille, J. Perez and G.P. Johari, *Macromolecules*, 1992, **25**, 5108.

Uptake and Transport of Gas in Microstructured Amorphous Matrices

A. Schoonman, J.B. Ubbink,* W.M. MacInnes and H. J. Watzke

NESTLÉ RESEARCH CENTER, VERS-CHEZ-LES-BLANC,
CH-1000 LAUSANNE 26, SWITZERLAND
*Corresponding author

1 Introduction

Amorphous carbohydrates are often used as encapsulation and coating materials. They are good barrier materials for many organic compounds and low molecular weight gases.[1-10,77-80]

The sorption and transport properties of amorphous carbohydrates depend strongly on the nature of the permeant, and on the structure and physical state of the amorphous matrix.[11-13] The barrier properties of carbohydrate matrices are strongly dependent on environmental conditions, particularly relative humidity and temperature. Many of these properties are directly related to the glass transition temperature (T_g) of the amorphous matrix.[14-17,77,78] Processes and properties in food science, which are affected by T_g, have been reviewed in detail elsewhere.[14,18-20,54]

Transport of low molecular weight compounds in amorphous materials is thought to take place *via* an activated diffusion process. A permeant moves through a matrix by a series of consecutive jumps, due to a redistribution of matrix free volume induced by thermal fluctuations.[21-25] In heterogeneous matrices, transport of gases is strongly dependent on the microstructure of the matrix. The effect of microstructure on the rate of transport of low molecular compounds has been extensively studied for many differently structured materials, such as closed-cell polymer foams,[26] (edible) dense films,[16,27-29] fractal pore structures,[30-32] and permselective membranes for gas separation, food packaging and protective coatings.[33]

In this paper, we describe our study of the uptake and transport of nitrogen in microstructured amorphous matrices consisting of maltodextrin and sodium caseinate in varying proportions. The work was divided, essentially, into two parts. First, we sought to understand the effect of material properties and environmental conditions on the uptake of gas by the amorphous matrix. Then, we studied the effect of sample microstructure, particularly porosity, on gas transport properties, by following the release of entrapped nitrogen from the matrices over time.

The paper is divided into six sections. Sections 2 and 3 contain brief introductory discussions of the sorption and transport of gases in amorphous polymers and microstructured matrices. Section 4 deals with materials and methods. In Section 5, results on gas uptake by, and release from, maltodextrin/sodium caseinate matrices are discussed. Finally, conclusions are presented in Section 6.

2 Sorption of Gases in Amorphous Polymers

The sorption of gases has been studied in great detail for a large number of synthetic and chemically modified polymers and polymeric composites.[11,21,24,34–37] A key concept underlying the quantification of the sorption of gases is the free volume of the polymer matrix.[22,23,33,38–40] Typically, sorption characteristics are strongly dependent on the nature of the polymer and permeant, and on the physical state of the polymer. Significant amounts of gas can be absorbed in amorphous materials because of their relatively large free volume. At least two different sorption regimes are ordinarily distinguished: sorption by glassy and by rubbery polymers.[41–44]

In the rubbery state, a polymer matrix behaves, in effect, as a viscoelastic liquid with finite relaxation times. In this case, sorption of gases is very similar to gas sorption by other liquids, and Henry's law behaviour is often observed for gas pressures that are not too high.[45]

In the glassy state, the mechanism of sorption is different, as matrix relaxation is much slower than the timeframe of the sorption experiment. To describe sorption by glassy polymers, a large number of models of varying degrees of sophistication have been put forward. Gas adsorption in glassy polymers is often heuristically explained in terms of the solubility of a gas in a polymer matrix, following either Henry's law or Flory–Huggins theory and Langmuir adsorption of gas in microscopic voids (*e.g.* see Krykin *et al.*[45] for a recent discussion of these so-called dual-mode sorption models). More fundamental approaches, in which gas sorption has been related to physical properties of a polymer matrix, have also been developed.[46–48] However, those approaches remain semi-quantitative and lack, in particular, an unequivocal explanation of the nature of the distribution of a penetrant in a matrix.

Although the free-volume concept has been applied to the sorption of gases by polymers for many years,[22] it is only recently that advanced experimental and computational techniques, such as positron-annihilation spectroscopy (for a recent overview, see Petrick[49]) and computer simulations,[50,51] have provided insight to the distribution of free volume on molecular- and nano-scales.

In the literature, the sorption of gases and other low molecular weight compounds by bio- and food polymers has received relatively little attention. Gas permeability has been determined for a large number of edible films (see Krochta and Baldwin[10] for a recent discussion), but, because a value of permeability coefficient is the product of the solubility of a gas and its diffusion coefficient,[52] this parameter does not provide information on the inherent sorption and diffusion processes. The diffusion of water and volatile organic compounds has been studied for a number of amorphous carbohydrates, but mainly in their rubbery state.[6,17,28,53] Water can be considered as a special case, because it acts as a plasticizer for bio- and food polymers,[14,54–56] and causes increased mobility of the matrix molecules, particularly above their T_g.[7,14,54]

3 Gas Transport in Microstructured Matrices

Diffusion of gases in polymers often follows Fick's law of diffusion, in both the rubbery and glassy states of a matrix.[12,57–59] Transport of gas through solid matrices has been explained by either of two different phenomenological models, the free-volume model or the dual-mode transport model. For systems in the glassy state, the dual-mode transport model is commonly used to describe gas permeation.[41–43] On the other hand, permeation of gas in a rubbery system is usually explained in terms of the free-volume model.[21–24,36,60] The free-volume model can also be applied in a modified form to glassy polymers.[24]

The microstructure of matrices influences gas transport properties to a great extent. Important structural factors influencing gas transport include matrix heterogeneity (*e.g.* presence of crystallites, phase separation),[61] existence of cracks and voids,[37] and size, shape and connectivity of pores.[62–67]

4 Materials and Methods

4.1 Preparation of Samples

Foamed powders composed of 12-DE maltodextrin (Sugro AG, Switzerland) and varying percentages of sodium caseinate (10, 20 and 30% w/w) (Säntis AG, Switzerland) were prepared according to a procedure outlined elsewhere.[68] The powders were prepared at three levels of overrun (50, 100 and 150%), by varying the amount of nitrogen injected into the dispersion. Overrun is defined as follows:

$$\text{Overrun} = \frac{\rho_s - \rho_f}{\rho_f} \times 100\% \tag{1}$$

where ρ_s is the density (g cm^{-3}) of the solution, and ρ_f is the density of the foam.

The solid foam was dried under controlled conditions in a vacuum dryer (Secfroid FCV 600 equipment, France), milled using a Frevitt mill (Frevitt, Switzerland) and sieved. The particle size fraction used for this study had a diameter of 0.4–0.9 mm.

The powder particles were used 'as is' or compressed into tablets. The use of maltodextrins for producing improved excipients for tabletting, and as direct tablet excipients, has been described in technical literature,[69] including patents. The tablets (diameter 38 mm, height 2 mm) were compressed at an estimated tabletting pressure of 260 Mpa, using a conventional workshop press (PRM 60 PHP, Rassant, France).

Samples (foamed powders and tablets) were analysed at various 'water activities' (a_w = 0.11, 0.23, 0.32 and 0.53; T = 25 ± 1 °C). The different a_ws were produced by equilibrating the powders in the presence of standard saturated salt solutions prior to tabletting.[70]

Dense granules were prepared by mixing 90% w/w 12-DE maltodextrin and 10% w/w sodium caseinate with water (70% w/w total solids) under vacuum for 12 hours. After vacuum-drying the resulting solution at 60 °C for 72 hours (Heraeus Instruments, type VT 6130P, Germany), the solid material was milled. The fraction with mean diameter of 1.9 mm was equilibrated at a_w = 0.23 (T = 25 ± 1 °C) and then used for gas uptake and release measurements.

4.2 Gas Uptake in Samples

The uptake of nitrogen by the samples was achieved by pressurizing them at room temperature in an autoclave, followed by heating them under pressure, usually to temperatures above T_g. By means of rapid cooling of the samples immediately after loading, the gas was retained within the matrix. This procedure has been described in detail elsewhere[71] and is illustrated in Figure 1.

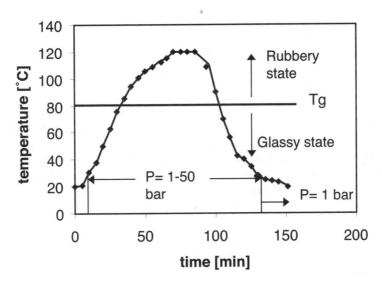

Figure 1 *Schematic illustration of experimental procedure for gas-loading of sample matrix. Temperature–time path is shown in relation to matrix physical state: state 1 – below* T_g, *glassy matrix, slow diffusion of gas; state 2 – above* T_g, *rubbery state, fast diffusion of gas; state 3 – far above* T_g, *rapid collapse of matrix*

4.3 Physical and Structural Analysis of Samples

4.3.1 Density and Porosity. The apparent density of samples was determined using an Accupyc 1330 pycnometer (Micromeritics, USA) at room temperature. Helium gas was used as the displacement gas, and the mean value of five runs per sample was determined. An equilibration rate of 0.4 kPag min^{-1} was used and found to yield sufficiently accurate results.

The closed porosity, ε_c, was calculated from the matrix density, ρ_m (g cm^{-3}), and the apparent density, ρ_a (g cm^{-3}):

$$\varepsilon_c = \frac{\rho_m - \rho_a}{\rho_m} \qquad (2)$$

The density of the matrix was dependent on its composition. To a good approximation, matrix density, ρ_m, was the weighted average of the density of maltodextrin (ρ_{MD}) and sodium caseinate (ρ_{SC}):[71]

$$\rho_m = \phi_{MD}\rho_{MD} + \phi_{SC}\rho_{SC} \qquad (3)$$

$$= \phi_{MD}\rho_{MD} + (1 - \phi_{MD})\rho_{SC}$$

where ϕ_{MD} and ϕ_{SC} are the volume fractions of maltodextrin and sodium caseinate, respectively. Matrix density was determined self-consistently from gas sorption experiments, as described below.

The open porosity, ε_0, was calculated as:

$$\varepsilon_0 = \frac{\rho_a - \rho_e}{\rho_a} \qquad (4)$$

where ρ_e is the envelope density (g cm^{-3}).

It should be noted that closed porosity was expressed as a fraction of the volume of a matrix and closed pores (thus excluding open porosity), whereas open porosity was expressed as a fraction of the geometric volume of a sample (*i.e.* the sample volume including open pores).

4.3.2 Thermal Analysis. Calorimetric measurements were carried out using a TA 8000/DSC 820 (Mettler-Toledo, Switzerland), equipped with an auto sampler. Results were recorded and analysed using a Mettler-GraphWare TA 72.2/5 software package. Temperature and heat flow were calibrated using the melting point (429.8 K) and melting enthalpy (28.45 J g^{-1}) of indium. Medium-pressure crucibles (ME-29990, Mettler-Toledo) were filled with 25 to 30 mg of sample and hermetically sealed. An empty aluminum pan was used as a reference. Measurements were carried out at a heating rate of 5 °C min^{-1} and a cooling rate of 20 °C min^{-1}. T_g was determined as the onset of the change in heat flow in a second heating scan.

4.4 Determination and Analysis of Gas Uptake and Release

Gas uptake or release by samples was determined both immediately after gas-loading and at regular intervals during storage. After loading, small amounts of sample (typically, 1–2 g) were stored in 20 ml glass vials that were hermetically sealed with an aseptic seal. Gas release was determined, at STP (standard temperature (298.15 K) and pressure (100 kPa)), by collecting the excess gas, accumulated in the headspace of a vial, in a water-filled column with ml-scaling.[71] Then, by the same method, the amount of gas retained in samples was determined by first dissolving the sample in a vial, by injecting a small volume of water, and then collecting the volume of gas thus liberated from the sample matrix.[71]

A convenient variable to use for expressing the capacity of a material to absorb gas is the specific volume of entrapped gas per unit volume of material, V^* (cm^{-3} cm^{-3}). We anticipated a direct relationship between matrix porosity and V^*:[71]

$$\bar{V}^* = c_f \left(1 - \frac{\rho_a}{\rho_m} \right) \tag{5}$$

where c_f is a coefficient of proportionality.

The partial molar volume of a solute in a solvent is defined as:[72]

$$\Phi_m = \left(\frac{\partial V}{\partial n} \right)_{P,T,n_s} \tag{6a}$$

where V is the volume of the system, n is the number of moles of solute, n_s is the number of moles of solvent, P is pressure and T is temperature.

Anticipating that, for high apparent densities, the partial molar volume of nitrogen entrapped in a matrix would be independent of density, we may write:

$$\Phi_m = \frac{10^6 V_m \varepsilon_c}{\bar{V}^*} \tag{6b}$$

where V_m is the molar volume (2.24×10^{-2} m^3 mol^{-1} at STP).

The release of nitrogen gas from tablets and granules, as a function of time, was quantified using Fick's laws of diffusion. For tablets, use was made of the solution to the diffusion equation for the geometry of an infinite slab:[73]

$$M(t) = M(0) \left(1 - \sum_{n=0}^{\infty} \frac{8}{(2n+1)^2 \pi^2} \times \exp\left(\frac{-D(2n+1)^2 \cdot \pi^2}{4l^2} t \right) \right) \tag{7}$$

where $M(t)/M(0)$ is the fraction of gas released at time t (s), $M(0)$ (cm^3 cm^{-3} or cm^3 g^{-1}) is the total amount of gas originally present in a tablet, l is the thickness of a tablet (m) and D is the diffusion coefficient (m^2 s^{-1}).

For analysis of gas release from powders, it was assumed that granules were sufficiently spherical to warrant use of the solution to the diffusion equation in spherical coordinates:[73]

$$M(t) = M(0) \left(1 - \frac{6}{\pi^2} \sum_{n=1}^{\infty} \frac{1}{n^2} \times \exp\left(-\frac{n^2 \pi^2 D t}{r^2} \right) \right) \tag{8}$$

where r is the radius of a granule (m).

Fitting of release curves using Equations 7 and 8 yielded an effective diffusion coefficient for gas release, denoted as D_e.

5 Results and Discussion

5.1 Gas Uptake

We first studied the uptake of nitrogen by dense samples (tablets and dense granules). This provided us with material properties used for the interpretation of gas uptake by foamed powders.

Generally, in the study of gas sorption by materials such as polymers, care is taken that a matrix is either able to equilibrate under the influence of an externally applied pressure, or that it fully maintains its microscopic and molecular structure during gas-loading experiments. However, as we were interested in gas uptake by fragile foamed structures (see Figure 2 for a representative image of the foam structure),[74] the temperature needed to be sufficiently high (close to or above T_g) to increase the rate of gas uptake and, in addition, to decrease the brittleness of the powder. However, the temperature needed to be not so high that the foam structure would collapse during the time frame of a gas-loading experiment. Because of those temperature and time constraints, our interest was

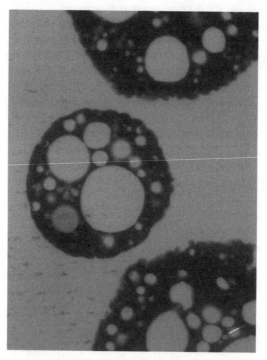

Figure 2 *Representative micrograph showing the internal structure of a solid foam. The image was obtained by phase-contrast microscopy of a thin slice of sample, embedded in resin and then stained*[71]

in the dynamic behaviour of matrix materials under the influence of an externally applied pressure, rather than in static sorption data.

In Figure 3, the volume of gas taken up by dense granules is shown as a function time, for different gas-loading temperatures. There was a clear difference between the results at 50 or 70 °C and those at 80 or 90 °C, with gas-loading being much faster for the two higher temperatures. This finding correlated well with T_g of the matrix, which was 92.5 ± 3.4 °C at $a_w = 0.23$. Within the time frame of these experiments, it was not clear whether the gas uptake capacity of the dense granules was lower at 50 and 70 °C than at 80 and 90 °C, or whether only the kinetics of gas sorption were much slower.

For both tablets and dense powders, within experimental accuracy, a linear relationship was found between specific volume of entrapped gas, V^* (cm^3 cm^{-3}), and matrix apparent density, as illustrated in Figure 4, which shows gas-loading data for a tablet sample. This linear relationship allowed us to obtain a self-consistent value for matrix density, by extrapolating the gas uptake curve to zero gas content. This method, which, as far as we are aware, is novel, is one of very few unambiguous ways to obtain consistent values for the matrix density of amorphous materials.[71] Generally, precise values for this property are difficult to obtain, because of the variability of free volume.

An important indicator of the state of nitrogen in dense matrices was the partial molar volume (Table 1). Partial molar volume in the entrapped state turned out to be much higher than that in the gaseous state at a given loading pressure. This was indicative of a dissolved state of nitrogen in the matrix. Also, the coefficient, c_f (in Equation 5), should have been numerically equal to the loading pressure (in bar), if entrapped nitrogen was in a gaseous state, but was actually much higher (loading pressures varied between 10 and 50 bar). In the dissolved state, nitrogen was either molecularly dispersed or entrapped in the

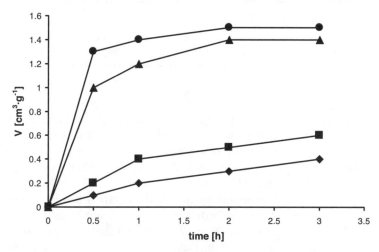

Figure 3 *Volume of gas taken up by dense granules* [$a_w = 0.23$, $T_g \sim 90$ °C, *loading pressure 50 bar*] *at different gas-loading temperatures.* ◆: T = 50 °C; ■: T = 70 °C; ▲: T = 80 °C; ●: T = 90 °C

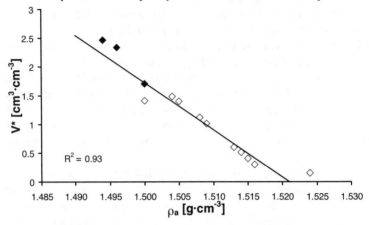

Figure 4 *Specific volume of entrapped gas, V*, as a function of apparent density, for a tablet composition consisting of 70% w/w 12-DE maltodextrin and 30% w/w sodium caseinate, after gas-loading. Open symbols: loading temperature of 70 °C; closed symbols: loading temperature of 90 °C*

Table 1 *Coefficient of proportionality, c_f, of gas-loading in tablets and granules, partial molar volume of dissolved gas, Φ_p, and extrapolated matrix density, ρ_m, as functions of tablet composition. R^2 is the correlation coefficient for the fit of the experimental data to Equation 5. Loading pressures for tablets varied between 10 and 30 bar. Loading pressure for granules was 50 bar*

Matrix composition	c_f	$\rho_m\,(g\,cm^{-3})$	$\Phi_p\,(cm^3\,mol^{-1})$	R^2
90/10	129	1.546	173	0.63
80/20	235	1.536	95.3	0.80
70/30	125	1.521	179	0.94
Granule, 90/10	237	1.538	94.5	0.91

form of very small inclusions, likely to be of nm size. In SEM images, large inclusions (micrometer-scale or larger) were virtually absent (data not shown).

Gas uptake by foamed powders proceeded by a different mechanism. The amount of gas taken up was much greater than that for dense granules and tablets, because large amounts of gas could be entrapped in the larger pores and bubbles (Figure 2) that acted like pressurized reservoirs. The total amount of gas taken up was well-approximated by the ideal gas law: when the pressure was doubled, the amount of gas also virtually doubled. A small fraction of the total gas would be taken up by the matrix. We expected that this quantity could be well-estimated from the gas-loading data for tablets and dense granules. In all powder compositions that showed a very high porosity, the capacity to entrap gas was reduced, especially at higher loading pressures. This may have been caused by lamellae surrounding the foam bubbles, which were too thin and thus could not withstand the forces produced by the pressure difference between the interior and exterior of the particles after gas-loading.

5.2 Gas Release

Data for gas release from foamed amorphous powders were fitted using Fick's law of diffusion, modified for appropriate geometry: a sphere in the case of powders and a slab in the case of tablets. This provided effective values for the diffusion coefficient, D_e. An example of fitted gas release data is shown in Figure 5. The effective D_e for gas release depended on both structural and physical parameters. The most important physical parameters were temperature and a_w; D_e increased with both a_w and T (data not shown). No details are given here, as our focus was on the effect of sample structure on the rate of gas release.

In Table 2, it can be seen that, under similar storage conditions (temperature and a_w), the values of D_e differ by several orders of magnitude for the variously structured samples. In particular, the values of D_e found for tablet samples were much higher than those for the other two structures. This finding hinted at an effect of microstructure of the tablet. If we assumed that the actual diffusion coefficient for nitrogen in maltodextrin/sodium caseinate matrices was well-approximated by the D_e for gas release from dense granules (which, of all samples studied, had the most homogeneous structure), then we could estimate the effective diffusion length scale for diffusional release of gas from tablets as:

$$a \approx \sqrt{\frac{D_e(\text{matrix})}{D_e(\text{tablets})}} \times l \approx 6 \times 10^{-5} \text{ m} \tag{9}$$

where l is tablet thickness (2 mm).

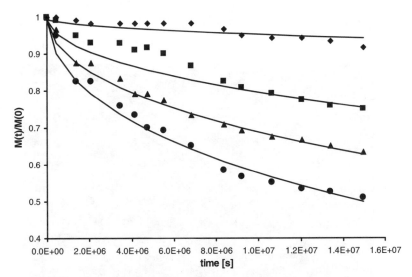

Figure 5 *An example of data for gas release from tablets [composition 90% w/w 12-DE maltodextrin, 10% w/w sodium caseinate; $a_w = 0.11$], fitted using Equation 5.* ♦: $T = 5 \degree C$; ■: $T = 30 \degree C$; ▲: $T = 50 \degree C$; ●: $T = 70 \degree C$

Table 2 *Effective diffusion coefficients, D_e, for gas release from various matrices composed of 12-DE maltodextrin and sodium caseinate*

Amorphous matrix	D_e at 5 °C [m s^{-2}]	D_e at 30 °C [m s^{-2}]
Dense granule	6.5×10^{-17}	1.9×10^{-15}
Foamed powder, $\varepsilon_c = 12\%$	3.1×10^{-17}	2.6×10^{-16}
Foamed powder, $\varepsilon_c = 30\%$	3.4×10^{-17}	2.8×10^{-16}
Compressed tablet	1.5×10^{-13}	7.1×10^{-13}

This effective diffusion length scale corresponded well with the size of grains visible in an SEM image (Figure 6) of the interior of a tablet (after gas-loading). Thus, our interpretation of the gas release data for tablets was that the rates of gas release from tablets were not determined by the overall dimensions of a tablet, but largely by the size of the grains within a tablet. Such an interpretation was also supported by the high open porosity of the tablets (around 40%, data not shown).

The effective diffusion of gas out of foamed powders increased only slowly with increasing closed porosity, for values of closed porosity below about 30% (Table 2, Figure 7). Only for values of closed porosity above about 30% was a significant increase in D_e observed. A detailed theoretical treatment of gas release was not justified, because the scatter in the values of D_e was rather large. This scatter was probably caused by a multitude of factors, such as sample structure, variations in gas pressure in the foam bubbles and heterogeneity of powder particles.

For convenience, in Figure 7, the prediction based on the Maxwell equation[75] is shown:

$$D_e = \frac{D_m \lfloor D_g + 2D_m - 2\varepsilon_c(D_m - D_g) \rfloor}{D_g + 2D_m + \varepsilon_c(D_m - D_g)} \qquad (10)$$

where D_m is the diffusion coefficient for nitrogen in the matrix and D_g is the

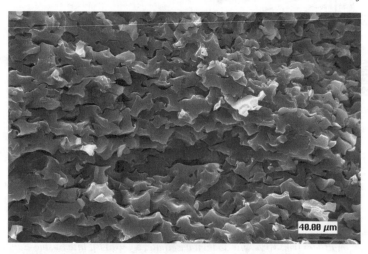

Figure 6 *SEM image of the interior of a tablet after gas-loading*

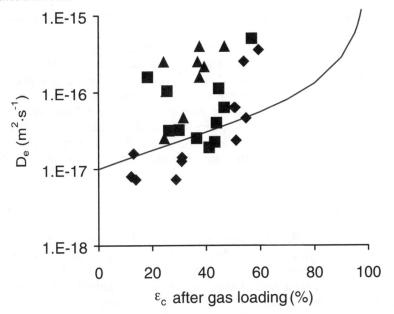

Figure 7 D_e *for gas release from porous powders, as a function of closed porosity after gas-loading.* ♦: *90% w/w 12-DE maltodextrin, 10% w/w sodium caseinate;* ■: *80% w/w maltodextrin, 20% w/w sodium caseinate;* ▲: *70% w/w maltodextrin, 30% w/w sodium caseinate; line: Maxwell equation (Equation 10)*

diffusion coefficient for nitrogen gas. For D_m, a value of 1×10^{-17} m^2 s^{-1} was used (corresponding to zero closed porosity), which was the mean value for gas release from dense matrices at $T_g - T_{storage} = 90\,°C - 5\,°C = 85\,°C$. For D_g, a value of 1×10^{-5} m^2 s^{-1} was used.[76]

The trend in D_e was reasonably well-predicted by the Maxwell equation. However, at levels of closed porosity approaching 60%, gas release from powders tended to increase much faster than was predicted by the Maxwell equation. As discussed earlier, this could have been caused by the fragility of the powders with very high overruns, leading to mechanical failure.

6 Conclusions

The primary parameter determining the gas uptake capacity of a maltodextrin/sodium caseinate matrix was the apparent density of that matrix. That density was variable within certain limits and could be influenced by processing, especially by thermal and pressure treatments.

For dense matrices, *i.e.* matrices for which the apparent density was close to the matrix density, gas uptake and release data reflected the material properties of the system, such as molecular mobility and free volume. It was found that nitrogen gas was sorbed in a relatively dense state, either molecularly dispersed or in nano-scale inclusions. Based on the use of gas sorption data, a method was developed to self-consistently determine the matrix density of amorphous materials.

By suitable processing, well-defined microstructures could be created, which dramatically influenced gas uptake and release characteristics. Two examples were studied: foamed powders and compressed tablets. In the case of foamed powders, high levels of closed porosity were found to act as reservoirs in which nitrogen could be entrapped in the gaseous state. Rates of gas release from foamed powders increased with increasing closed porosity, especially at high values of closed porosity. In the case of compressed tablets, rates of gas release were strongly influenced by the open porosity and granular interior of the tablets.

Acknowledgements

Dr Chris Bisperink, Prof. Martine Le Meste (Université de Bourgogne, Dijon, France) and Prof. Marcus Karel (MIT, Boston, USA) are thanked for continuous support of the work. We thank Marie-Lise Dillmann and Marie-France Clerc for the phase-contrast and SEM microscopy, respectively. We are indebted to the management of Nestec Ltd for granting permission to publish this work.

References

1. M. Rosenberg, I.J. Kopelman and Y. Talmon, *J. Agric. Food Chem.*, 1990, **38**, 1288.
2. M. Rosenberg and T.Y. Sheu, *Int. Dairy J.*, 1996, **6**, 273.
3. I. Goubet, J.L. Le Quere and A.J. Voilley, *J. Agric. Food Chem.*, 1998, **46**, 1981.
4. A. Voilley and M. Le Meste, 'Properties of Water in Foods', D. Simatos and J.L. Multon, (eds.), Martinus Nijhoff, Dordrecht, 1985, 357.
5. G.A. Reineccius, 'Encapsulation and Controlled Release of Food Ingredients', S.J. Risch and G.A. Reineccius, (eds.), ACS Symp. Ser., 1995, American Chemical Society, Washington.
6. L.C. Menting, B. Hoogstad and H.A.C. Thijssen, *J. Food Technol.*, 1970, **5**, 111.
7. G. Levi and M. Karel, *J. Food Eng.*, 1995, **24**, 1.
8. N. Gontard, R. Thibault, B. Cuq and S. Guilbert, *J. Agric. Food Chem.*, 1996, **44**, 1064.
9. M.P. Haak, Spray drying method and apparatus for concurrent particle coating, European Patent Specification, 1989, No. 0344375 B1.
10. J.M. Krochta and E.A. Baldwin, 'Edible Coatings and Films to Improve Food Quality', Nisperos-Carriedo, (ed.), Technomic, Lancaster, 1994.
11. A.Y. Houde and S.A. Stern, *J. Membr. Sci.*, 1997, **127**, 171.
12. J. Crank and G.S. Park, 'Diffusion in Polymers', Academic Press, London, 1968.
13. C.E. Rogers, 'Polymer Permeability', J. Comyn, (ed.), Elsevier Applied Science, London, 1985, Chapter 2.
14. H. Levine and L. Slade, *Water Sci. Rev.*, 1988, **3**, 79.
15. M. Karel and I. Saguy, 'Water Relationships in Foods', H. Levine and L. Slade, (eds.), Plenum Press, New York, 1991, 157.
16. I. Arvanitoyannis, M. Kalichevsky and J.M.V. Blanshard, *Carbohydr. Polym.*, 1994, **24**, 1.
17. R. Parker and S.G. Ring, *Carbohydr. Res.*, 1995, **273**, 147.
18. T.W. Schenz, *Food Hydrocoll.*, 1994, **9**, 307.

19. V.B. Tolstoguzov, *Nahrung*, 2000, 44, 76.
20. D. Champion, M. Le Meste and D. Simatos, *Trends Food Sci. Technol.*, 2000, **11**, 41.
21. R.M. Barrer, 'Diffusion in and Through Solids', E.K. Rideal, (ed.), Cambridge University Press, Cambridge, 1951.
22. M.H. Cohen and D. Turnbull, *J. Chem. Phys.*, 1959, **31**, 1164.
23. H. Fujita, A. Kishimoto and K. Matsumoto, *Trans. Faraday Soc.*, 1960, **56**, 424.
24. J.S. Vrentas and J.L. Duda, *J. Appl. Polym. Sci.*, 1978, 22, 2325.
25. W. Paul, H. Weber and K. Binder, *Ann. Phys. (Leipzig)*, 1998, **5–6**, 554.
26. S.E. Bochenkov and A.V. Pertsov, *Colloid J.*, 1997, **59**, 147.
27. I. Arvanitoyannis, E. Psomiadou and A. Nakayama, *Carbohydr. Polym.*, 1996, **31**, 179.
28. H.J. Park and M.S. Chinnan, *J. Food Eng.*, 1995, **25**, 497.
29. K.S. Miller and J.M. Krochta, *Trends Food Sci. Technol.*, 1997, **8**, 228.
30. R. Muralidhar and D. Ramkrishna, *Transport Porous Media*, 1993, **13**, 79.
31. J.C. Kimball and H.L. Frisch, *Phys. Rev. A*, 1991, **43**, 1840.
32. J.F. Gouyet, 'Diffusion in Materials', A.L Laskar *et al.*, (eds.), Kluwer Academic Publishers, The Netherlands, 1990.
33. A. Thran, G. Kroll and F. Faupel, *J. Appl. Polym. Sci.*, 1999, **37**, 3344.
34. H. Fujita, *Fortschr. Hochpolym. Forsch.*, 1961, **32**, 1.
35. S.A. Stern, G.R. Mauze and H.L. Frisch, *J. Polym. Sci., Polym. Phys. Ed.*, 1983, **21**, 1275.
36. M. Litt, *J. Rheol.*, 1986, **30**, 853.
37. M.I. Beck and I. Tomka, *J. Polym. Sci., Polym. Phys. Ed.*, 1997, **35**, 639.
38. K. Haraya and S.T. Hwang, *J. Membr. Sci.*, 1992, **71**, 13.
39. T.A. Barbari, *Phys. Ed.*, 1993, **31**, 501.
40. Y.P. Yampolskii, Y. Kamiya and A.Y. Alentiev, *J. Appl. Polym. Sci.*, 2000, **76**, 1691.
41. D.R. Paul and W.J. Koros, *J. Polym. Sci., Polym. Phys. Ed.*, 1976, **14**, 675.
42. P. Meares, *Spec. Publ. – R. Soc. Chem.*, 1986, **62**, 1.
43. H. Kumazawa and S.Y. Bae, *J. Appl. Polym. Sci.*, 1996, **60**, 115.
44. H. Odani and T. Uyeda, *Polym. J.*, 1991, **23**, 467.
45. M.A. Krykin, V.I. Bondar, M. Kukharsky and A.V. Tarasov, *J. Polym. Sci., Polym. Phys. Ed.*, 1997, 1339.
46. G.G. Libscomb, *AIChE J.*, 1990, **36**, 1505.
47. J.S. Vrentas and C.M. Vrentas, *Macromolecules*, 1991, **24**, 2404.
48. L. Leibler and K. Sekimoto, *Macromolecules*, 1993, **26**, 6937.
49. R.A. Petrick, *Prog. Polym. Sci.*, 1997, **22**, 1.
50. F. Lançon, L. Bilard, W. Chambron and A. Chamberod, *J. Phys. F: Met. Phys.*, 1985, **15**, 1485.
51. J. Han and R.H. Boyd, *Polymer*, 1996, **37**, 1797.
52. J. Crank, 'The Mathematics of Diffusion', 2nd ed., Clarendon Press, Oxford, 1975.
53. M. Räderer, A. Besson and K. Sommer, *Proc. IDS*, Leuvenhorst, The Netherlands, 2000.
54. H. Levine and L. Slade, *Carbohydr. Polym.*, 1986, **6**, 213.
55. D. Simatos, G. Blond and M. Le Meste, *Cryo-Lett.*, 1989, **10**, 77.
56. M. Karel, M.P. Buera and Y.H. Roos. 'The Glassy State in Foods', J.M.V. Blanshard and P.J. Lillford, (eds.), Nottingham University Press, Nottingham, 1993, 13.
57. F.J. Norton, *J. Cell Plast.*, 1982, **18**, 300.
58. M. Loncin, *Food Process Eng.*, 1980, 354.
59. M.C. Page and L.R. Glicksman, *J. Cell. Plast.*, 1992, **28**, 268.
60. D. Turnbull and M.M. Cohen, *J. Chem. Phys.*, 1961, **34**, 120.
61. D.F. Baldwin, N.P. Suh and M. Shimbo, *Cell. Polym.*, 1992, **38**, 109.

62. R.B. Leslie, P.J. Carrillo, T.Y. Chung, S.G. Gilbert, K. Hayakawa, S. Marousis, G.D. Saravacos and M. Solberg, 'Water Relationships in Foods', H. Levine and L. Slade, (eds.), Plenum Press, New York, 1991, 365.
63. C.H. Tong and D.B. Lund, *Biotechnol. Prog.*, 1990, **6**, 67.
64. J.D. Mellor, 'Preconcentration and Drying of Food Materials', S. Bruin, (ed.), Elsevier Science Publishers, Amsterdam, 1988.
65. G.D. Saravacos, 'Engineering Properties of Foods', M.A. Rao and S.S.H. Rizvi, (eds.), Marcel Dekker, New York, 1995.
66. P.L. Drzal, A.F. Halasa and P. Kofinas, *Polymer*, 2000, 41, **12**, 4671.
67. B.J. Briscoe and T. Savvas, *Adv. Polym. Technol.*, 1998, **17**, 87.
68. F. Witschi, Dissertation ETH No. 13336, 1999, Swiss Federal Institute of Technology, Zürich.
69. R.J. Alexander, 'Starch Hydrolysis Products', F.W. Schenck and R.E. Hebeda, (eds.), VCH Publishers, Weinheim, 1992, Chapter 8, 233.
70. L. Greenspan, *J. Res. Natl. Bur. Stand. (US) – Part A. Phys. and Chem.*, 1977, **81A**, 89.
71. A. Schoonman, J. Ubbink, C. Bisperink, M. Le Meste and M. Karel, *Biotechnol. Prog.*, 2001, in press.
72. P.W. Atkins, 'Physical Chemistry', 2nd ed., Oxford University Press, Oxford, 1982.
73. J. Crank, 'The Mathematics of Diffusion', Clarendon Press, Oxford, 1956.
74. A. Schoonman, G. Mayor, M.L. Dillmann, C. Bisperink and J. Ubbink, *Food Res. Int.*, 2001, **34**, 913.
75. G.K. Vagenas and V.T. Karathanos, *Biotechnol. Prog.*, 1991, **7**, 419.
76. 'Handbook of Chemistry and Physics', 81st ed., CRC Press, London, 2000.
77. H. Levine, L. Slade, B. Van Lengerich and J.G. Pickup, US Patent 5009900, 1991.
78. H. Levine, L. Slade, B. Van Lengerich and J.G. Pickup, US Patent 5087461, 1992.
79. H.E. Swisher, US Patent 2809895, 1957.
80. H.E. Swisher, US Patent 3041180, 1962.

Theories of Unstable Aqueous Systems: How Can They Help the Technologist?

Recent Developments in the Theory of Amorphous Aqueous Sytems

Pablo G. Debenedetti and Jeffrey R. Errington

DEPARTMENT OF CHEMICAL ENGINEERING,
PRINCETON UNIVERSITY

1 Computer Simulation of Carbohydrate–Water Systems

There has been much progress in this area since the previous BioUpdate conference on solid aqueous solutions in 1995, *e.g.* refs. 1–13. All-atom models are now routinely used, and the force fields account realistically for dispersion interactions (generally modeled with the Lennard–Jones potential), electrostatics (modeled by means of partial charges) and intramolecular forces (bond bending, bond stretching, torsion). The method of choice for carbohydrate–water simulations is generally molecular dynamics, in which the equations of motion are integrated in time. Both constant–pressure and constant–volume simulations have been used. The accurate prediction of solution density and diffusion coefficients (both water and carbohydrate), in systems as concentrated as 80 weight % carbohydrate, is by now standard.[2] Accurate melting curve calculations, on the other hand, are not yet feasible.[3] While glass transition estimates are possible,[3,4] they are inevitably based on extrapolation, and hence their accuracy depends on the method, theory and assumptions used to effect the extrapolation.

By far the most common application of molecular simulation in the study of carbohydrate–water mixtures is the investigation of microscopic structure and dynamics. Examples include the determination of site-specific solvation structure,[1,8] and the prediction of concentration-dependent intramolecular hydrogen bond formation[1] and changes in solvation structure.[2] Simulations can also yield useful insights into the evolution of hydrogen bond dynamics as a function of solution concentration.[7]

A large gap still exists between the time and length scales probed in computational studies of carbohydrate–water mixtures ($t \leq 100$ ns; $N \leq 50$ carbohydrate molecules; $L \approx 20$ Å) and the corresponding experimentally relevant scales, especially in highly viscous, concentrated solutions. Realistic calculations involving ternary systems that include biomolecules appear beyond the reach of present-day computational power. Nevertheless, simulations provide valuable molecular-level insight into solution structure and dynamics, and can play a useful role in elucidating the relative effectiveness of various carbohydrates as cryo- and lyoprotectants, at least insofar as their effectiveness is related to hydrogen bond dynamics and the details of carbohydrate–water interactions.

2 Systems with Attainable Kauzmann Temperatures

In his classic review, which remains as insightful today as when it was published, more than half a century ago, Kauzmann called attention to an 'apparent paradox'. This refers to the fact that, liquid heat capacities being higher than those of the corresponding crystal phase, the entropy of melting is consumed rapidly upon supercooling.[14] For many substances, the entropy of the super-cooled liquid is in fact on a collision course with that of the stable crystal, by the time the glass transition intervenes. Modest extrapolation beyond the glass transition predicts an entropy crossing to occur at a temperature T_K, below which the liquid would have lower entropy than the solid. Since the crystal approaches zero entropy as T tends to 0, this state of affairs, were it to continue to arbitrarily low temperature, would represent a violation of the Third Law of Thermodynamics and has come to be known as the Kauzmann paradox.

For the last fifty years, Kauzmann's paradox has been central to the formulation of theories of the glassy state. The prevailing view is that the experimentally observed glass transition is a kinetically controlled manifestation of an underlying thermodynamic transition that the supercooled liquid would have to undergo in order to avoid an entropy crisis. This transition is commonly referred to as an ideal glass transition, and it would require that the liquid settle into the unique, deepest, potential energy minimum corresponding to a non-crystalline arrangement of its constituent molecules.[15] Although the rigorous theoretical basis underlying the notion of an ideal glass transition has been questioned,[16] the Kauzmann temperature has played a key role in glass studies, experimental as well as theoretical. Underlying this ubiquity is the observation that for many liquids, in particular those that fall in the 'fragile' extreme of Angell's useful strong/fragile classification, the Kauzmann temperature (obtained by mild extrapolation of calorimetric measurements) is very close to the Vogel–Tammann–Fulcher (VTF) temperature of structural arrest, T_o, obtained by fitting relaxation data to the VTF equation,

$$\tau = \tau_o \exp\left(\frac{B}{T - T_o}\right) \tag{1}$$

In this equation, τ is a relaxation time (or, equivalently, a shear viscosity) and B is a temperature-independent constant. In other words, $T_K \approx T_o$ (see ref. 17 for an excellent compilation of experimental data), in agreement with the notion of structural arrest brought about by a system being trapped in a unique potential energy minimum. Thus, the Kauzmann temperature has come to be viewed as an absolute limit, unattainable in practice because of the extreme slow-down of structural relaxation that inevitably accompanies its approach. This picture is further reinforced by the theory of Adam and Gibbs,[18] according to which:

$$\tau = \tau_o \exp\left(\frac{C}{Ts_{conf}}\right) \qquad (2)$$

In this equation, C is a temperature-independent constant, and s_{conf} is the configurational entropy, commonly associated with the existence of multiple potential energy minima of the same depth.[15] The Adam–Gibbs equation has proved remarkably successful and versatile in the correlation and extrapolation of relaxation kinetics data (*e.g.* dielectric relaxation, viscosity). At an ideal glass transition, s_{conf} vanishes, as the system settles into a unique configuration. Note, however, that it is the vanishing of the liquid's configurational entropy, rather than the crossing of the liquid and crystalline total entropy curves, that brings about structural arrest.

At this point, it is useful to invoke experimental data. The melting curves of He^3 and He^4 exhibit pressure minima in the (P,T) plane.[19] The melting curve of the polymer poly(4-methylpentene – 1) (P4MP1) exhibits a pressure maximum in the (P,T) plane.[20-22] In these systems, therefore, the following condition is satisfied:

$$\frac{dp_m}{dT} = 0 \qquad (3)$$

where p_m denotes the melting pressure at a given temperature. For the helium isotopes, the volume change on melting is positive at this point, whereas for P4MP1, it is negative. Invoking the Clausius–Clapeyron equation:

$$\frac{dp_m}{dT} = \frac{\Delta s}{\Delta v} \qquad (4)$$

where Δs and Δv denote the changes in entropy and volume upon freezing (or melting), we can see that the point along the melting curve, where Equation 3 is satisfied, is necessarily a Kauzmann point, since the entropies of the liquid and crystal phases are equal at the given temperature and pressure.

There exist systems with attainable Kauzmann temperatures. This implies that there is no logical connection between the possibility of a Kauzmann locus[23] along which $\Delta s = 0$ and the possible existence of a positive-temperature singularity, at which the configurational entropy of the supercooled liquid vanishes, and at which relaxation times and shear viscosity diverge. This important distinction, hitherto not considered in theoretical or experimental studies of glass-forming liquids, must be incorporated in future investigation of relaxation and thermodynamics of viscous liquids and glasses.

3 The Quantification of Structure

Glasses possess processing history-dependent structure. The precise (quantitative) description of structure in systems lacking long-range crystalline order is therefore an important question. Progress in this area should lead to improved techniques for relating glass structure, processing history and physical properties. Recently, fresh insights have resulted from the introduction of structural order parameters.[24,25] For spherically symmetric systems, such as sphere packings, translational and bond-orientational order parameters were introduced, which can easily be measured in the course of a computer simulation. The translational order parameter measures the spatial persistence of the tendency of pairs of molecules to adopt preferential separations. The bond-orientational order parameter measures the distribution of 'bond' angles formed by a central molecule with pairs of nearest-neighbors.

By mapping configurations onto an 'order parameter plane', whose coordinate axes are the above-mentioned translational and bond-orientational measures of structure, an important result was obtained. All non-equilibrium configurations, and in particular, glasses formed by rapid compression and jamming of hard spheres, were found to lie in a distinct region of the order parameter plane, one not occupied by stable equilibrium states. This means that purely geometric measures can distinguish equilibrium from non-equilibrium states. It was further found that glasses generated using different compression rates have easily distinguishable measures of structural order, and that a single curve is formed in the order parameter plane by all glasses so generated. Both measures of order were found to increase monotonically as the compression rate decreased.[24]

More recently, these ideas have been extended to water,[26] using the same measure of translational order, and a water-specific measure of orientational order.[27] The latter is a simple measure of local tetrahedrality around individual molecules. Using these two order parameters, Debenedetti and Errington[26] showed that water's anomalies constitute a cascade, each triggered as the system becomes progressively more ordered. Structural anomalies, whereby both measures of structural order decrease upon compression, occur over the broadest range of temperatures and densities. Kinetic anomalies, whereby the diffusion coefficient increases upon compression, are located entirely within the region of structural anomalies. Thermodynamic anomalies, whereby the liquid expands upon constant-pressure cooling, are in turn contained inside the region of diffusive anomalies. It was further found that in the entire region of the temperature-density plane, where water exhibits anomalous behavior in its transport or thermodynamic properties, translational and orientational order are strictly correlated; the average separation between molecules cannot be changed without simultaneously changing their mutual orientation. This strict correlation makes it possible to rank states according to their degree of structural order. By doing this, Errington and Debenedetti[26] were able to find the minimum amount of structural order required to trigger structural, kinetic and thermodynamic anomalies.

Although the above-mentioned results for spheres and water are promising, much remains to be done for systems of practical interest in biomolecule preservation. Appropriate order parameters must be identified, which take into account the geometric degrees of freedom of industrially relevant matrix molecules (*e.g.* disaccharides). It is also necessary to extend these ideas to mixtures, possibly through the introduction of suitable combining rules for the various order parameters. These and other challenges will have to be tackled by theoreticians and simulators, if practical tools for the quantification of structure in pharmaceutical glasses are to result.

4 Kinetic and Thermodynamic Fragility

Angell proposed a useful classification of liquids along a fragile-to-strong scale.[28] Fragile liquids exhibit marked deviations from Arrhenius behavior in the temperature dependence of their relaxation times; when plotted in scaled Arrhenius fashion [log (relaxation time) *vs.* T_g/T], this functionality shows positive curvature, indicating an effective activation energy that increases markedly upon cooling (super-Arrhenius behavior). In contrast, strong liquids exhibit Arrhenius functionality (linear behavior in Arrhenius coordinates) over appreciable ranges of temperature. Silica is often mentioned as the prototypical strong liquid, whereas *o*-terphenyl is the canonical fragile glass-former. In general, strong liquids, such as the network oxides, SiO_2 and GeO_2, have tetrahedrally coordinated structures, whereas the molecules of fragile liquids experience non-directional, dispersive forces.

Various measures of fragility have been introduced. A particularly useful one, because of its experimental accessibility, is $F_{1/2}$,[29] defined as:

$$F_{1/2} = 2(T_g/T_{1/2} - 1/2) \tag{5}$$

where $T_{1/2}$ is the temperature at which the logarithm of the relaxation time is intermediate between the glass transition [log t (sec) $= 2$] and high-temperature limits [log t (sec) $= -14$]. At $T_{1/2}$, in other words, the relaxation time is 10^{-6} sec. For strong liquids, $F_{1/2} = 0$. The maximum possible fragility, on the other hand, corresponds to the limit $F_{1/2} = 1$. Recently, Martinez and Angell[30] showed that there exists a remarkable correlation between fragility and a corresponding thermodynamic measure. Specifically, Martinez and Angell[30] considered the temperature dependence of the ratio $\Delta S (T_g)/\Delta S(T)$, that is to say, the ratio of differences between liquid and crystal entropies at T_g and at the temperature of interest (that the temperature-dependent entropy difference should be in the denominator, not the numerator, follows from the Adam-Gibbs equation). Defining a thermodynamic fragility as

$$F_{3/4} = 4(T_g/T_{3/4} - 3/4) \tag{6}$$

we again have a quantity that varies between 0 (strong limit) and 1 (fragile limit). Martinez and Angell[30] showed convincingly that the kinetic fragility measure based on the temperature dependence of the relaxation time was linearly related to the thermodynamic fragility (based on the rate at which the entropy of fusion is consumed on supercooling). Their extensive compilation of data for more than twenty liquids of all classes (including organic, ionic and network-forming examples) suggests that the relationship between kinetic and thermodynamic measures of fragility is fundamental and broadly applicable. While this connection between dynamics and thermodynamics is useful for practical purposes (data correlation and extrapolation), its molecular origin remains poorly understood.[31]

References

1. P.B. Conrad and J.J. dePablo, *J. Phys. Chem. A*, 1999, **103**, 4049.
2. N.C. Ekdawi-Sever, P.B. Conrad and J.J. dePablo, *J. Phys. Chem. A*, 2001, **105**, 734.
3. E.R. Caffarena and J.R. Grigera, *J. Chem. Soc. Faraday Trans.*, 1996, **92**, 2285.
4. E.R. Caffarena and J.R. Grigera, *Carbohydr. Res.*, 1997, **300**, 51.
5. M. Carlevaro, E.R. Caffarena and J.R. Grigera, *Int. J. Biol. Macromol.*, 1998, **23**, 149.
6. E.R. Caffarena and J.R. Grigera, *Carbohydr. Res.*, 1999, **315**, 63.
7. C.J. Roberts and P.G. Debenedetti, *J. Phys. Chem. B*, 1999, **103**, 7308.
8. J. Behler, D.W. Price and M.G.B. Drew, *Phys. Chem. Chem. Phys.*, 2001, **3**, 588.
9. Q. Liu and J.W. Brady, *J. Am. Chem. Soc.*, 1996, **118**, 12276.
10. Q. Liu, R.K. Schmidt, B. Teo, P.A. Karplus and J.W. Brady, *J. Am. Chem. Soc.*, 1997, **119**, 7851.
11. Q. Liu and J.W. Brady, *J. Phys. Chem. B*, 1997, **101**, 1317.
12. K.J. Naidoo and J.W. Brady, *Chem. Phys.*, 1997, **224**, 263.
13. B. Leroux, H. Bizot, J.W. Brady and V. Tran, *Chem. Phys.*, 1997, **216**, 349.
14. W. Kauzmann, *Chem. Rev.*, 1948, **43**, 219.
15. P.G. Debenedetti and F.H. Stillinger, *Nature*, 2001, **410**, 259.
16. F.H. Stillinger, *J. Chem. Phys.*, 1988, **88**, 7818.
17. C.A. Angell, *J. Res. NIST*, 1997, **102**, 171.
18. G. Adam and J.H. Gibbs, *J. Chem. Phys.*, 1965, **43**, 139.
19. J. Wilks, 'The Properties of Liquid and Solid Helium', Clarendon Press, Oxford, 1967.
20. S. Rastogi, M. Newman and A. Keller, *Nature*, 1991, **353**, 55.
21. S. Rastogi, M. Newman and A. Keller, *J. Polym. Sci. (Polym. Phys.) B*, 1993, **31**, 125.
22. A.L. Greer, *Nature*, 2000, **404**, 134.
23. F.H. Stillinger, P.G. Debenedetti and T.M. Truskett, *J. Phys. Chem. B*, 2001, **105**, 11809.
24. T.M. Truskett, S. Torquato and P.G. Debenedetti, *Phys. Rev. E*, 2000, **62**, 993.
25. S. Torquato, T.M. Truskett and P.G. Debenedetti, *Phys. Rev. Lett.*, 2000, **84**, 2064.
26. J.R. Errington and P.G. Debenedetti, *Nature*, 2001, **409**, 318.
27. P.L. Chau and A.J. Hardwick, *Mol. Phys.*, 1998, **93**, 511.
28. C.A. Angell, *J. Non-Cryst. Sol.*, 1991, **131–133**, 13.
29. J.L. Green, K. Ito, K. Xu and C.A. Angell, *J. Phys. Chem. B*, 1999, **103**, 3991.
30. L.M. Martinez and C.A. Angell, *Nature*, 2001, **410**, 663.
31. F.H. Stillinger and P.G. Debenedetti, *J. Chem. Phys.*, 2002, **116**, 3353.

Studies on Raffinose Hydrates

Kazuhito Kajiwara,[1]* Akihito Motegi,[1] Masashi Sugie,[1] Felix Franks,[2] Sigeru Munekawa,[3] Toshio Igarashi[3] and Akira Kishi[3]

[1]DEPARTMENT OF BIOSCIENCES, TEIKYO UNIVERSITY OF SCIENCE AND TECHNOLOGY, 2525 YATSUSAWA UENOHARA-MACHI, KITATSURU-GUN, YAMANASHI, 409-0193 JAPAN
[2]BIOUPDATE FOUNDATION, 25 THE FOUNTAINS, 229 BALLARDS LANE, LONDON N3 1NL, UK
[3]X-RAY RESEARCH LABORATORY, RIGAKU CORPORATION, 9-12, MATSUBARA-CHO, 3-CHOME, AKISHIMA-SHI, TOKYO, 196-8666 JAPAN

1 Introduction

After having been largely ignored for several decades, the physical and thermomechanical properties of solid polyhydroxy compounds (PHCs), and particularly of oligomeric sugars, are receiving increasing attention, mainly because of new insights into their varied applications in food and pharmaceutical product and process technology.[1] Because of a general lack of physicochemical information, PHC solid/solid and solid/liquid phase behaviour is still largely unexplored. This has led to some 'surprising' discoveries, e.g. the formation of a metastable crystalline mannitol hydrate, which can form under certain conditions when the alditol is dried from an aqueous solution.[2] An elucidation of the eutectic, peritectic and polymorphic behaviour of PHCs is bedevilled by the general observation that phase transitions tend to be slow and sometimes not measurable at all in real time. In part this phenomenon of slow phase transitions must be due to the extremely complex crystal structures, basically akin to that of ice, in which individual molecules are linked by hydrogen bonds to several neighbouring molecules, giving rise to chains and complex, infinite, 3D networks.[3] Such complex molecular arrangements are likely to give rise to low nucleation rates of solid daughter phases within an ordered mother phase. Another consequence of the extensively hydrogen bonded, supramolecular domains is the propensity of PHCs to display highly non-Newtonian rheologies. The technological exploitation of thixotropy, shear thinning, plastic flow, gel and glass formation, etc. are many and varied.

In the past, existing information on the viscometric behaviour of PHCs in solution was apparently sufficient in its scope for their technical applications as thickeners, gelling agents and stabilisers by the food and healthcare product industries. Their purposeful uses in pharmaceutical, and particularly biopharmaceutical processing, however, require a much better understanding of phase transitions, interactions with residual water, and possible physical and/or chemical changes with long relaxation times, such as might lead to product deterioration during storage.

Until recently the boundary to practical knowledge was set by the glass transition, T_g, on the assumption that beyond or below T_g a solid system is 'stable', at least for periods of practical interest.[4-6] In food and pharmaceutical product technology such periods came to be identified as shelf lives, although the so-called 'shelf' was frequently to be found in a refrigerator, in order to allow for commercially adequate storage periods. It has since come to be realised that physical and chemical changes persist at temperatures well beyond and below T_g, although this fact had already been pointed out by Kauzmann in his seminal paper of 1948![7] Although most such changes are considered to be detrimental, resulting in a deterioration of product quality, on occasions they can also be beneficial. Thus, in the slow crystallisation of PHC hydrates from the amorphous state, in the presence of residual water, the PHC acts as a 'chemical desiccant'.[8] This phenomenon is of practical interest in the drying of complex mixtures from aqueous solution, or the removal of traces of water vapour that have inadvertently found their way into a product during long-term storage.

While most sugars crystallise in the anhydrous form, some hydrates have been identified, among them glucose·H_2O, α,α-trehalose·$2H_2O$, β,β-trehalose·$4H_2O$ and raffinose·$5H_2O$, as well as others. The above examples have been chosen for discussion, because their crystal structures are known. By analogy with salt hydrates, a controlled PHC dehydration at constant temperature might be expected to reveal distinct intermediate hydrates and, finally, the crystalline, anhydrous form. The crystalline anhydrous form of α,α-trehalose can indeed be prepared by careful removal of water from the dihydrate, followed by heating and suitable annealing.[9] However, no intermediate hydrates have yet been detected. The reverse process, *i.e.* the crystallisation of the dihydrate from the amorphous sugar, containing a low amount of water, has also been studied. Its rate of formation can be increased dramatically by annealing the solid mixture at temperatures above T_g and just below the melting point, T_m, of the dihydrate.[8] Attempts to apply similar techniques to β,β-trehalose·$4H_2O$ have not led to any positive results. Although its T_m could be established, no further crystallisation processes could be detected, at least over a period of several days, and the search for a crystalline anhydrous form also proved unsuccessful.[10]

The raffinose–water system is of particular interest, not only because it appears to be the only known sugar pentahydrate, but because there might be several possibilities of intermediate crystalline hydrate states, although no such structures have actually been positively identified. Past evidence in support of intermediate hydrates has been based on sorption/desorption isotherms and the interpretation of powder X-ray diffraction data and DSC scans.[11,12] The search for a crystalline anhydrous raffinose (AR) has also been unsuccessful up to now.

The main aim of the investigations here reported was to obtain more conclusive structural and thermodynamic information about possible intermediate hydrates R4W, R3W, R2W and R1W. A secondary aim was a re-examination of the R5W crystal structure, which might shed new light on changes in the sugar–sugar and sugar–water hydrogen bonding patterns during the controlled removal of water.

2 Experimental

2.1 Thermochemical Measurements

Raffinose pentahydrate, α-D-galactopyranosyl-$(1\rightarrow6)$-O-α-D-glucopyranosyl-$(1\rightarrow2)$-β-D-fructofuranoside (ex Sigma, >99% pure), was used without further purification. Distilled water was used in all experiments. Samples containing raffinose were prepared as follows, unless otherwise described: Powdered raffinose pentahydrate (R5W) was held for 3–10 days at 30 °C in DSC pans in a desiccator over solid NaOH or a saturated NaOH solution. Each sample was weighed periodically, until its water content had been reduced to the desired value. Residual water contents were determined by mass loss during evaporation.

Thermal measurements were performed in a Shimadzu DSC-50, helium being used as purge gas. The instrument had previously been calibrated with water and indium to within ±1 °C at a scanning rate of 5 °C min^{-1}. For determinations of enthalpies of dehydration, samples were analysed in open DSC pans and at heating rates of between 0.1 and 1.0 °C min^{-1}, in steps of 0.1 °C min^{-1}, and finally at 3.0 and 5.0 °C min^{-1}. All scans were performed over the temperature range 15 to 150 °C. For melting point and heat of fusion measurements, samples were hermetically sealed in DSC pans and heated from room temperature to 120 °C at a rate of 5 °C min^{-1}.

2.2 Single-crystal X-ray Diffraction

A colourless block crystal of R5W was prepared as follows: powdered R5W was dissolved in distilled water to give a saturated solution which was then stored at room temperature until the water had evaporated. R5W was recrystallised and a single crystal grown by several exchanges of the saturated solution. The crystal, having approximate dimensions of $0.50 \times 0.45 \times 0.25$ mm was mounted on a glass fibre.

All measurements were made on a Rigaku RAXIS-RAPID Imaging Plate diffractometer, using graphite monochromated MoKα = 0.71 Å radiation (50 kV, 100 mA). Indexing was performed from two oscillations with exposure times of 1 min. Data were collected at 23 ± 1 °C, to a maximum 2θ value of 54.7°. A total of 44 images, corresponding to 220.0° oscillation angles were collected with two different goniometer settings and exposure times of 0.40 min deg^{-1}. The camera radius was 127.40 mm. Readout was performed in the 0.10 mm pixel size, and the data processing was performed with the PROCESS-AUTO programme package. Of the 25 200 reflections which were collected, 3388 were unique; equivalent reflections were merged.

2.3 XRD-DSC

XRD-DSC measurements were performed with a Rigaku XRD-DSC II. Diffraction patterns were obtained with a 50 kV, 40 mA wavelength of Cu–Kα = 1.54 Å. R5W powder, which had previously been ground to 45 micrometers mesh, was loaded into a sample holder and scanned at a rate of 20° (2θ) min^{-1}, while the sample was being heated from room temperature to 130 °C at a rate of 10 °C h^{-1}.

3 Results and Discussion

3.1 Single-crystal X-ray structure of R5W

Crystallographic data (atomic coordinates, anisotropic displacement parameters, bond lengths and angles, *etc.*), and refinement techniques will be reported elsewhere. Essentially the structural details of R5W are in agreement with those reported by Jeffrey and Huang,[3] but of a higher precision. Figure 1 shows the positions of the five water molecules with respect to the three sugar residues,

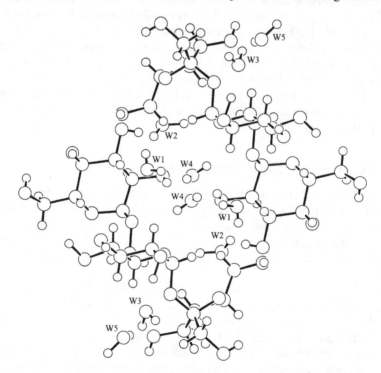

Figure 1 *Single-crystal X-ray diffraction for raffinose pentahydrate (R5W), showing the location and numbering of the various water molecules and drawn from data reported by Jeffrey and Huang*[3]

and the stereopictures in Figure 2 graphically illustrate the tunnel-like structure of the hydrogen-bonded network.

3.2 Thermodynamics of Fusion and Dehydration

The DSC dehydration scans in Figure 3 reveal three symmetrical endotherms of which only two depend on the heating rate. It should be emphasised here that a heating rate of $0.2\,°C\,min^{-1}$ may be considered as low when compared to most reported DSC studies on solids. Nevertheless, for very slow relaxation processes such temperature scanning rates may still not be low enough to prevent the appearance of instrumental artefacts.

Immediately after completion of the dehydration measurements the samples were returned to room temperature and examined for physical changes and/or discoloration. In every case they had remained white and powdery. The observed transitions were therefore assigned to distinct hydrates. Enthalpies corresponding to each of the two lower endotherms corresponded to 20% of the total enthalpy change. They were therefore assigned to R4W and R3W, respectively. The third endotherm, which could not be detected at heating rates of $0.2\,°C$ min^{-1}, was constant at $80\,°C$ but increased in intensity with increasing heating rate. No evidence could be gained for the existence of lower hydrates or for a crystalline anhydrate.

Melting points were measured for preparations, dried from R5W powder, in the composition range $0.3 < x < 0.7$, where x is the raffinose mol fraction. The measured melting point of R5W, $85\,°C$, is in good agreement with various previously reported values. For preparations with $x > 0.5$, melting temperatures reach a constant value of $90\,°C$. A well-formed minimum is also observed at $x = 0.35$. In our previous studies, based on the dehydration by storing prepara-

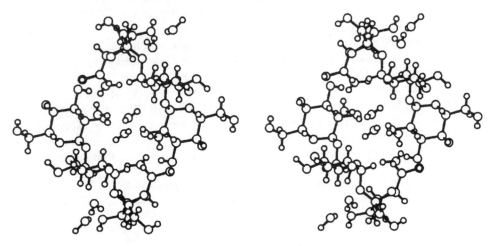

Figure 2 *Stereopictures of raffinose pentahydrate, which graphically illustrate the tunnel-like structure of the hydrogen-bonded network*

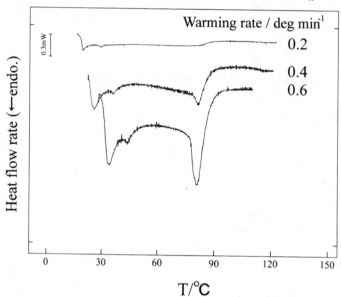

Figure 3 *DSC measurements of heat flow during dehydration of raffinose hydrate. Warming rates were 0.2, 0.4 and 0.6 °C min⁻¹, respectively. For determinations of enthalpies of dehydration, samples were analysed in open DSC pans*

tions in a vacuum oven overnight, the observed melting points rose as x exceeded 0.5.[13] The current results suggest, however, that this was not a sufficiently long period for the establishment of equilibrium. In the series of experiments here reported, preparations were dried for periods of 5–10 days, and this treatment produced constant melting points for mixtures with $x > 0.5$. It is therefore possible that such preparations consisted of mixtures of R1W and an anhydrous amorphous phase. Similar results, *i.e.* a constant melting point of R5W, had been observed for mixtures of R5W and amorphous raffinose.

Melting points and enthalpies of fusion of the crystalline species are summarised in Table 1. The enthalpies decrease with a decreasing mol ratio water:raffinose. The removal of the first water molecule from R5W produced a difference in the fusion enthalpies between R5W and R4W of 16.1 kJ mol⁻¹. Similarly, further stepwise dehydration from R4W to R3W and from R3W to R1W produced $\Delta(\Delta H)$ values of 25.1 and (18.6×2) kJ mol⁻¹, respectively. Therefore, when one water molecule is dehydrated, the difference of fusion enthalpies is approximately 20 kJmol⁻¹. The hypothetical fusion enthalpy of

Table 1 *Melting points and enthalpies of fusion of raffinose hydrates*

	T_m/K	$\Delta H/kJ\ mol^{-1}$
R5W[13]	352.7 ± 0.1	88.8 ± 0.4
R4W	357.3 ± 0.3	72.7 ± 2.2
R3W[13]	358.4	47.6
R1W	358.2 ± 0.3	10.4 ± 0.9

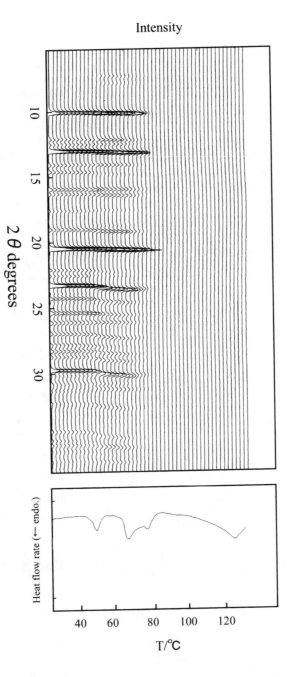

Figure 4 *Measurements by simultaneous powder X-ray diffraction-differential scanning calorimetry (XRD-DSC). Figure at left is an XRD pattern and at right is a DSC scan. Raffinose pentahydrate powder was scanned at a rate of $20\,^{\circ}C\,(2\theta)\,min^{-1}$, while the sample was being heated from room temperature to $130\,^{\circ}C$ at a rate of $10\,^{\circ}C\,h^{-1}$*

(crystalline) anhydrous raffinose (AR) is thus estimated to be approximately equal to zero, indicating that AR in the preparations used was in the amorphous state.

3.3 XRD-DSC

Figure 4 shows the thermodynamic and structural changes occurring during a complete dehydration of R5W. On the removal of the first water molecule, to yield a R4W stoichiometry, small peaks appear at *ca.* 8° (2θ), which do not appear in the R5W diffractogram.[11,12] Further dehydration is characterised by the appearance of the next two, only partly resolved, endotherms and leads to complete amorphisation. It is accompanied by gradual shifts in the XRD positions of some other minor peaks and an eventual disappearance of all crystalline diffraction signals. Although yet another endotherm is visible at 125 °C, its origin is unclear, but it cannot be associated with the evaporation of water.

In order to obtain some additional information about possible existence of intermediate hydrates, a simulation was performed on the effects of removing each one of the five water molecules in turn from their different locations in R5W.

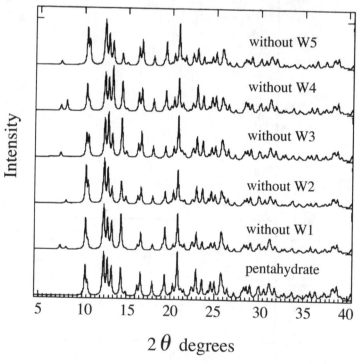

Figure 5 *A simulation was performed on the effects of removing each one of the five water molecules in turn from their different locations in R5W. The simulation was based on the assumption that the hydrogen-bonded, crystalline raffinose framework is structurally unaffected by the removal of any one of the water molecules at a time. The bottom line is the initial R5W diffractogram*

The simulation is based on the assumption that the hydrogen-bonded, crystal-line raffinose framework is structurally unaffected by the removal of any one of the water molecules at a time. The results are shown in Figure 5, where the bottom line is the initial R5W diffractogram. The picture obtained by the removal of either W1 or W4, corresponds to the experimental results shown in Figure 4, *i.e.* the two minor peaks appear. The X-ray structure analysis indicates that of all the five water molecules, W1 is the only one that is linked to the network by three, rather than four hydrogen bonds.[3] We therefore suggest that the R4W crystal structure, identified in our study, is obtained by the removal of W1 from R5W.

The complete R5W X-ray analysis also shows that the hydrogen-bonded raffinose molecule chains in the crystal are linked by water molecules. There are no direct hydrogen bond links between sugar residues in adjacent chains. It appears, therefore, that upon the removal of all five water molecules, the raffinose framework collapses, and no recrystallisation of AR takes place, as it actually does in the case of anhydrous α,α-trehalose. It must, however, be borne in mind, that the apparent inability of the hydrogen bond network to repair itself in the absence of water may be a kinetic artefact, related to the high T_g of AR. Such a conclusion can be drawn from an inspection of Figure 6, which shows the state diagram of the raffinose–water system. While the melting point curve is nearly constant, changing only from 352 to 360 K over the whole composition range, T_g changes from 135 K (for pure water)[14] to 376 K for AR. The T_g curve converges

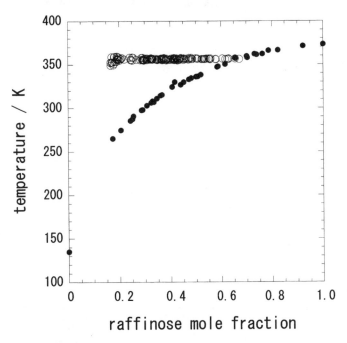

Figure 6 *A state diagram raffinose–water system. Open circles are the melting points, and closed circles are the glass transition temperatures*

with the liquidus at $x \approx 0.7$. T_g/T_m ratios for carbohydrates tend to lie in the neighbourhood of 0.75,[4] which suggests that, the structure of any crystalline AR, if it could ever be prepared in real time, would have to be quite different from that of R5W, and its melting point would be substantially higher than the above relationship suggests.

References

1. J.M.V. Blanshard and P.J. Lillford (eds.), 'The Glassy State in Foods', Nottingham University Press, 1993.
2. Y. Lian, N. Milton, E.G. Groleau, D.S. Mishra and R.E. Vansickle, *J. Pharm. Sci.*, 1999, **88**, 196.
3. G.A. Jefffrey and D. Huang, *Carbohydrate Res.*, 1990, **206**, 173.
4. L. Slade and H. Levine, *Crit. Rev. Food Sci. Nutrition* 1991, **30**, 115.
5. J.L. Green and C.A. Angell, *J. Phys. Chem.*, 1989, **93**, 2880.
6. F. Franks, R.H.H. Hatley and S.F. Mathias, *Pharm. Technol. Int.*, 1991, **3**, 24.
7. W. Kauzmann, *Chem. Rev.*, 1948, **43**, 219.
8. B.J. Aldous, A.D. Auffret and F. Franks, *Cryo-Letters* 1995, **16**, 181.
9. F. Shafizader and R.A. Sussot, *J. Org. Chem.* 1973, **38**, 3710.
10. C.J. Roberts and F. Franks, *J. Chem. Soc. Faraday Trans.*, 1996, **92**, 1337.
11. A. Saleki-Gerhardt and G. Zografi, *J. Pharm. Sci.*, 1995, **84**, 318.
12. K. Kajiwara, F. Franks, P. Echlin and A.L. Greer, *Pharm. Res.*, 1999, **16**, 1441.
13. K. Kajiwara and F. Franks, *J. Chem. Soc. Faraday Trans.*, 1997, **93**, 1779.
14. M. Sugisaki, H. Suga and S. Seki, *Bull. Chem. Soc. Japan* 1968, **41**, 2591.

Comparison Between WLF and VTF Expressions and Related Physical Meaning

Alberto Schiraldi

DISTAM, UNIVERSITÀ DI MILANO, VIA CELORIA 2, 20133 ITALY

The equations proposed by Williams, Landel and Ferry (WLF)[1] and Vogel, Tamman and Fulcher (VTF)[2] have been widely used to describe the behaviour of liquids and amorphous solids, including glasses of either polymer or non-polymer nature. The two expressions are shown in the box below.

WLF

$$a_T = \frac{\tau(T)}{\tau(T_g)} = \exp\left[\frac{C1(T-T_g)}{C2+(T-T_g)}\right]$$

$$\log a_T = \frac{C1(T-T_g)}{C2+(T-T_g)}$$

VTF

$$\tau = \tau(\infty)\exp\left[\frac{B}{(T-T_o)}\right]$$

$$\log\frac{\tau}{\tau(\infty)} = \frac{B}{(T-T_o)}$$

'a_T' is the ratio of any mechanical relaxation time, τ, at the temperature, T, with respect to its value at a reference temperature, such as T_g, and primarily reflects the temperature dependence of a segmental friction coefficient or mobility, on which the rates of all configurational rearrangements depend. 'T_o' is the temperature at which the relaxation time relevant to molecular displacements becomes infinite. '$\tau(\infty)$' is the relaxation time for $T\to\infty$.

A few simple algebraic relationships hold:

$$\tau(T) = \tau(T_g)\exp\left[\frac{C1(T-T_g)}{C2+(T-T_g)}\right] = \tau(\infty)\exp\left[\frac{B}{(T-T_o)}\right] \qquad (1)$$

For $T = T_g$

$$\tau(T_g) = \tau(\infty)\exp\left[\frac{B}{(T_g-T_o)}\right] \qquad (2)$$

which can be replaced in Equation 1 to give:

$$\exp\left[\frac{B}{(T_g - T_o)}\right] \times \exp\left[\frac{C1(T - T_g)}{C2 + (T - T_g)}\right] = \exp\left[\frac{B}{(T - T_o)}\right] \tag{3}$$

or

$$\left[\frac{B}{(T_g - T_o)}\right] + \left[\frac{C1(T - T_g)}{C2 + (T - T_g)}\right] = \left[\frac{B}{(T - T_o)}\right]$$

$$\frac{(T - T_o)(T_g - T_o)}{B} = -\frac{C2 + (T - T_g)}{C1} \tag{4}$$

Putting $T = T_o$,

$$C2 = T_g - T_o$$

and

$$B = -C1C2$$

that is:

$$\tau(T_g) = \tau(\infty)\exp[-C1]$$

$$C1 = \log\left[\frac{\tau(\infty)}{\tau(T_g)}\right] \tag{5}$$

The two equations can therefore be formally related to each other,[3,4] although this does not imply the same meaning for either.

It can be of help to remember that the WLF equation holds for $(T_g + 100\ \text{K}) \geq T \geq T_g$ and therefore does not cover the (T_o, T_g) range. It was derived from Ferry's[5] expression:

$$a_T = \frac{\eta T_s \rho}{\eta_s T \rho} \tag{6}$$

where η and ρ are viscosity and density, respectively, and $T_s \approx (T_g + 50\ \text{K})$ is a suitable reference temperature. a_T primarily reflects the temperature dependence of a segmental friction coefficient or mobility, on which the rates of all configurational rearrangements depend. The equation was proposed to describe low molecular mass polymers, the viscosity of which depends on temperature (for $T > T_g$) in a way that can be empirically correlated with the effect of the molecular mass on the T_g value. In this perspective, the approximate (because of the uncertainty in the experimental determination of T_g and related quantities) expression shown in the above box fits many findings for $T > T_g$.

The VTF equation (see the above box) holds for $T > T_o$. Since:

$$\frac{\tau(\infty)}{\tau} \propto \frac{\eta(\infty)}{\eta} \tag{7}$$

then

$$\eta = \eta(\infty) \exp\left[\frac{B}{(T-T_o)}\right] = \eta(\infty) \exp\left[-C1\frac{T_g-T_o}{(T-T_o)}\right] \qquad (8)$$

which can be compared with:

$$\eta = \eta(\infty) \exp\left[\frac{C}{TSc}\right] \qquad (9)$$

which is the Adam-Gibbs expression for the viscosity,[6] where $\eta(\infty)$ is the viscosity at $T = \infty$, and Sc is the excess configurational entropy of the system with respect to a stable solid state. One can easily recognize that:

$$Sc = \frac{C}{B}\frac{T-T_o}{T} = Sc(\infty)\left(1 - \frac{T_o}{T}\right) \qquad (10)$$

where $Sc(\infty)$ is the excess configurational entropy at infinite temperature.

T_o is a meaningful reference point, in the sense that it defines a state where molecular displacements within a given undercooled liquid system would vanish, together with the excess configurational entropy. One can, in many cases, identify T_o with the Kauzmann temperature, T_K, where the entropy of the undercooled liquid becomes equal to that of the thermodynamically stable solid phase.[7] Equation 10 allows for a decrease of T_g, whenever a process produces a decrease of Sc, as when annealing occurs at $T_o < T < T_g$.

The meaning of $Sc(\infty)$ is more vague, as long as the idea of a superheated liquid, where the molecular mobility should be comparable with that in a gaseous phase, is tenable, only assuming a concomitant large increase of pressure, as in the case of supercritical fluids. This adds a variable that is not included in the VTF equation.

As long as a decrease in molecular mass within a given family of homologous polymers reduces T_o and T_g by approximately the same amount, the difference $(T_g - T_o)$ can be assumed to remain approximately constant. In this sense, $C2$ in the WLF equation can be given a 'universal value', namely, 51.6 K. As long as the ratio $[C/Sc(\infty)]$ may be given a 'universal' constant value, also the parameter B in the VTF equation and the parameter $C1$ in the WLF equation have 'universal' constant values, namely 900 K and -17.44 (when a_T is given in a $\log_{10}$ scale), respectively.[3,4] This implies that:

$$\frac{C}{Sc(\infty)} = -C1C2 = B \qquad (11)$$

which combines properties at T_g, T_o and T_∞.

Given that the relaxation time associated with the drop in heat capacity at T_g and detected with a calorimetric experiment is of the order of 10^2 s, the 'universal value' of $C1$ imposes an expectation of 10^{-15} s for $\tau(\infty)$. The relevant viscosity would encompass the range 10^{-5}–10^{12} Pa s (Figure 1). These extremes can be

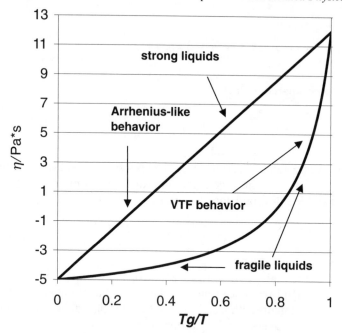

Figure 1 *Strong and fragile liquids, according to Angell* et al.[8]

connected with a straight line (Arrhenius-like behaviour) in the plane (log η, T_g/T) for some systems that are called 'strong liquids', whereas a curved line, which corresponds to Equation 8, describes the behaviour of systems called 'fragile liquids'.[8] In such a plot, the extremes of the temperature scale correspond to T_g and $T = \infty$, respectively, while T_o lies in the $T_g/T > 1$ range, where Arrhenius behaviour is expected for either strong or fragile liquids (Figure 2).

On passing across the $T_g/T = 1$ condition, the VTF trend can be expected to change into that of the corresponding tangent (Figure 2), namely:

$$-\log\left[\frac{\tau}{\tau(\infty)}\right] = -C1\frac{(x-a)}{(1-a)} \tag{12}$$

where x and a represent T_g/T and T_o/T_g, respectively. The parameter a is a measure of 'fragility' in the sense of Angell.[8] When $a = 0$, the maximum 'strong liquid' behaviour is observed, whereas 'fragile liquids' correspond to $0 < a < 1$ values. Relaxation time instead of viscosity appears in Equation 12, as the latter would be meaningless for a solid-like system, *i.e.* for $T < T_g$.

It should be noticed that $Sc(T_g) = Sc(\infty)$ for $a = 0$; in other words, maximum 'strong liquids' would keep the same excess entropy at temperatures well above their own T_g.

If the maximum 'strong liquid' can be assumed to go through the T_g threshold without any change of heat capacity, then the extrapolation of the Arrhenius-like trend of the relaxation time to temperatures below T_g could be a reasonable

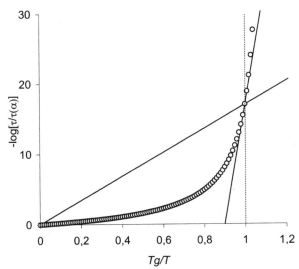

Figure 2 *Extension of Angell's plot toward the range where the system becomes solid-like.
The VTF trend (open circles) has been extrapolated with the straight line tangent
to the VTF curve at the point* $T_g/T = 1$. *The example shown corresponds to* a = 0.9
(see Equation 12)

approximation. At the same time, if such a trend is also reasonable for vitrified
'fragile' liquids, then a different trend should be envisaged across the T_g threshold, such as to replace a 'physical discontinuity' at a well-defined T_g with a glass
transition temperature range (Figure 3). Such a behaviour would not imply any
'universality' for the parameters $C1$ and $C2$.

The choice of a suitable reference temperature in either the WLF or VTF
equation allows for their use in the description of diffusion-limited processes. An
example could be the nucleation and growth of a crystal phase, which is maxi-
mized at some temperature above T_g and below the melting point, T_m.

This process is sustained by molecular displacements that allow the formation
of crystal nuclei; for this reason, the nucleation takes place above T_g. Migration
of other molecules toward nuclei sustains the formation of crystal embryos and
eventually the growth of macroscopic crystal phases. Taking into account that
the medium's viscosity decreases with increasing temperature, the growth rate
can be assumed to increase with increasing T. For $T > T_m$, no crystal would be
stable, whereas the driving force that sustains crystal growth increases on cooling
below T_m. Its effects are, however, diminished because of the concomitant
increase of the medium's viscosity. The rate of crystal formation is therefore the
result of the joint probability of forming nuclei and increasing their size up to a
macroscopic scale. The rate (indicated by k) of each single event can be described
with a VTF-like expression:

$$k_{overall} = k_{nucl} \times k_{growth} = k_{0,nucl} \exp\left(\frac{B_{nucl}}{T - T_{0,nucl}}\right) \times k_{0,growth} \exp\left(\frac{B_{growth}}{T_{0,growth} - T}\right)$$

(13)

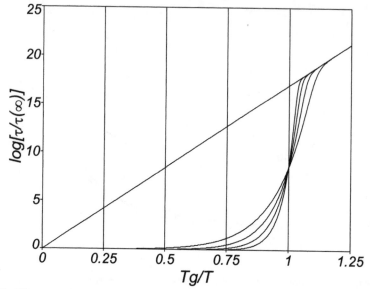

Figure 3 *Modification of Angell's plot for 'fragile liquids', with a sigmoid trend across a glass transition range. The examples shown correspond to different skew sigmoid functions that attain the straight line trend at different T_o values. The straight line corresponds to an ideal 'strong liquid' that does not undergo any heat capacity drop across the glass transition range*

In either exponential term, T_o is a suitable reference temperature at which the corresponding process rate vanishes. Within the temperature range of interest, $T_{0,nucl}$ and $T_{0,growth}$ are the lower and upper limits, respectively. They could be tentatively identified with the glass transition temperature, T_g, of the still amorphous system and the melting point, T_m, of the crystals, respectively.

When applied to starch retrogradation in pasta and rice,[9] Equation 13 satisfactorily fits the experimental findings, when T_m and T_g lie in the ranges of $50\,°C \pm 3\,°C$ and $-10 \pm 3\,°C$, respectively, which are in good agreement with the experimental values.[4]

References

1. M.L. Williams, R.F. Landel and J.D. Ferry, *J. Am. Chem. Soc.*, 1955, **77**, 3701.
2. H. Vogel, *Physik Z.*, 1921, **22**, 645; G. Tamman and W. Hesse, *Z. Anorg. Allgem. Chem.*, 1926, **156**, 245; G.S. Fulcher, *J. Am. Chem. Soc.*, 1925, **8**, 339.
3. C.A. Angell, *Polymer*, 1997, **38**, 6261.
4. L. Slade and H. Levine, 'The Glassy State in Foods', J.M.V. Blanshard and P.J. Lillford (eds.), Nottingham University Press, Loughborough, 1993, 35.
5. J.D. Ferry, *J. Am. Chem. Soc.*, 1950, **72**, 3746; J.D. Ferry and E.R. Fitzgerald, *J. Appl. Phys.*, 1953, **24**, 911.
6. G. Adam and J.H. Gibbs, *J. Chem. Phys.*, 1965, **43**, 139.
7. W. Kauzmann, *Chem. Rev.*, 1948, **43**, 219.
8. C.A. Angell, R.D. Bressel, J.L. Green, H. Kanno, M. Oguni and E.J. Sare, *J. Food Eng.*, 1994, **22**, 115.
9. M. Riva, D. Fessas and A. Schiraldi, *Cereal Chem.*, 2000, **77**, 433.

Progress in Food Processing and Storage

Progress in Food Processing and Storage, Based on Amorphous Product Technology

Louise Slade and Harry Levine

CEREAL SCIENCE GROUP, NABISCO BISCUIT & SNACKS R&D,
KRAFT FOODS NORTH AMERICA, EAST HANOVER, NJ 07936, USA

1 Introduction

The science of sugar glasses has been studied for over 70 years.[1-6] Much more recently (22 years ago), the 'food polymer science' (FPS) approach[7] was developed to study glasses and glass transitions, and their effects on processing, product quality and storage stability, in foods. Since 1980, this 'new' approach to food research – beyond that of the 'water activity' concept of moisture management – has been used to understand structure–function relationships, the effects of plasticization by water – as illustrated by means of state diagrams – on thermal, mechanical, rheological and textural properties, and physical (meta)stability in the non-equilibrium glassy solid state, *vs* instability in the rubbery or viscous liquid state.[8-12]

Blanshard[13] has stated that 'there is no doubt that the perspectives and new understanding of food processing, storage and stability provided by a recognition of the existence and characteristics of the glassy and rubbery states have been one of the most stimulating and pervasive developments in food research in the past half century'. We are proud to have been credited by Blanshard,[13] and similarly by Roos *et al.*,[14] with having played a pioneering role in this area of food research.

2 Recent Applications and Progress

Key concepts of the FPS approach, especially as it has been more widely applied in the 1990s, include a practical understanding of the significance of the glass transition temperature, T_g, and its temperature range, how T_g is defined for multi-component, aqueous amorphous blends, and how T_g relates to the relative mobilities of individual components, including water, in such blends.[15] The latter two aspects are illustrated in Figure 1.[15] For a given glass-forming solute–water blend, T_g is determined by the weight-average molecular weight (Mw) of that blend. A probe molecule of the same Mw would be immobilized at the T_g

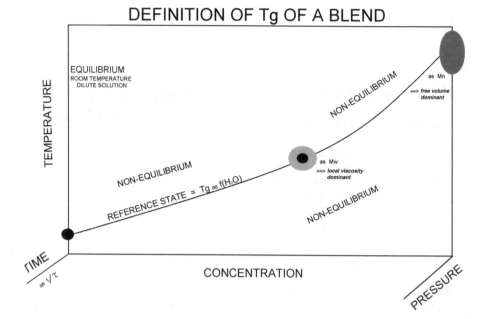

Figure 1 *Definition of* T_g *of a blend, illustrated by a glass curve plotted on a state diagram of temperature* vs *solute concentration*

of the blend, due to the dominance of local viscosity. In contrast, a solute molecule of higher molecular weight than Mw of the blend would be already immobilized at a higher T_g (determined by solute number-average molecular weight (Mn), due to the dominance of free volume), while water molecules, of lower molecular weight than Mw of the blend, would still be capable of showing some translational mobility (but much less than that of bulk liquid water) in the glassy blend at $T < T_g$ of the blend.[15]

Due to water's well-known plasticizing effect, T_g decreases monotonically (curvilinearly) with increasing moisture content, but the (inverse) linear relationship is between T_g and measured system relative humidity (% RH), rather than moisture content.[11,16] Like the glass transition itself, plasticization is a kinetic process, so the mere presence of water is not a guarantee that plasticization has already occurred.[11,17,18]

In response to the question 'what is the T_g of my system?', the following can be stated.[15] Across different glass-forming solute–water systems, T_g at a given RH increases with increasing solute(s) Mw, up to the so-called 'entanglement molecular weight'. This fact has been illustrated previously in terms of starch gelatinization temperatures and mold spore germination lag times.[9] If any ice can form in a system upon cooling, then the system is located above its T_g curve; i.e. Tsystem $> T_g = T_g'$, Wsystem $> W_g = W_g'$, and RHsystem > 0. For some solutes, W_g is more important than T_g, with respect to the location of the system relative to its T_g curve. Nevertheless, the controlling influence of the glass curve can extend to temperatures over 100 K above the glass curve. For example, the non-Arrhenius behavior of water itself extends to 155 K above T_g of water alone. For single solutes, control by the glass curve extends from T_g up to T_m and the ratio of T_m/T_g (in K) predicts the relative rigidity of a given glass.

An understanding of the effects of glass transition phenomena on freshness and shelf-life of various types of food systems starts with a description – in terms of a 'universal sorption isotherm'[19] (shown in Figure 2) – of the general relationship between moisture content and relative vapor pressure (RVP) or %RH. For example, for different baked goods and other cereal-based and related food ingredients and products, it has been shown that low-moisture cookies (viewed as candy glasses), bread dough and flour all exhibit typical ('universal') behavior, while baked bread does not.[19] Figure 2 illustrates the concept that multiple texture stabilization requires control of moisture content, sample RH, molecular T_g and network T_g. Portions of the glass curves for sorbitol, for a non-networked biopolymer and for a permanent network have been positioned relative to the 'universal isotherm curve' on the diagram, as a way of illustrating that molecular T_g controls water vapor migration, while network T_g controls bulk liquid water migration. Figure 2 also points out that, for thermosets such as cheese and baked bread, it is relatively easy to effect surface dehydration, but difficult to remove bulk water. In contrast, it is easier to remove bulk water from a raisin, which typically exhibits desorption hysteresis.

Selected examples of applications of the FPS approach to food processing and

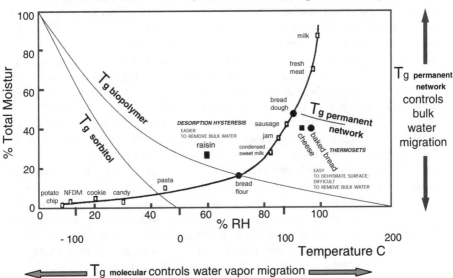

Figure 2 *% Relative humidity of common foods at room temperature and typical steady-state moisture contents, plotted as a 'universal sorption isotherm'*

to studies on product quality attributes – *e.g.* crispness as a textural aspect of freshness over shelf-life – can be described, which illustrate the significant progress made in recent years, based on amorphous product technology, as applied to methods for producing and then maintaining the freshness and improving the storage stability of foods. For example, an earlier description of whole-grain processing, based on the state diagram and glass curve for starch–water,[11] has led to a deeper understanding of the importance of T_g – which controls the initial expansion[20,21] and subsequent collapse and resulting shrinkage of expanded products[22–24] – in extrusion processing. Today, the FPS approach and state diagrams are explicitly used as a technological basis for describing pasta drying[25] and extrusion processing of cereal-based products.[26,27] [For completeness, we note that fresh orange juice is a good example of a food product for which T_g is not a relevant predictor of shelf-life, in contrast to frozen orange juice concentrate, for which T_g is the most relevant predictor.]

To consumers, texture is one of the most important quality attributes of food products. The creation of texture during product manufacture[28] and its stability during product storage[29,30] are subjects to which the FPS approach has been widely applied in the past five years. The direct relationship between product T_g – in comparison to the ambient temperature of product storage or consumption – and the textural property of crispness, which is a critical attribute of many amorphous snack food products, has become increasingly recognized,[11] as well as diagrammatically illustrated.[31] This recognition first arose from an intuitive realization and appreciation of the fundamental correlation between the glassy solid state (below T_g of the product matrix) and sensory crispness, and in contrast, between the rubbery liquid state (above T_g) and loss of crispness.[11,29] Intuitively, loss of crispness could be understood to often result from moisture uptake during product storage – allowed by inferior barrier packaging material – leading to plasticization (synonymous with 'softening') and a consequent decrease in product T_g.[15] For amorphous, starch-based extrudates,[31] it was shown that a sensory crispness score increased monotonically with increasing T_g. An early empirical correlation (inversely linear) between decreasing crispness intensity and increasing snack product %RH[32] was later explained on the basis of T_g and the effect of water uptake and resulting plasticization, to raise %RH, lower T_g and thereby reduce crispness.[33] Another early empirical finding confirmed the correlation between decreasing sensory crispness intensity and increasing %RH, resulting from moisture uptake, for commercial breakfast cereal products.[34] Later, this finding was explained as an effect of plasticization by water, to sharply reduce the mechanical modulus of the crisp, glassy solid matrix (at the 'glassy plateau') to that of the soggy, rubbery liquid matrix.[15] It had been previously shown – for a cereal-based solid food foam, equilibrated to various moisture contents, to achieve different extents of plasticization by water – that modulus and sensory hardness score were directly linearly correlated.[35] In a study on cookies during commercial baking, it was found that crispness developed (*i.e.* sensory crispness score increased) as the modulus of elasticity increased with decreasing cookie moisture content.[28] In baked cookies,[36] and in extruded amorphous model snack products,[37,38] recent experimental evidence has further confirmed the direct relationship among T_g, water plasticization, and sensory crispness. In other recent studies on extruded glassy cereal-based foods[39] and crispy puffed extrudates,[40] the effect of limited plasticization (resulting in anti-plasticization) by water – at moisture contents well below that at T_g – on sensory crispness, sensory hardness, and mechanical properties was suggested to result from a secondary physical transition, from brittle to ductile [definition: deform-able or plastic], which could occur in the glassy solid matrix at $T_\beta < T_g$, the possibility of which had been mentioned earlier.[11]

References

1. G.S. Parks, H.M. Huffman and F.R. Cattoir, *J. Phys. Chem.*, 1928, **32**, 1366.
2. G.S. Parks and W.A. Gilkey, *J. Phys. Chem.*, 1929, **33**, 1433.
3. G.S. Parks, S.B. Thomas and W.A. Gilkey, *J. Phys. Chem.*, 1930, **34**, 2028.
4. G.S. Parks, L.J. Snyder and F.R. Cattoir, *J. Chem. Phys.*, 1934, **2**, 595.
5. G.S. Parks and S.B. Thomas, *J. Am. Chem. Soc.*, 1934, **56**, 1423.
6. G.S. Parks, L.E. Barton, M.E. Spaght and J.W. Richardson, *Physics*, 1934, **5**, 193.
7. L. Slade and H. Levine, 'Industrial Polysaccharides – The Impact of Biotechnology and Advanced Methodologies', S.S. Stivala, V. Crescenzi and I.C.M. Dea (eds.), Gordon and Breach Science Publishers, New York, 1987, 387.
8. H. Levine and L. Slade, *Water Sci. Rev.*, 1988, **3**, 79.
9. L. Slade and H. Levine, *Pure Appl. Chem.*, 1988, **60**, 1841.
10. F. Franks, R.H.M. Hatley and S.F. Mathias, *BioPharm*, 1991, **4(9)**, 38.
11. L. Slade and H. Levine, *Crit. Rev. Food Sci. Nutr.*, 1991, **30**, 115.
12. H. Levine and L. Slade, 'Dough Rheology and Baked Product Texture: Theory and Practice', H. Faridi and J.M. Faubion (eds.), Van Nostrand Reinhold/AVI, New York, 1990, 157.
13. J.M.V. Blanshard, 'Physico-Chemical Aspects of Food Processing', S.T. Beckett (ed.), Chapman & Hall, London, 1995, 17.
14. Y.H. Roos, M. Karel and J.L. Kokini, *Food Technol.*, 1996, **50(11)**, 95 and 1997, **51(2)**, 31.
15. L. Slade and H. Levine, *Adv. Food Nutr. Res.*, 1995, **38**, 103.
16. H.W. Starkweather, 'Water in Polymers', S.P. Rowland (ed.), ACS Symposium Series 127, Washington, ACS, 1980, 433.
17. R.C. Hoseney, 'Principles of Cereal Science and Technology', Amer. Assoc. Cereal Chem., St. Paul, 1986.
18. J.K. Sears and J.R. Darby, 'The Technology of Plasticizers', Wiley-Interscience, New York, 1982.
19. L. Slade and H. Levine, 'Phase/State Transitions in Foods', M.A. Rao and R.W. Hartel (eds.), Marcel Dekker, NY, 1998, 87.
20. C.M. Chen and A.I. Yeh, *J. Cereal Sci.*, 2000, **32**, 137.
21. C. Boischot, C.I. Moraru and J.L. Kokini, *Cereal Chem.*, 2001, submitted.
22. J.R. Mitchell, J. Fan and J.M.V. Blanshard, *Extrusion Communique*, 1994, March, 10.
23. J. Fan, J.R. Mitchell and J.M.V. Blanshard, *Int. J. Food Sci. Technol.*, 1996, **31**, 55.
24. J. Fan, J.R. Mitchell and J.M.V. Blanshard, *Int. J. Food Sci. Technol.*, 1996, **31**, 67.
25. C. Zweifel, B. Conde-Petit and F. Escher, *Cereal Chem.*, 2000, **77**, 645.
26. B. Strahm, *Cereal Foods World*, 1998, **43**, 621.
27. B. Strahm, B. Plattner, G. Huber and G. Rokey, *Cereal Foods World*, 2000, **45**, 300.
28. L. Piazza and P. Masi, *Cereal Chem.*, 1997, **74**, 135.

29. Y.H. Roos, 'Phase/State Transitions in Foods', M.A. Rao and R.W. Hartel (eds.), Marcel Dekker, NY, 1998, 57.
30. Y.H. Roos, K. Jouppila and E.S. Soderholm, 'Water Management in the Design and Distribution of Quality Foods', Y.H. Roos, R.B. Leslie and P.J. Lillford (eds.), Technomic, Lancaster, PA, 1999, 429.
31. K.F. Breslauer and G. Kaletunc. CAFT progress report, Rutgers University, NJ, 1991.
32. E.E. Katz and T.P. Labuza, *J. Food Sci.*, 1981, **46**, 403.
33. L. Slade and H. Levine, 'The Glassy State in Foods', J.M. V. Blanshard and P.J. Lillford (eds.), Nottingham University Press, Loughborough, 1993, 35.
34. F. Sauvageot and G. Blond, *J. Texture Stud.*, 1991, **22**, 423.
35. P.J. Lillford. 'Foams: Physics, Chemistry and Structure', A.J. Wilson (ed.), Springer-Verlag, London, 1990, 149.
36. C.C. Seow, C.K. Vasanti Nair and B.S. Lee, 'Food Preservation by Moisture Control', G.V. Barbosa-Canovas and J. Welti-Chanes (eds.), Technomic, Lancaster, PA, 1995, 697.
37. M.K. Karki, Y.H. Roos and H. Tuorila, IFT 1994 Annual Meeting, Atlanta, June 29.
38. Y.H. Roos, K. Roininen, K. Jouppila and H. Tuorila, *Int. J. Food Prop.*, 1998, **1**, 163.
39. G. Roudaut, C. Dacremont and M. Le Meste, *J. Texture Stud.*, 1998, **29**, 199.
40. E. Van Hecke, K. Allaf and J.M. Bouvier, *J. Texture Stud.*, 1998, **29**, 617.

The Effect of Microstructure on the Complex Glass Transition Occurring in Frozen Sucrose Model Systems and Foods

H.D. Goff, K. Montoya and M.E. Sahagian

DEPT. OF FOOD SCIENCE, UNIVERSITY OF GUELPH, GUELPH, ON, N1G 2W1, CANADA

1 Introduction

Many frozen foods contain high sugar contents (*e.g.* fruits, dessert products, ice cream). The presence of sugar depresses the initial freezing point and, due to subsequent freeze-concentration of the sugars, results in a temperature-dependent equilibrium between the ice crystals and the unfrozen solution surrounding them.[1] At sufficiently low temperature, this freeze-concentrated unfrozen phase reaches sufficient viscosity to become glassy. Many studies have been conducted using sucrose solutions to model these frozen foods.[2-7] Although the state diagram has been modeled and published by many people,[8-11] one difficulty has always been reconciling experimental data with the state diagram.[12] On warming a maximally freeze-concentrated sugar solution, one that has been properly annealed at $T > T_g$ to promote maximal freeze-concentration, in a differential scanning calorimeter (DSC), two baseline shifts in the heat flow curve are evident.[2,3,6,10,11] Assigning these two events on the state diagram has been controversial.[11-13] The most widely accepted belief at present seems to be that the lower temperature transition, at around $-42\,^{\circ}\mathrm{C}$ in properly annealed systems, is T_g', the T_g of the maximally freeze-concentrated glass,[8,9] while the warmer temperature transition, at around $-35\,^{\circ}\mathrm{C}$, represents the onset of melting of ice crystals.[10,14,15] This warmer temperature transition is often denoted T_m', and results from delayed melting of ice into the highly viscous unfrozen phase.[10,14,15]

Ice cream is a frozen food containing a high concentration of sugar. It is unique from other such systems, in that it is frozen in a scraped-surface heat exchanger under very high shear (dynamic) conditions. Nucleation of ice occurs at start-up at the wall, followed by growth of ice at the wall, and with each rotation of the scraper blade, ice fragments are sheared off and deposited into the bulk of the unfrozen liquid.[16] The ice crystals that have been deposited in the bulk solution and the remaining ice fragments at the wall both grow, due to the rapid removal of heat from the system. After initial start-up of a scraped-surface freezer, little, if any, nucleation occurs either at the wall or in the bulk, and the crystallization process is dominated by growth.[17] The resulting ice crystals from this process are typically discrete and spherical in morphology.[18] This is in contrast to the freezing of a food system or model solution under static (quiescent) conditions, where undercooling of the solution by several degrees is common, followed by rapid nucleation under conditions very far from equilibrium. In such a system, nucleation is expected to completely dominate growth in establishing ice crystal morphology.[19]

The objectives of the research reported in this paper were to examine the effect of freezing procedure (quiescent *versus* dynamic) and storage conditions (time and temperature fluctuations) on thermal transitions and microstructure of frozen sucrose solutions and model food systems, and to interpret the nature and source of the thermal transitions, based on the microstructure of the samples.

2 Materials and Methods

2.1 Experimental Variables

Sucrose solutions of 40% concentration were used in the first part of this study. Model aqueous systems meant to emulate the solution phase of ice cream were also used. They were comprised of 13.6% skim milk powder, 11.3% sucrose and 5.6% starch hydrolysate (42 DE). Both systems were quiescently frozen in small containers (either DSC pans or 2 ml aluminum, covered pans) to $-35\,°C$ and held for 24 h prior to analyses. Both systems were also dynamically frozen to $-5\,°C$ in a scraped-surface freezer (Taylor Freezers, Rockton, IL) in 8 minutes, then quiescently frozen as above to $-35\,°C$ and held for 24 h prior to analyses. All four treatments (quiescently and dynamically frozen, sucrose solutions and model systems), after 24 h, were subjected to temperature cycling at $-15\pm5\,°C$, with a 4 h ramp and 4 h hold cycle, for repeated cycles. Treatments were also stored at either $-35\,°C$ or $-30\,°C$ for up to 10 weeks.

2.2 Response Variables

A DSC with modulated-temperature (MT) capabilities (DSC2910, TA Instruments, New Castle, DE) was used for thermal evaluation. Heat-flow calibration was performed with gallium, temperature calibration was performed with gallium and indium, and heat-capacity calibration was performed with sapphire. Samples at $-35\,°C$ were placed into the DSC, cooled to $-60\,°C$ and scanned from $-60\,°C$ to $5\,°C$ at $1\,°C\ min^{-1}$, with heat-only modulation, amplitude of $\pm0.159\,°C$, period of 60 s.

Microstructure was evaluated by cryo-scanning electron microscopy (SEM) with an EMScope (Kent, UK) SP2000 cryo-system, on a Hitachi S-570 SEM (Tokyo, Japan) equipped with a Voyageur Image Acquisition system (Noran Instruments, Middleton, WI).[20] From the micrographs obtained, equivalent

diameter based on cross-sectional area of the ice crystals was determined, and a
median diameter (X_{50}) was determined from the cumulative distribution function of a logistic dose-response model.[20]

3 Results and Discussion

3.1 Thermal Transitions in 40% Sucrose Solutions under Quiescent
Freezing Conditions

Two transitions were clearly evident from the MT-DSC analyses, both in the
total heat flow curve and in the complex heat capacity $(C_p{}^*)$ curve in quiescently
frozen 40% sucrose solutions (Figure 1). For the purposes of this discussion, we
refer to them as Tr1, the lower temperature transition, and Tr2, the warmer
temperature transition. A small endotherm can be seen in the non-reversing
curve at both transitions. When data were collected during cooling, the $C_p{}^*$
cooling and warming curves were completely overlapping.[21] When the period of
the modulation was varied, both transitions also showed frequency dependence,[21] as has also been shown by dielectric analysis.[21] The Tr2 was dependent
on the presence of ice in the sample; it was not present in an 82% sucrose solution
$(\text{conc.} = C_g{}')$.[3,6,10,21,22] Upon warming of solutions to $-25\,°C$, followed by
cooling to $-60\,°C$ and rewarming, both transitions were completely overlapping, hence both were reversible transitions.[21] All of the above led to the
conclusion that in quiescently frozen and annealed solutions, both transitions
exhibited glass transition-like behavior.

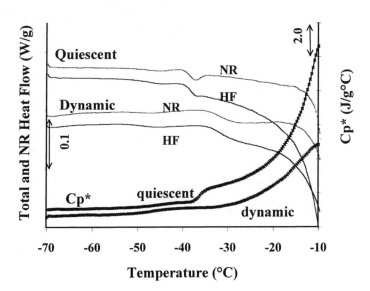

Figure 1 *Total heat flow (HF), non-reversing (NR) heat flow and complex heat capacity*
($C_p{}^*$) curves for quiescently and dynamically frozen 40% sucrose solutions

3.2 Thermal Transitions in 40% Sucrose Solutions under Dynamic Freezing Conditions

After sucrose solutions had been dynamically frozen in the scraped-surface heat exchanger and subsequently hardened, their thermal behaviour was much different. Although still present in the total heat flow curve, the Tr2 had completely disappeared from the C_p^* curve (Figure 1), where reversing events such as glass transitions should be evident. The endotherm in the non-reversing curve had become much broader, representative of a baseline shift leading directly into the melt (Figure 1). These dynamically frozen samples were also warmed in the DSC to $-5\,°C$ and recooled, and they showed exactly the same effect, *i.e.* the transitions were completely reversible up to a few degrees below the melting temperature.[21] If samples were warmed to $5\,°C$ and recooled, they again exhibited similar behaviour to the quiescently frozen samples.[21] The above results implied that the Tr2 no longer exhibited glass transition-like behavior after dynamic freezing.

3.3 Thermal Transitions in 40% Sucrose Solutions as affected by Storage Conditions

3.3.1 Quiescent Freezing. The Tr2 in the C_p^* curve was found to disappear with repeated temperature cycling at $-15\pm5\,°C$, while the endotherm in the non-reversing curve became a baseline shift (Figure 2). This implied that the quiescently frozen sample took on dynamic-like characteristics after temperature cycling. With storage of quiescent-frozen samples at $-35\,°C$, a diminishing of Tr2 in the C_p^* curve was seen between 5 and 7 weeks, although the change in slope of C_p^* at Tr2 was still evident (Figure 3). With storage of quiescent-frozen

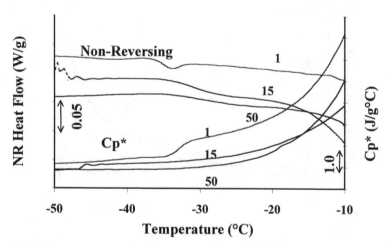

Figure 2 *Non-reversing (NR) heat flow and complex heat capacity (C_p^*) curves for quiescently frozen 40% sucrose solutions after 1, 15 and 50 temperature fluctuation cycles at T > Tr2*

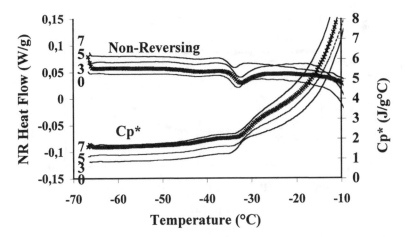

Figure 3 *Non-reversing (NR) heat flow and complex heat capacity (C_p^*) curves for quiescently frozen 40% sucrose solutions stored at $-35\,°C$ for 0, 3, 5 and 7 weeks*

samples at $-30\,°C$, a complete loss of Tr2 in the C_p^* curve was seen within 3 weeks (Figure 4). The non-reversing curve showed the baseline shift at Tr2 after 3 weeks, characteristic of the dynamically frozen samples. This again implied that frozen storage at $T > T_g$ was sufficient to convert the thermal behavior of these samples from that of their original quiescent condition to one similar to that seen in the dynamically frozen samples.

3.3.2 Dynamic Freezing. The dynamically frozen samples showed no change in thermal behaviour as a result of temperature cycling. The only baseline shift that was evident in the C_p^* curve was at Tr1 (Figure 5).

3.4 Microstructure of 40% Sucrose Solutions

Microstructural analysis revealed that the quiescently frozen samples exhibited very ordered, interconnected, dendritic-type patterns, with considerable unfrozen solution constrained within the structure of the crystal resulting from undercooling and rapid nucleation (Figure 6A). However, the dynamically frozen samples showed discrete, rounded crystals, resulting from secondary nucleation in the scraped-surface freezer, followed by crystal growth during hardening (Figure 6C). After temperature cycling, the area of the ice/unfrozen interfaces decreased due to recrystallization in both systems. Recrystallization in the quiescently frozen samples resulted in a loss of dendritic structure and formation of discrete, rounded crystals (Figure 6B). Recrystallization in the dynamically frozen samples resulted in crystal growth but little change in overall morphology (Figure 6D). Interfacial area may have played an important role in dictating thermal behaviour; however, there appeared to be other factors involved. Table 1 shows size distribution analyses for the quiescently frozen

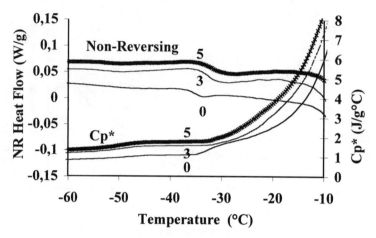

Figure 4 *Non-reversing (NR) heat flow and complex heat capacity (C_p^*) curves for quiescently frozen 40% sucrose solutions stored at $-30\,°C$ for 0, 3 and 5 weeks*

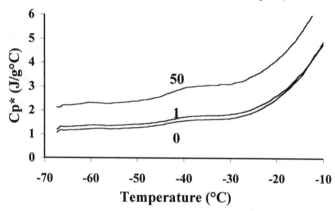

Figure 5 *Complex heat capacity (C_p^*) curves for dynamically frozen 40% sucrose solutions before (0) and after 1 and 50 temperature fluctuation cycles at $T > Tr2$*

Table 1 *Median diameter (X_{50}) and slope of the cumulative ice crystal size distribution function in 40% sucrose solutions frozen quiescently and held at either -35 or $-30\,°C$*

Time	$-35\,°C$		$-30\,°C$	
	X_{50} (μm)	Slope (%/μm)	X_{50} (μm)	Slope (%/μm)
1 day	14.0 a	11.3 a	14.4 a	17.3 a
3 weeks	13.4 a	11.9 a	17.1 a	10.7 ab
5 weeks	14.5 a	13.1 a	20.9 a	6.6 b

Values followed by the same letter in a column do not differ ($p < 0.05$)

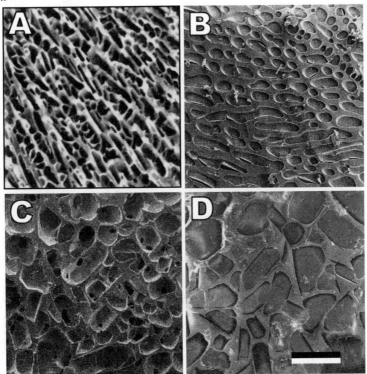

Figure 6 *Microstructure of quiescently frozen 40% sucrose solutions before (A) and after (B) 50× temperature cycling at T > Tr2, and dynamically frozen 40% sucrose solutions before (C) and after (D) 50× temperature cycling at T > Tr2. Bar = 50 μm for A–C and 80 μm for D*

samples stored at −35 °C and −30 °C. Although the thermal behaviour of the −30 °C samples after storage was remarkably different from that of either the fresh samples or those stored at −35 °C, the median diameter among all these samples was not significantly different. However, a significant broadening of the distribution occurred, as seen from a decrease of the slope values of the cumulative distribution function, and reflective of recrystallization. Thus, although there was some correlation between ice crystal size (hence, surface area) and thermal behavior, thermal behavior seemed to be related to other factors as well, notably crystal morphology.

3.5 Model Systems

The model systems behaved exactly the same as the sucrose solutions. Both showed a baseline shift at Tr2 in the total heat flow curve. Quiescently frozen samples showed a Tr2 in the C_p^* curve and an endotherm in the non-reversing curve (Figure 7). Dynamically frozen samples did not show a Tr2 in the C_p^* curve, but rather showed a baseline shift in the non-reversing curve (Figure 7). Quiescently frozen solutions, after temperature cycling at $T > T_g$, did not show a

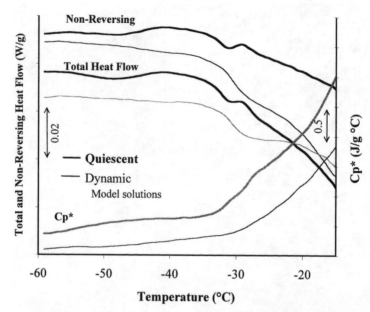

Figure 7 *Total and non-reversing heat flow and complex heat capacity* (C_p^*) *curves for quiescently and dynamically frozen model solutions emulating the unfrozen phase of an ice cream mix*

Tr2 in the C_p^* curve, but did show a baseline shift in the non-reversing curve (data not shown). Hence, as before, after temperature cycling, the thermal behavior of the quiescently frozen samples converted to that similar to the dynamically frozen samples. Dynamically frozen solutions, after temperature cycling, did not change (data not shown).

Microstructural analysis showed an ordered, dendritic pattern in quiescently frozen samples, and discrete rounded crystals in either dynamically frozen solutions or quiescent samples after temperature cycling (Figure 8). The change in morphology of the quiescently frozen samples, as a result of temperature cycling, was clearly evident in these model solutions.

3.6 Discussion and a Further Test of Our Hypothesis

It may be that heterogeneity exists in the unfrozen phase of quiescently frozen solutions, leading to another phase more concentrated in sucrose than is the bulk phase. This sucrose phase would only be formed in quiescent systems, as a result of rapid nucleation. It is possible that the diffusion of water to the surface of a crystal at low temperatures, from the neighborhood of the crystal, was not compensated by diffusion from the bulk phase. This would have produced a concentrated, non-equilibrated phase around the crystal, which would be detected as a glass transition in the reversing component (at Tr2), during warming of the frozen solution in the MTDSC. Such a phase would disappear after tempera-

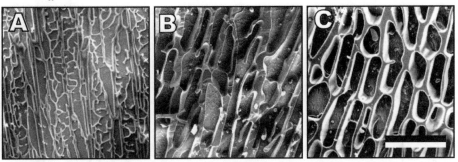

Figure 8 *Microstructure of quiescently frozen model solutions before temperature cycling (A) and after 15 × (B) and 50 × (C) temperature cycling at* T > Tr2. *Bar = 50 μm*

ture cycling, because of the mobility of the unfrozen phase at temperatures higher than T_g', leading to perfection of the surfaces of crystals and homogeneity in the unfrozen phase. In dynamic systems, the nature of the crystallization process would have prevented its formation at a detectable level.

To test this hypothesis, sucrose solutions were slow-cooled (2 °C min^{-1}) from 25 °C to −60 °C (nucleation occurred at −20 °C), annealed for 60 min at −35 °C, recooled and warmed at 2 °C min^{-1} from −60 °C to −5 °C to melt most (but not all) of the crystals (1st warm), cooled again at 2 °C min^{-1} to −60 °C, annealed for another 60 min at −35 °C, and then recooled and scanned at 2 °C min^{-1} from −60 °C to 5 °C (2nd warm). Temperature was modulated during the heating ramps at an amplitude of ±0.318 °C min^{-1}, 60 s period (heat-iso modulation). Although the total heat flow curves looked similar, showing both transitions in both warming curves, there was virtually no Tr2 in the C_p^* curve in the 2nd warm (Figure 9). The baseline shift in the total heat flow curve at Tr2 in the 2nd warm came from the non-reversing curve, not the reversing curve. This temperature treatment resulted in thermal behavior similar to that for the samples from the dynamic freezing process. When data from cooling curves were collected (temperature modulation also during cooling) and analysed, a baseline shift in the total heat flow curve at Tr2 occurred in the 1st cool, but not in the 2nd cool. If the Tr2 during warming resulted from a delayed onset of melting, it would not be expected in the cooling curve. Likewise, in the C_p^* curve, the Tr2 was evident in the 1st cool, but not in the 2nd cool (data not shown).

Micrographs of the structure throughout this process illustrated the change in ice crystal morphology from the 1st warm to the 2nd warm (Figure 10). At the resolution of the light microscope, nucleation during the 1st cool resulted in a cloud of crystals too small to be detectable. It is likely that this microstructure led to a heterogeneous distribution of sucrose and ice crystals, which would lead to a complex relaxation of the glass. However, once the crystals were ripened through the 1st warm and 2nd cool, they became discrete and easily resolved. It would thus appear that the microstructure from rapidly nucleated crystals resulted in Tr2 as a glass transition, while a maturation of these crystals led to a loss of Tr2 as a glass transition. Crystal morphology and unfrozen phase structure were also

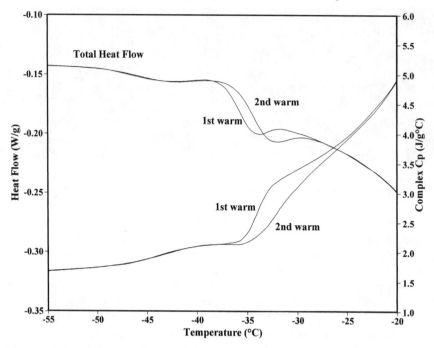

Figure 9 *Total heat flow (top) and complex heat capacity (bottom) curves for a 40% sucrose solution that had been quiescently frozen to $-60\,°C$, warmed to $-5\,°C$ (1st warm), recooled to $-60\,°C$, and rewarmed to 5 $°C$ (2nd warm)*

examined by cryo-SEM (Figure 11). The ice crystals at $-35\,°C$ during the 1st warm were very dendritic, with many thin lamellae of unfrozen phase within the crystals and thick lamellae of unfrozen phase between the crystals, whereas the microstructure at $-35\,°C$ during the 2nd warm showed a much more homogeneous distribution of unfrozen phase surrounding discrete, rounded crystals.

Three recent papers have used MT-DSC to look at the glass transition in frozen sucrose solutions.[13,22,23] Aubuchon *et al.*[13] followed a 'slow freeze and anneal' method and showed the same results as those of our quiescently frozen samples or in the 1st warm from the experiment above. They concluded, also from quasi-isothermal experimental results, that Tr2 is a glass transition (or at least appears glass transition-like). Knopp *et al.*[23] cooled their sucrose solutions to $-25\,°C$, then warmed to $-10\,°C$ and held 20 min, then cooled to $T < T_g$ and rewarmed. They showed the same results as ours in the 2nd warm above, and those from the dynamically frozen samples, the quiescently frozen and temperature-cycled samples, and the stored samples, *viz.*, no Tr2 in the C_p^* curve, although it was seen in the total heat flow curve. They concluded that Tr2 was not a glass transition, but the onset of melting. Thus, it appears that the freezing protocol has a major impact on the nature of Tr2.

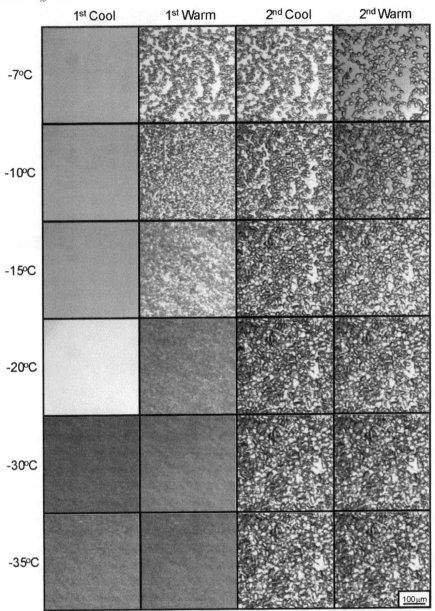

Figure 10 *Images of 40% sucrose solutions on a cold-stage of a light microscope, following the same temperature regime as the thermograms in Figure 9. The solution was first cooled to −60 °C (left column, from top to bottom; nucleation occurred at −21 °C), then warmed to −5 °C (2nd column from left, bottom to top), cooled again to −60 °C (2nd column from right, top to bottom), and finally warmed from −60 °C to 5 °C (right column, bottom to top)*

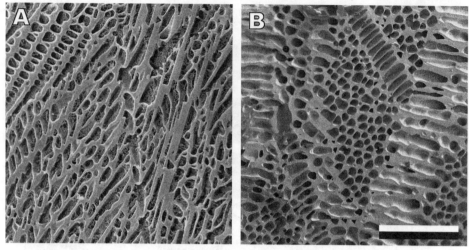

Figure 11 *Microstructure, as determined by cryo-SEM, of quiescently frozen 40% sucrose solutions at −35 °C during the first warm (A) and during the second warm (B), following the same freezing protocol as for the images in Figure 10. Bar = 60μm*

4 Conclusions

Freeze-concentrated sucrose solutions, after quiescent freezing, had two glass-like transitions. However, the nature of Tr2, the higher temperature relaxation, was changed by either dynamic freezing, or by temperature-cycling quiescently frozen samples, or by storing quiescently frozen samples at $T > Tr2$, or by allowing ice crystals to mature after nucleation at $T < T_f$, the final melting temperature. Microstructural differences revealed that ice crystal size or surface area may have had an effect, but so too did ice crystal morphology. Quiescent freezing resulted in considerable undercooling, followed by rapid nucleation and very small, dendritic, crystals. The glass transition-like nature of Tr2 may have resulted from the formation of a sucrose-rich, unequilibrated phase trapped around or within the rapidly nucleated ice crystals, and/or solute inclusions within the crystals themselves. Such a phase may have arisen from a concentration gradient and limited diffusion at the crystal interface. This phase relaxed at a higher temperature than did the bulk unfrozen phase, resulting in the two-step glass transition. When this phase disappeared, from maturation or perfection of the ice crystals or equilibration of the unfrozen phase, the transition seen at Tr2 in the total heat flow curve lost its glass transition-like characteristics, and then appeared to have been due solely to the onset of delayed melting.

Acknowledgements

The authors wish to thank the Natural Sciences and Engineering Research Council of Canada (NSERC) for financial support of this research.

References

1. H.D. Goff, *Food Res. Int.*, 1992, **25**, 317.
2. G. Blond and D. Simatos, *Thermochim. Acta*, 1991, **175**, 239.
3. S. Ablett, M.J. Izzard and P.J. Lillford, *J. Chem. Soc. Faraday Trans.*, 1992, **88**, 789.
4. M. Le Meste and V. Huang, *J. Food Sci.*, 1992, **57**, 1230.
5. W.M. MacInnes, 'The Glassy State in Foods', J.M.V. Blanshard and P.J. Lillford, (eds.), Nottingham University Press, Loughborough, UK, 1993, 223.
6. M.E. Sahagian and H.D. Goff, *Thermochim. Acta*, 1994, **246**, 271.
7. H.D. Goff and M.E. Sahagian, *Thermochim. Acta*, 1996, **280**, 449.
8. H. Levine and L. Slade, *J. Chem. Soc. Faraday Trans. I*, 1988, **84**, 2619.
9. H. Levine and L. Slade, *Cryo-Lett.*, 1988, **9**, 21.
10. Y. Roos and M. Karel, *Int. J. Food Sci. Technol.*, 1991, **26**, 553.
11. S. Ablett, A.H. Clark, M.J. Izzard and P.J. Lillford, *J. Chem. Soc. Faraday Trans.*, 1992, **88**, 795.
12. H.D. Goff, *Pure Appl. Chem.*, 1995, **67**, 1801.
13. S.R. Aubuchon, L.C. Thomas, W. Theuerl and H. Renner, *J. Thermal Anal.*, 1998, **52**, 53.
14. E.Y. Shalaev and F. Franks, *J. Chem. Soc. Faraday Trans.*, 1995, **91**, 1511.
15. W.Q. Sun, *Cryo-Lett.*, 1997, **18**, 99.
16. R.W. Hartel, *Trends Food Sci. Technol.*, 1996, **7**, 315.
17. A.B. Russell, P.E. Cheney and S.D. Wantling, *J. Food Eng.*, 1999, **39**, 179.
18. K.B. Caldwell, H.D. Goff and D.W. Stanley, *Food Struct.*, 1992, **11**, 1.
19. D.S. Reid, W. Kerr and J. Hsu, *J. Food Eng.*, 1994, **22**, 483.
20. A.A. Flores and H.D. Goff, *J. Dairy Sci.*, 1999, **82**, 1399.
21. M.E. Sahagian, PhD Thesis, University of Guelph, 2001.
22. M.J. Izzard, S. Ablett, P.J. Lillford, V.L. Hill and I.F. Groves, *J. Thermal Anal.*, 1996, **47**, 1407.
23. S.A. Knopp, C. Chongprasert and S.L. Nail, *J. Thermal Anal.*, 1998, **54**, 659.

Relationship Between the Glass Transition, Molecular Structure and Functional Stability of Hydrolyzed Soy Proteins

Vanesa Zylberman[1] and Ana M.R. Pilosof[2]

DEPARTAMENTO DE INDUSTRIAS, FACULTAD DE CIENCIAS EXACTAS Y NATURALES, UNIVERSITY OF BUENOS AIRES, (1428) BUENOS AIRES, ARGENTINA

[1]Research Fellow, Universidad de Buenos Aires, República Argentina
[2]Member of Consejo Nacional de Investigaciones Científicas y Técnicas de la República Argentina

1 Introduction

The main objective in spray- or freeze-drying commercial preparations of proteins is to lower molecular mobility to such an extent that chemical and physical changes are restricted. The amorphous glassy state has been shown to be an important factor in the physical stability of proteins used in foods.[1,2] However, abundant experimental evidence has shown that the glassy state is necessary but not sufficient to ensure the chemical stability of dry proteins.[3–6] Restricted mobility would make difficult an unfolding process, thus delaying denaturation of proteins, but reactions such as hydrophobic aggregation, deamidation, cross-linking, racemization and β-elimination could easily occur in low-moisture proteins because of the proximity of reactants.[7,8] Some of these reactions could occur, not only in the rubbery state but also in the glassy state, because they are not diffusion-dependent.[7,9] In particular, it has been shown that, due to a relatively high effective protein concentration in the solid state, intermolecular processes such as aggregation are more prevalent than intramolecular processes.[10,11]

Soy proteins are used for the production of many food products because of their broad range of functional properties. The 7S and 11S globulins, which account for about 70% of total proteins, largely determine the emulsion, foaming and gelling behavior of soy proteins.

The glass transition temperature (T_g) values for native 7S and 11S globulins, as a function of moisture, have been determined.[12] Both globulins were shown to be highly plasticizable by water, with 11S having a higher T_g than 7S.

Specific functionality of soy proteins is achieved by modifying protein structure by physical or chemical methods. However, there is little information available on the impact of modification of soy protein structure on dry T_g or on how that relates to storage stability. Transglutaminase treatment of 7S and 11S soy globulins, which generated crosslinks, was shown to lower T_g.[13]

Limited enzymatic hydrolysis is a common procedure to improve foaming and emulsifying properties of soy proteins, as well as solubility across a wide range of pH. Such a hydrolysis process generates smaller polypeptides with molecular weight distributions that vary according to hydrolysis conditions and enzyme specificity. There have been few studies reported on the T_g of protein hydrolysates and their physical/chemical stability. For example, the glass transition and caking of a fish protein hydrolysate were studied by Aguilera *et al.*,[2] and the effect of relative humidity (RH) on caking of soy sauce powder was studied by Hamano and Sugimoto.[14]

In the present work, we analysed the effect of enzymatic hydrolysis on molecular structure and T_g of soy proteins and its influence on the stability (of foaming properties and color) of the dry system. The specificity of proteases was also evaluated.

2 Materials and Methods

2.1 Sample Preparation

A commercial soy protein isolate (SPI, 91.5% protein, 0.8% lipid) from Protein Technologies International (St. Louis, MO, USA) was hydrolyzed with neutral proteases from *Aspergillus oryzae* (F) and *Bacillus subtilis* (B) (400 000 HUT g^{-1} and 120 222 NPU g^{-1}, respectively), obtained from Quest Argentina.

SPI (500 ml, 6 g 100 ml^{-1}) was hydrolyzed batch-wise for 1 h at 50 °C by treatment at the following enzyme-to-substrate ratios (E/S, %): 0.05, 0.2, 0.3, 0.5 and 1.5. Hydrolysis was stopped by heating at 80 °C for 10 min. The variation in pH was very small (*i.e.* maximum decrease was 0.3 pH units) and was adjusted back to the original value with diluted NaOH.

The entire protein system was lyophilized in a Stokes Model 21 freeze-drier, which operated at −40 °C under vacuum of less than 100 μm Hg.

The degree of hydrolysis (DH), defined as the percentage of peptide bonds cleaved, was calculated from a determination of free amino groups by the method of Church *et al.*,[15] using *o*-phthaldialdehyde.

Lyophilized samples were equilibrated for one week at 25 °C over a saturated sodium acetate solution (RH = 22%). Moisture content of equilibrated samples was determined in triplicate in a vacuum oven at 70 °C for 48 h over desiccant.

Following equilibration to 22% RH, vials containing protein were hermetically sealed and stored in ovens at 60 or 90 °C for 92 h. Another vial was maintained at 25 °C.

2.2 Differential Scanning Calorimetry (DSC)

Equilibrated samples were hermetically sealed in 40 μl aluminium pans and run on a Mettler TA 4000 Thermal Analysis System. DSC runs, at least in duplicate, were performed from 5 to 120 °C at 5 °C min^{-1}. Thermograms were analysed, using Mettler TA 72 software, to determine T_g from the midpoint of the glass transition. DSC calibration was performed using ice and indium fusion thermograms, and an empty aluminium pan was used as a reference.

2.3 Gel Electrophoresis

Samples of untreated and hydrolyzed SPI, as well as a native protein isolate obtained in our laboratory, were analysed by polyacrylamide gel electrophoresis (PAGE), using a Mini-Protean II dual-slab cell system (Bio-Rad Laboratories), by the method of Laemmli.[16]

Samples (20 μl, 2 μg μl^{-1}) were dissolved in 0.5 M Tris-HCl (pH 6.8) buffer with 2% SDS and 5% β-mercaptoethanol. Resolving and stacking gels contained 12 and 4% acrylamide, respectively.

2.4 Physicochemical Properties

Foaming capacity and kinetics of liquid drainage from foams were determined by the method of Carp *et al.*[17] Foam expansion, after whipping 30 ml of 3% protein (w/w) in distilled water, was calculated as (foam volume – 30 ml) x 100/30 ml. The rate of liquid drainage from foams (K_{dr}) was calculated from a kinetic model proposed by Carp *et al.*[17]

Color was determined using BYK-GARDENER Color View 9000 equipment with D65 illuminant and 10° vision angle. The change in sample color was expressed as the color difference, Δ_{Lab} (using the L, a, b colorimetry system), between treated sample and untreated SPI.

3 Results and Discussion

3.1 Characterization of Protein Samples

Figure 1 shows the SDS-PAGE pattern for the commercial SPI used as substrate for hydrolysis (lane 3), in comparison with a native SPI (lanes 1 and 2). Comparison of bands indicates, for commercial SPI, an absence of the acidic–basic (AB) – 11S subunit that was present in the native protein (lane 1). Lane 2, corresponding to native SPI under reducing conditions, shows that when the disulfide bond between acidic (A) and basic (B) polypeptides was reduced, the resulting electrophoretic pattern was similar to that exhibited by commercial SPI, confirming that the AB subunit was reduced during the manufacturing process. Upon heating, the disulfide bonds linking acidic–basic polypeptides are cleaved.[18] Furthermore, the absence of any endotherm in DSC thermograms (not shown) indicated that the commercial isolate was denatured. The commercial SPI also contained protein aggregates that did not enter the SDS-PAGE gel.

Similar DH values were obtained with both enzymes (Figure 2). However, subunit degradation, analysed by SDS-PAGE, was different. Since equal

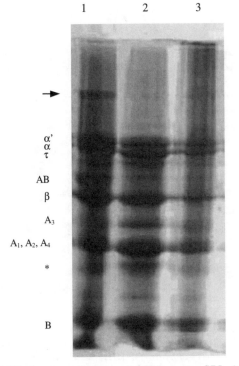

Figure 1 *SDS-PAGE of native and commercial SPIs: native SPI without 2-mercaptoethanol (lane 1); native SPI with 2-mercaptoethanol (lane 2); and commercial SPI without 2-mercaptoethanol (lane 3). Arrow indicates high molecular weight aggregates*

amounts of protein were loaded on the gels, bands of darker intensity indicated that the corresponding protein fractions were more resistant to enzymatic breakdown. As shown in Figure 2, the bacterial protease caused a more pronounced degradation of protein fractions. The protein bands between 200 and 30 kDa (α', α, β, A-polypeptides) in the bacterial hydrolysate were unnoticeable after 2% DH. The band corresponding to B-polypeptide was more resistant to enzymatic breakdown and was still noticeable at 5% DH (mainly in fungal hydrolysates). Similar behavior was reported by Kim *et al.*,[19] who used several commercial proteases. The more compact conformation of B-polypeptide, due to its higher hydrophobicity, would make it less accessible to enzyme attack.

3.2 Foaming Properties and Color

Foaming capacity was greatly improved by hydrolysis (from 100% to about 240%), reaching a plateau at 4% DH (bacterial protease) or 6% DH (fungal protease). Below the plateau value, the bacterial-modified SPI exhibited higher foaming capacity than the fungal-modified SPI (data not shown), as a consequence of more pronounced protein degradation. Nevertheless, the rate of liquid drainage was increased from 0.02 ml min^{-1} to 0.06 ml min)$^{-1}$ for DH = 5%, regardless of the enzyme. Improved foaming capacity and decreased foam stabil-

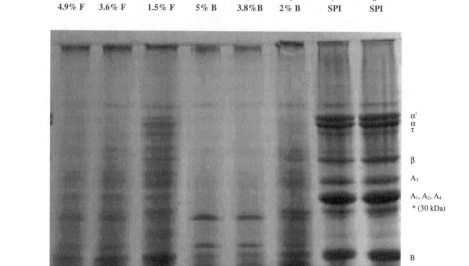

1	2	3	4	5	6	7	8
4.9% F	3.6% F	1.5% F	5% B	3.8%B	2% B	SPI	SPI

Figure 2 *SDS-PAGE (with 2-mercaptoethanol) of commercial SPIs hydrolyzed with fungal protease (lanes 1–3), with bacterial protease (lanes 4–6), and non-hydrolyzed SPI (lanes 7–8). The degrees of hydrolysis (DH, %) are indicated at the top of the gel*

ity of SPI, due to limited hydrolysis with different enzymes, has been reported previously.[20–23] The color of SPI was not modified by hydrolysis.

3.3 Glass Transition Temperature

The commercial SPI showed two different T_gs, which are marked by arrows in Figure 3 (SPI, untreated). Morales and Kokini[12] suggested that the lower T_g (T_g1) corresponded to 7S globulin and the higher T_g (T_g2) to 11S globulin. The T_g values we obtained for the commercial SPI (moisture = $5.8 \pm 0.3\%$, d.b.) were 49 and 86.4 °C, respectively. These values are higher than those reported[12] for isolated 7S and 11S globulins at a similar water content. The differences could be accounted for by the existence of protein aggregates in our commercial SPI, differences in plasticization originating from modification of protein structure throughout the manufacturing process, and/or to thermodynamic incompatibility between the two globulins, which might raise the transition temperatures. Indeed, the existence of two glass transitions in the whole SPI suggests that the two globulins are immiscible and separate into different amorphous phases. Also, we must consider the difficulty in comparing T_g values within the range of extremely high plasticization of soy proteins.

The equilibrium water content of SPI (at 22% RH) was not significantly affected by DH or the enzyme used for hydrolysis. This behavior may be ascribed to the limited hydrolysis carried out (maximum DH achieved was 17%), and to a

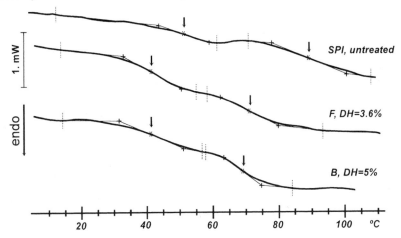

Figure 3 *DSC thermograms showing the glass transitions of SPI and hydrolyzed SPI*

balance between increased polar and hydrophobic groups resulting from hydrolysis. The average equilibrium water content, for fungal-modified SPIs of different DH, was 5.72 ± 0.69, and for bacterial-modified SPIs, $6.1 \pm 0.45\%$ (d.b.). Thus, sample water contents were essentially equivalent.

DSC thermograms for hydrolyzed SPIs (F and B in Figure 3) also showed two different glass transitions, indicating that the hydrolysis products from the polypeptides comprising the 7S and 11S globulins were also incompatible. Changes in heat capacity ranged from 0.2 to 0.4 J g^{-1} K. Extensive degradation of the polypeptides comprising 7S globulin (illustrated in Figure 1) did not significantly modify the T_g of this component. In contrast, enzyme treatment was more effective in decreasing the T_g corresponding to the polypeptides comprising 11S globulin, which were less extensively degraded by enzymes.

When DH increased to 5%, T_g2 decreased from 86.5 °C (for commercial SPI) to 71.9 °C (for fungal-modified SPI) or 69.5 °C (for bacterial-modified SPI) (Figure 4). SPI further hydrolyzed to DH = 13% with fungal protease still showed two T_gs (at 47.8 and 62.8 °C), but at DH = 17%, it showed only one T_g at 51 °C.

The 'additive contribution theory' for prediction of T_g of proteins[24] may explain most of the observed results. The main point of this approach is that the chemical structure of the repeating unit (amino acid) of a protein provides basic information about its physical properties. Thus, among the chemical characteristics of a protein (amino acid composition, disulfide bonds, molecular weight (MW) and quaternary structure), the amino acid composition is sufficient to permit estimation of T_g. T_g depends mainly on a protein's chemical structure, but not on MW, when the latter exceeds a critical degree of polymerization (DP).

The following observations could be explained:

1. For the commercial SPI, T_g1 (7S) $< T_g2$ (11S), and the globulins had lost their quaternary structure. The higher T_g of native 11S globulin has been ex-

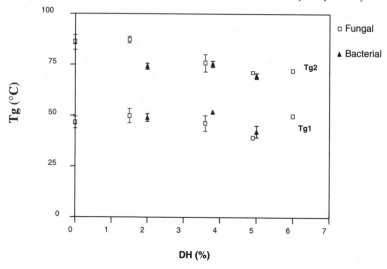

Figure 4 T_g *values as a function of DH for hydrolyzed SPIs*

plained[12] on the basis of its higher MW (350 kDa), in comparison to that of 7S globulin (150–210 kDa). In the commercial SPI, the 11S polypeptides A and B were found not to be linked (Figure 1), and their MWs were lower than those of the α', α, and β polypeptides comprising 7S globulin. This finding suggested that T_g did not depend on MW of the protein fractions, but rather on their chemical structure. According to the additive contribution theory,[24] the higher proportion of glycine and alanine (which increases T_g) and lower proportion of leucine (which decreases T_g) in 11S polypeptides would account for the higher T_g of 11S globulin.

2. The significant decrease in $T_g 2$ (11S) with DH, but not in $T_g 1$ (7S):

Owing to the higher DP of 7S (450) in comparison to 11S (311) globulin,[24] the components of 11S globulin would reach, at a lower DH, a critical DP, below which T_g decreases with MW.

For homologous series of amorphous synthetic polymers or starch hydrolysis products, T_g increases with increasing MW (or DP), up to a plateau limit for the region of entanglement coupling, where T_g levels off with further increases in MW.[25] Below the entanglement limit, there is a linear relationship between increasing T_g and decreasing inverse MW.

3.4 Storage Stabilty

Storage temperatures were selected on the basis of $T_g 1$ and $T_g 2$ values. Samples were stored at (a) 25 °C, where $(T - T_g)$ values were < zero (*i.e.* in the glassy solid state), (b) 60 °C, a temperature in between $T_g 1$ and $T_g 2$, and (c) 90 °C, where samples were totally rubbery (Table 1).

For samples in the glassy state or slightly above T_g1 (*i.e.* at 25 or 60 °C, respectively), no significant changes were observed in foaming properties (Figures 5 and 6) after 92 h of storage. However, after sample storage in the rubbery liquid state (at 90 °C), where $(T - T_g1)$ and $(T - T_g2) > 0$, foaming capacity increased at low DH values (Figure 5), and the rate of liquid drainage from foams also increased (Figure 6). Increased foam expansion, upon storage in the rubbery state at 90 °C, could be due to the formation of soluble complexes with good interfacial properties *via* Maillard reactions.[26]

Browning of enzyme-modified SPI occurred only after storage at 90 °C. Browning of model systems has also been shown to occur in the rubbery state, where such systems had enhanced molecular mobility.[27,28] The degree of browning was greatly enhanced with increasing DH (Figure 7), and when SPI

Table 1 *Differences between storage temperatures* (T) *and glass transition temperatures* (T_g1, T_g2) *for SPIs of different DH. F: hydrolyzed with fungal protease; B: bacterial protease.*

DH (%)	T = 60 °C		T = 90 °C	
	$T - T_g1$ (°C)	$T - T_g2$ (°C)	$T - T_g1$ (°C)	$T - T_g2$ (°C)
1.5% F	10.1	−23.7	40.1	6.3
3.6% F	13.8	−15.75	43.8	14.25
4.9% F	20.9	−11.1	50.9	18.9
6% F	1.4	−11.9	40.4	18.1
2% B	11.05	−14.25	41.05	15.75
3.8% B	8.2	−15.35	38.2	14.65
5% B	17.85	−9.5	47.85	20.5

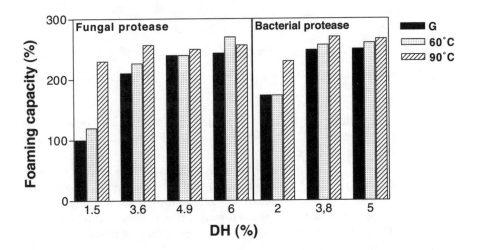

Figure 5 *Foaming capacity of hydrolyzed SPIs stored at 25 °C (G = glassy), or after storage at 60 or 90 °C for 92 h*

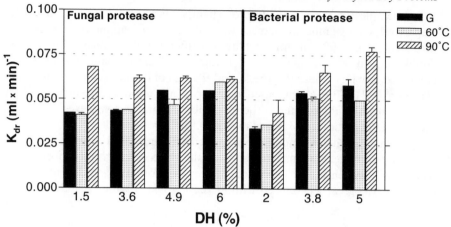

Figure 6 *Rate constants for liquid drainage (K_{dr}) from hydrolyzed SPI foams stored at 25 °C (G = glassy), or after storage at 60 or 90 °C for 92 h*

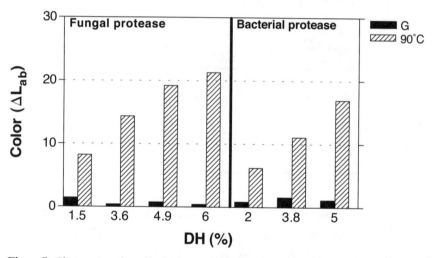

Figure 7 *Changes in color of hydrolyzed SPIs stored at 25 °C (G = glassy), or after storage at 90 °C for 92 h*

was hydrolysed with fungal protease. No relationship was found between the degree of color development and $(T - T_g)$ values, but DH accounted for 95% of the variations in sample color.

Previous studies[29,30] on physicochemical changes in SPI and dry wheat gluten in the vicinity of their T_gs have shown that little change occurred in the glassy state. Above T_g, changes occurred at accelerating rates; foaming properties decreased and proteins browned. However, it was demonstrated that, in addition to the glass transition, the interdependent effects of moisture content and temperature were also very important.

4 Conclusions

Modification of the molecular structure of soy protein isolates, resulting from limited hydrolysis, did not affect the lower of two T_gs but decreased the upper T_g. Structural modification greatly influenced the foaming properties, but not the color, of SPIs.

Although the glass–rubber transition was shown to impact the functional stability of enzyme-modified SPI, T_g of the dry proteins could not fully account for most of the changes caused by increased DH. The MW and type of peptides produced by specific enzymatic action, as well as storage temperature itself, also helped to account for the observed changes. Color and foam stability were better preserved through use of bacterial rather than fungal protease.

Acknowledgements

The authors acknowledge financial support from the Universidad de Buenos Aires and the Consejo Nacional de Investigaciones Científicas y Técnicas y Agencia Nacional de Promoción Científica y Tecnológica de la República Argentina.

References

1. L. Slade and H. Levine, *Crit. Rev. Food Sci. Nutr.*, 1991, **30**, 115.
2. J.M. Aguilera, G. Levi and M. Karel, *Biotechnol. Prog.*, 1993, **9**, 651.
3. Y.H. Chen, J.L. Aull and L.N. Bell, *Food Res. Int.*, 1999, **32**, 467.
4. C. Schebor, M.P. Buera and J. Chirife, *J. Food Eng.*, 1996, **30**, 269.
5. V.M. Taragano and A.M.R. Pilosof, *Biotechnol. Prog.*, 2001, **17**, 775.
6. M.R. Terebiznik, M.P. Buera and A.M.R. Pilosof, *Lebensm Wiss. Technol.*, 1998, **31**, 143.
7. H.R. Costantino, R. Langer and A.M. Klibanov, *J. Pharm. Sci.*, 1994, **83**, 1662.
8. V.V. Mozhaev, *TIBTECH*, 1993, **11**, 88.
9. A.I. Liapis, M.J. Pikal and R. Bruttini, *Drying Technol.*, 1996, **14**, 1265.
10. L.N. Bell, M.J. Hageman and J.M. Bauer, *Biopolymers*, 1995, **35**, 201.
11. S.P. Duddu and P.R. Dal Monte, *Pharm. Res.*, 1997, **14**, 591.
12. D.A. Morales and J.L. Kokini, 'Phase/State Transitions in Foods', M.A. Rao and R.W. Hartel (eds.), Marcel Dekker, NY, 1998, Chapter 10, 273.
13. A. Mizuno, M. Mitsuiki and M. Motoki, *J. Agric. Food Chem.*, 2000, **48**, 3286.
14. M. Hamano and H.J. Sugimoto, *J. Food Proc. Preserv.*, 1978, **2**, 185.
15. F.C. Church, H.E. Swaisgood, D.H. Porter and G.L. Catignani, *J. Dairy Sci.*, 1983, **66**, 1219.
16. A. Laemmli, *Nature*, 1970, **227**, 680.
17. D.J. Carp, G.B. Bartholomai and A.M.R. Pilosof, *Lebensm. Wiss. Technol.*, 1997, **30**, 253.
18. S. Utsumi and J.E. Kinsella, *J. Agric. Food Chem.*, 1985, **33**, 297.
19. S.Y. Kim, P.S.W. Park and K.C. Rhee, *J. Agric. Food Chem.*, 1990, **38**, 651.
20. J. Adler-Nissen and H.S. Olsen, 'Functionality and Protein Structure', A. Pour-El (ed.), American Chemical Society, Washington DC, 1979.
21. L.S. Bernardi Don, A.M.R. Pilosof and G.B. Bartholomai, *JAOCS*, 1991, **68**, 102.

22. A.M. Calderón De La Barca, R.A. Ruiz-Salazar and M.E. Jara-Marini, *Food Chem. Toxicology*, 2000, **65**, 246.

23. G. Pusky, *Cereal Chem.*, 1975, **52**, 655.

24. Y.I. Matveev, V.Y. Grinberg and V.B. Tolstoguzov, *Food Hydrocolloids*, 1997, **11**, 125.

25. H. Levine and L. Slade, 'Physical Chemistry of Foods', H. Schwartzberg and R. Hartel (eds.), Marcel
    Dekker, New York, 1992, 83.

26. J.M. Ames, 'Biochemistry of Food Proteins', B.J.F. Hudson (ed.), Elsevier Applied Science, London, 1992, 99.

27. L.N. Bell, D.E. Touma, K.L. White and Y.H. Chen, *J. Food Sci.*, 1998, **63**, 625.

28. R. Karmas, M.P. Buera and M. Karel, *J. Agric. Food Chem.*, 1992, **40**, 873.

29. B.E. Elizalde and A.M.R. Pilosof, 'Proceedings 3rd European Conf. Grain Legumes', Valladolid, Spain, 1998, 54.

30. B.E. Elizalde and A.M.R. Pilosof, *J. Food Eng.*, 1999, **42**, 97.

A Study of Vitrification of Australian Honeys at Different Moisture Contents

P.A. Sopade,[1,2] B. Bhandari,[1] B. D'Arcy,[1] P. Halley[2] and N. Caffin[1]

[1]FOOD SCIENCE & TECHNOLOGY GROUP, SCHOOL OF LAND AND FOOD SCIENCES, AND
[2]DEPARTMENT OF CHEMICAL ENGINEERING, UNIVERSITY OF QUEENSLAND,
ST. LUCIA QLD 4072, AUSTRALIA

Nomenclature

A	constant
B	constant
C	constant
ΔC_{pi}	change in specific heat capacity of component i
D	mean relative deviation modulus
K	constant
n	number of experimental data
q	constant
r	correlation coefficient
T_g	glass transition temperature of a mixture
T_{ge}	experimental glass transition temperature
T_{gi}	glass transition temperature of component i ($i=1$ for water and 2 for anhydrous honey)
T_{gp}	predicted glass transition temperature
W_i	weight fraction of component i ($i=1$ for water and 2 for anhydrous honey)

1 Introduction

The glass transition phenomenon in food systems, with its various influences on processing, storage and handling, is receiving extensive attention.[1-6] Consequently, food components, ingredients and additives are being studied, as is the role of food operations and processes on the glass transition. Sugars and their mixtures have been studied,[5,7-10] but the situation in naturally occurring sugar

169

mixtures such as honeys is not well known. Apart from different sources of honeys, the uses to which honeys are put in the food industry are varied and diverse. Since glass transition data provide a valuable complementary tool for the physical characterisation of foods, such studies on honeys are important. Rubin et al.[11] and Kantor et al.[12] have studied the glass transition of some honeys, and although they studied the influence of moisture content, neither group applied any glass transition model to assist in prediction of glassy behaviour. We have observed a dearth of information on the physical properties of Australian honeys, and only recently reported[13] on the glass transition behaviour of undiluted samples. Although the studied honeys varied in moisture content, the influence of water could not be properly assessed because of differences in sugar content. The glass transition of a given honey, at different moisture contents, is expected to provide a better examination of the effect of water on these honeys. Honeys can be incorporated in formulations that vary in overall moisture content, in the design of products with different characteristics.

Moreover, certain characteristics of foods, particularly those that depend on the molecular structure of a food, have been related to the glass transition. Stickiness, caking, collapse and crispness are notable examples.[1,5,14] For extrusion cooking, Fan et al.[15] obtained a direct relationship between specific mechanical energy and glass transition temperature (T_g), for which Chen and Yeh[16] reported a strong correlation with expansion temperature. Roos[1] showed a linear relationship between T_g and collapse temperature for maltodextrins of different dextrose equivalents (10–25). The relationship between moisture sorption properties and glass transition has also been discussed.[5,17,18] With respect to honeys, Bhandari et al.[6] and Sopade et al.[13] have speculated that crystallisation tendency can be inferred from T_gs. In light of this, knowledge of vitrification in honeys could provide a fundamental basis for understanding their physical characteristics and those of their products. Therefore, the present study was conducted to investigate the glass transition as a function of moisture content in liquid Australian honeys, and to apply appropriate glass transition models to describe the dependence.

2 Materials and Methods

2.1 Materials

Ten varieties of liquid Australian honey, obtained from Capilano Honey Ltd, Richlands, were studied. The sample treatment and characteristics of the honeys have been described previously.[17] Moisture content was determined using a refractometer (RFM 340 Refractometer, Bellingham & Stanley, Turnbridge Wells, UK), and four moisture levels (≈ 25, 30, 35 and 40%) were prepared for each honey type, by adding calculated (material balance) amounts of ultrapure water. The dilution level was based on the likely maximum hygroscopicity of honeys.[19] All the diluted samples were thoroughly mixed before analysis, and no sample was equilibrated for less than two days prior to T_g measurements. To minimise deterioration (at high water levels), all samples were refrigerated throughout the study, but equilibrated at room temperature before measurement.

2.2 Glass Transition Analysis

A differential scanning calorimeter (TA 2920 Modulated DSC, TA Instruments, New Castle, DE) was used. It was calibrated with indium ($T_m = 156.6\,°C$, $\Delta H = 28.5$ J g^{-1}), mercury ($T_m = -38.9\,°C$, $\Delta H = 11.4$ J g^{-1}) and *n*-heptane ($T_m = -90.6\,°C$). Unannealed samples (10–20 mg) were scanned from -130 to $50\,°C$ at $10\,°C$ min^{-1}, while annealed samples (10–20 mg) were scanned from -130 to $-50\,°C$, held for 30 min, then cooled to $-130\,°C$, before scanning to $50\,°C$, as described elsewhere.[13]

2.3 Statistical Analysis

Standard statistical packages (Statistica™, Excel97™ and Minitab™) were used for regression and analysis of variance (ANOVA). The following models were investigated for describing the moisture dependence of T_g.
  Gordon–Taylor:

$$T_g = \frac{W_1 T_{g1} + K W_2 T_{g2}}{W_1 + K W_2} \tag{1}$$

This is one of the earliest models, but its use has been typically confined to binary systems, and this is regarded as its major limitation.[1] However, it is possible to extend it, in the following form, to multi-component systems:[1]

$$T_g = \frac{\sum\limits_{i=1}^{n} K_{(i-1)} W_i T_{gi}}{\sum\limits_{i=1}^{n} W_i K_{(i-1)}} \tag{1a}$$

Couchman–Karasz: The exact form of this model is reported[1] to be a better predictor:

$$\ln T_g = \frac{\sum\limits_{i=1}^{n} W_i \Delta C_{pi} \ \ln \ T_{gi}}{\sum\limits_{i=1}^{n} W_i \Delta C_{pi}} \tag{2}$$

Kwei:

$$T_g = \frac{W_1 T_{g1} + K W_2 T_{g2}}{W_1 + K W_2} + q W_1 W_2 \tag{3}$$

Fox:

$$\frac{1}{T_g} = \frac{W_1}{T_{g1}} + \frac{W_2}{T_{g2}} \tag{4}$$

or

$$T_g = \frac{T_{g1}T_{g2}}{W_i T_{g2} + W_2 T_{g1}}$$

Fried:

$$\ln\frac{T_g}{T_{g1}} = \frac{W_2 \ln\dfrac{T_{g2}}{T_{g1}}}{W_1 \ln\dfrac{T_{g2}}{T_{g1}} + W_2} \tag{5}$$

Pochan–Beatty–Hinman:

$$\ln T_g = W_1 \ln T_{g1} + W_2 \ln T_{g2} \tag{6}$$

Huang:

$$T_g = \left[\frac{W_1 \Delta C_{p1}(T_{g1} + T_{g2}) + 2W_2 \Delta C_{p2} T_{g2}}{W_1 \Delta C_{p1}(T_{g1} + T_{g2}) + 2W_2 \Delta C_{p2} T_{g1}}\right] T_{g1} \tag{7}$$

Linear:

$$T_g = W_1 T_{g1} + W_2 T_{g2} \tag{8}$$

The characteristics and derivations of these models are available elsewhere, as are also other models that have been proposed.[1,4,16,20] However, the above models are among the most widely used and were focused on during this study.

3 Results and Discussion

3.1 Evidence of Glass Transition

Defining a shift in heat capacity as an indication of transition, we observed that all the honeys exhibited such a shift for all levels of dilution studied. Typical thermograms are shown in Figure 1. Generally, an 'endotherm' **a** was found for moisture contents up to 25%, and two or more 'endo-' and/or 'exotherms' **b** were found at moisture contents greater than 25%. As reported previously,[13] 'endotherm' **a** was more pronounced in undiluted annealed samples. Since annealing was done at a temperature within the glassy state of these samples, 'endotherm' **a** may have originated from molecular relaxation.[10,21] The depth of 'endotherm' **a** was essentially negligible, when the honeys were not annealed,[13] and for diluted samples, the 'endotherm' was not only nonexistent, but the thermograms of unannealed samples coincided with the pre-annealed sections of the corresponding sample thermograms (as shown in Figure 2). The overlap could be attributed to the absence of any relaxation, because the temperature at

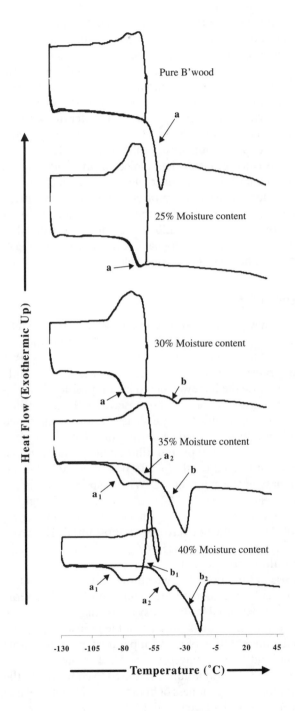

Figure 1 *DSC thermograms for a typical honey (B'wood), at the indicated moisture levels, showing the following different configurations: a = glass transition, a_1 = first glass transition (pre-anneal), a_2 = second glass transition (post-anneal), b = melting of ice crystals, b_1 = crystallisation during annealing, and b_2 = melting of ice crystals after annealing*

which annealing was done was higher than that of the glass transition region for the diluted samples. It can be observed from Figures 1 and 2 that no significant endo- or exotherm was evident before the glass transition region, to indicate any physicochemical changes in samples during cooling to the starting temperature ($-130\,^\circ$C). Hence, it can be inferred that quench-cooling (by liquid nitrogen) of the samples to the starting temperature prevented both sugar and water crystallisation. Even when some diluted and undiluted samples were cooled, at either 10 or 20 $^\circ$C min^{-1}, to this temperature, crystallisation of any type was also absent.

The effectiveness of rapid cooling in preventing freezing of water during glass transition studies has been reported by Chang *et al.*,[22] while Hartel and Shastry[23] noted that rapid reduction in temperature will prevent nucleation and formation of a crystal lattice, because of insufficient molecular mobility. Crystal growth is always dependent on nucleation, and since the latter is a kinetic event, a certain time scale is required for orientation of molecules into a crystal lattice. For honeys, that could take weeks or months,[24] so the absence of any endo- or exotherm during cooling of our samples to $-130\,^\circ$C was not surprising. The situation with endotherm **b** (Figure 1), which may have resulted from melting of the crystallised component in the honeys, is discussed below.

3.2 Moisture Dependence of T_g

Table 1 shows T_g results for honeys at moisture levels up to 40%, while Figure 3 illustrates the typical trend of T_g *vs* moisture content. Our T_gs are identical to previously published values[11,12] for other honeys. Generally, one-way ANOVA showed that the onset, glass transition and end temperatures were significantly ($p < 0.05$) different, possibly because the honeys differed in composition.[13] However, those compositional differences did not reflect themselves in the ΔC_p values for the transitions. While no differences in ΔC_p were measured for the undiluted honeys,[13] the diluted honeys effectively yielded significant ($p < 0.05$) differences. A two-way ANOVA revealed that honey type and dilution, individually (main effect) or collectively (interaction), significantly ($p < 0.05$) affected ΔC_p, but no trend was apparent for honey as the main effect. With respect to moisture content of honeys, an inverse relationship with ΔC_p was evident, even though the coefficient of variation (CV) was about 10%. With such a low CV, the differences among honeys, in terms of ΔC_p, could not have been substantial. However, Roos[1] and Khalloufi *et al.*[18] reported an increase in ΔC_p with increasing weight fraction of water, but our results contradict that.

Many models, such as those in Equations 1–8 which were investigated in this study, have been used to describe the moisture dependence of T_g. The Gordon–Taylor model is the most widely used for binary systems, and we show G–T regression parameters for our data in Table 2. In view of the industrial importance of honeys, we preferred a presentation of relevant parameters in such table form, for ease of applications and future analysis of results. The mean relative deviation modulus, D, defined[25,26] as in Equation 9, was used to complement the regression parameters in examining the goodness of fit of the models, and a value less than 10% indicates acceptability of a given model.[27]

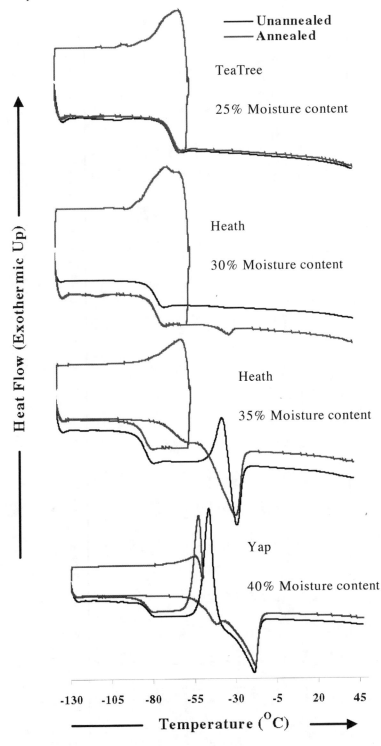

Figure 2 *DSC thermograms for typical diluted honeys, showing the influence of annealing*

Table 1 *Glass transition parameters for Australian honeys[a] at different levels of dilution*

Honey	Undiluted[b]				25%				30%				35%				40%			
	Onset °C	T_g °C	End °C	ΔC_p J°C⁻¹g⁻¹	Onset °C	T_g °C	End °C	ΔC_p J°C⁻¹g⁻¹	Onset °C	T_g °C	End °C	ΔC_p J°C⁻¹g⁻¹	Onset °C	T_g °C	End °C	ΔC_p J°C⁻¹g⁻¹	Onset °C	T_g °C	End °C	ΔC_p J°C⁻¹g⁻¹
BTIB	-45.6± 0.10[c]	-41.4± 0.06	-38.6± 0.04	0.9± 0.02	-61.1± 0.06	-58.2± 0.21	-55.5± 0.11	0.7± 0.03	-70.9± 0.21	-67.9± 0.04	-64.7± 0.21	0.8± 0.02	-80.0± 0.28	-76.6± 0.25	-73.3± 0.06	0.8± 0.00	-86.9± 0.47	-84.0± 0.31	-80.7± 0.20	0.7± 0.06
B'wood	-45.8± 0.09	-41.7± 0.08	-38.8± 0.09	0.9± 0.07	-63.5± 0.22	-60.6± 0.08	-57.6± 0.04	0.7± 0.04	-73.1± 0.64	-70.3± 0.33	-67.0± 0.04	0.7± 0.06	-79.6± 0.24	-76.7± 0.10	-73.4± 0.16	0.7± 0.03	-86.9± 0.11	-84.3± 0.14	-81.0± 0.13	0.8± 0.04
Gumtop	-46.5± 0.37	-42.2± 0.14	-39.3± 0.10	0.9± 0.05	-63.2± 0.84	-60.1± 0.08	-57.1± 0.05	0.8± 0.08	-71.8± 0.13	-68.7± 0.07	-65.6± 0.07	0.8± 0.04	-79.4± 0.17	-76.3± 0.15	-72.9± 0.6	0.8± 0.06	-85.9± 0.18	-84.1± 0.20	-81.1± 0.10	0.6± 0.07
Heath	-47.0± 0.31	-43.0± 0.33	-39.9± 0.21	0.8± 0.06	-64.3± 0.76	-60.9± 0.28	-57.9± 0.38	0.8± 0.02	-73.3± 0.09	-69.5± 0.04	-66.2± 0.01	0.9± 0.01	-80.1± 0.13	-76.6± 0.34	-73.4± 0.55	0.8± 0.10	-87.9± 0.44	-84.4± 0.51	-81.5± 0.30	0.9± 0.05
L'wood	-42.3± 0.10	-38.4± 0.08	-35.7± 0.01	0.9± 0.07	-63.9± 0.18	-60.4± 0.07	-57.5± 0.05	0.9± 0.01	-72.6± 0.05	-69.4± 0.00	-66.3± 0.08	0.8± 0.02	-79.2± 0.26	-76.4± 0.02	-73.0± 0.16	0.8± 0.06	-86.6± 0.28	-83.9± 0.40	-80.8± 0.13	0.7± 0.05
NLIB	-46.5± 0.24	-42.4± 0.14	-39.5± 0.06	0.9± 0.07	-61.9± 0.14	-58.7± 0.22	-56.0± 0.04	0.8± 0.00	-72.6± 0.18	-69.5± 0.05	-66.5± 0.20	0.8± 0.00	-78.8± 0.31	-75.5± 0.08	-72.2± 0.19	0.8± 0.04	-86.1± 0.22	-83.4± 0.23	-80.2± 0.19	0.7± 0.03
STBark	-50.3± 1.01	-45.9± 0.95	-42.6± 0.72	0.9± 0.05	-62.7± 0.17	-59.7± 0.01	-56.6± 0.04	0.8± 0.05	-71.9± 0.07	-68.3± 0.06	-64.7± 0.15	0.8± 0.01	-80.1± 0.12	-76.8± 0.22	-73.2± 0.08	0.8± 0.02	-86.5± 0.17	-84.0± 0.45	-80.8± 0.21	0.7± 0.03
Teatree	-49.8± 0.06	-45.4± 0.02	-42.2± 0.05	0.9± 0.02	-62.5± 0.49	-59.5± 0.47	-56.8± 0.52	0.8± 0.01	-73.6± 0.01	-70.7± 0.03	-67.6± 0.08	0.8± 0.00	-82.4± 0.37	-79.2± 0.40	-75.8± 0.33	0.8± 0.05	-87.7± 0.30	-85.0± 0.21	-82.1± 0.14	0.7± 0.06
Yap	-44.3± 0.09	-40.2± 0.32	-37.6± 0.01	1.0± 0.12	-63.2± 0.09	-60.4± 0.08	-57.6± 0.08	0.8± 0.01	-72.0± 0.38	-69.1± 0.35	-66.2± 0.07	0.8± 0.01	-79.4± 0.13	-76.6± 0.28	-73.4± 0.36	0.7± 0.04	-86.5± 0.19	-84.2± 0.15	-81.0± 0.31	0.7± 0.02
Y'Box	-45.7± 0.28	-41.4± 0.09	-38.3± 0.08	0.9± 0.03	-64.1± 0.16	-61.3± 0.13	-58.3± 0.05	0.8± 0.02	-70.7± 0.16	-67.9± 0.11	-64.8± 0.14	0.7± 0.02	-79.6± 0.61	-76.8± 0.12	-73.5± 0.08	0.8± 0.06	-87.8± 0.22	-84.7± 0.17	-81.8± 0.50	0.7± 0.01

[a] BTIB=Blue top iron bark, B'wood=Bloodwood, L'wood=Leatherwood, NLIB=Narrow leafed iron bark, STBark=Stringybark, Yap=Yapunyah and Y'box=Yellowbox.

[b] Adapted from Sopade et al.[13]

[c] Figures are means ± standard deviations.

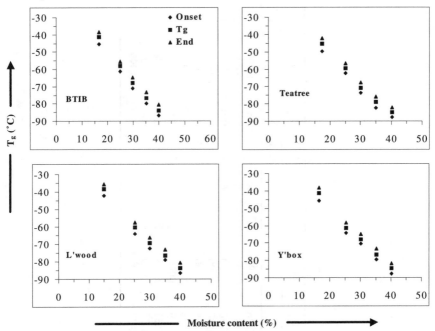

Figure 3 *Typical relationship between* T_g *and moisture content for honeys prior to annealing*

$$D = \frac{100}{n} \sum_{i=1}^{n} \frac{|T_{ge} - T_{gp}|}{T_{ge}} \tag{9}$$

The T_g of amorphous water, T_{g1}, has been reported[1] to range from -134 to $-144\,°C$, but -135 and $-139\,°C$ are widely accepted values. When a non-linear regression was done without fixing T_{g1}, the statistics (not shown) obtained revealed that all the models except Fox and Huang gave very low D ($\leq 3\%$) and high r (≥ 0.995). Although the Fox model gave a fairly high r (≥ 0.952), its D value was $\geq 6\%$. Interestingly, the T_g of water, predicted by the Gordon–Taylor (-129 to $-147\,°C$) and Fried (-132 to $-134\,°C$) models, were within the values reported previously.[1] Generally, when T_g for water was fixed at either -135 or $-139\,°C$ for the regression, all the models were either slightly worse or totally unsuitable in describing the moisture dependence of T_g for honeys. In addition, the T_g values predicted for anhydrous honeys (T_{g2}), by the Gordon–Taylor (-3 to $13\,°C$), Fried (7 to $18\,°C$) and Huang (3 to $12\,°C$) models, were comparable to the T_g values for the anhydrous form of the sugars (fructose, glucose, sucrose and maltose)[1,3,20] that make up honeys.

The T_g values for these anhydrous sugars can be used, in the Couchman–Karasz model (Equation 2) and, as we proposed, in the 'generalised' Gordon–Taylor model (Equation 1a), to predict the T_g of honeys. Although high r (> 0.998) and low D ($< 2\%$) values were obtained, some of the predicted parameters were not in line with known values for the sugars. When literature[1] values ($T_{gwater} = -135\,°C$, $T_{gfructose} = 10\,°C$, $T_{gglucose} = 36\,°C$, $T_{gsucrose} = 67\,°C$,

Table 2 *Parameters of the Gordon–Taylor model for honeys*

| Honey | Tg1 not fixed | | | | | Tg1 = −135 °C | | | | Tg1 = −139 °C | | | |
	Tg1(°C)	Tg2(°C)	K	r	D	Tg2(°C)	K	r	D	Tg2(°C)	K	r	D
BTIB	−138	9	0.38	0.999	1	12	0.36	0.999	1	8	0.40	0.999	1
B'wood	−139	3	0.42	0.999	1	7	0.38	0.999	1	4	0.42	0.999	1
Gumtop	−133	11	0.35	1.000	1	10	0.36	1.000	1	6	0.40	1.000	1
Heath	−129	13	0.31	1.000	0	7	0.37	1.000	1	3	0.41	0.999	1
L'wood	−131	10	0.34	0.999	1	6	0.38	1.000	1	4	0.42	1.000	1
NLIB	−132	11	0.34	1.000	0	8	0.38	0.999	1	4	0.42	0.999	1
STBark	−147	−3	0.51	0.998	2	5	0.39	1.000	1	2	0.43	1.000	0
Teatree	−141	4	0.41	1.000	0	10	0.35	0.998	2	6	0.39	0.998	2
Yap	−132	10	0.34	1.000	0	7	0.37	1.000	0	4	0.41	1.000	1
Y'Box	−140	5	0.41	0.999	1	8	0.37	0.999	1	5	0.41	0.999	1

$T_{\text{gmaltose}} = 92\,°\text{C}$) were used for the constituents, identical r and D values were obtained, but negative K ($= \Delta C_{\text{pcomponent}}/\Delta C_{\text{pwater}}$) values were predicted, as before, for some components. Perhaps, in a multi-component system such as honey, component–component interactions may be pronounced. The use of T_{g} values for pure components, with no compensation for likely interactions, may be unrealistic. The Kwei model has been proposed to take into account such interactions,[20] but its predictive ability was not particularly good for honeys. In addition, monosaccharides have been reported[1,5] to generally affect the glass transition of di- and oligosaccharides. Possibly, another source of uncertainty in literature values for these sugars is in sample preparation, whereby sugars typically are melted to produce the amorphous form for T_{g} analysis.[5,7,28] Melting involves a phase change, with its attendant molecular reorganisation. In a mixture of sugars with different melting points, there is a distinct possibility that, while heating a lower-melting one to a higher temperature, pyrolysis might occur. The T_{g} of such a mixture might be incorrectly evaluated. For instance, Roos and Karel[9] freeze-dried their sugar samples and obtained a value of about 62 °C (midpoint of onset and end temperatures) for T_{g} of sucrose, while Orford *et al.*[7] used a melting technique and reported a value of 70 °C. Thus, apart from differences in equipment, the preparation of samples for T_{g} analysis will affect results. As recognition of the importance of the glass transition increases in food research,[5] such discrepancies need to be minimised, and this underlines the need for standardisation of procedures, in order to facilitate comparisons among laboratories.

In view of our observations with the Couchman–Karasz and 'generalised' Gordon–Taylor models, honeys are better treated as binary systems. Hence, the Fried, Huang and Gordon–Taylor models are appropriate for describing the moisture dependence of T_{g} for honeys. These models have been used for sugar-containing systems, and their suitability for honeys is not surprising, as honeys are concentrated sugar solutions. While the Fried model does not have a heat capacity parameter, both the Huang and Gordon–Taylor models do have one that is based on the K parameter as defined earlier. Assuming that ΔC_{p} for water is 1.94 $\text{J}\,°\text{C}^{-1}\text{g}^{-1}$, as used by Kalichevsky *et al.*,[20] the predicted ΔC_{p} values (0.60–0.99 $\text{J}\,°\text{C}^{-1}\text{g}^{-1}$) for honeys, from the Gordon–Taylor model, were close to the experimental values. This was true, regardless of the way the parameter, which represents T_{g} of water in the model, was handled. Although there are other values for ΔC_{p} of water in the literature,[1] their use would have under-predicted ΔC_{p}, because they are less than the assumed value (1.94 $\text{J}\,°\text{C}^{-1}\text{g}^{-1}$). From our studies, it appears that 1.94 $\text{J}\,°\text{C}^{-1}\text{g}^{-1}$ is a more realistic value for ΔC_{p} of water. Consequently, the Gordon–Taylor model can be regarded as the most appropriate one for honeys, and its K value varied to a certain extent with the honeys (as shown in Table 2), but generally less than for the K values reported for concentrated solutions of sugars.[1] If 'strong' liquids exhibit only small or undetectable changes in heat capacity at T_{g},[29] then honeys are relatively less fragile than sugars. The fragility of glass-forming liquids, based on their K-values, has consequences for the dependence of viscosity on temperature.[29] It can be observed in Table 2 that the predicted T_{g}s for anhydrous honeys (T_{g2}) varied with the honey, and in line with previous observations by both Roos[1] and Rahman,[3] they were linearly related to K:

$$K = 0.458 - 0.011 \, T_{g2} \tag{10}$$

$$r = 0.949$$

3.3 Post-anneal Transition

As shown in Figures 1 and 2, annealed samples exhibited two glass transitions: one before and another after annealing. The pre-anneal one was analysed as described above, while the post-anneal one could be seen to have occurred after part of the water had crystallised during annealing (isothermal at $-50\,°C$ for 30 min). Consequently, the post-anneal T_gs for honeys at some levels of dilution were higher than the pre-anneal T_gs. It appeared that a critical moisture content was reached (as illustrated in Figure 4), before freezable water crystallised out, leaving the diluted honeys more concentrated (maximally freeze-concentrated[5]). It can be inferred that, for moisture levels below this critical value, the annealing temperature was essentially in the glassy solid zone, where molecular motion of any sort, which could have initiated nucleation and crystallisation, was stopped or greatly impeded. However, above the critical moisture level, the annealing temperature was within the rubbery or viscous liquid zone, and molecules were relatively freer to associate and crystallise. It is recognised that pure water freezes at $0\,°C$, but solutes, such as those in honeys, depress the freezing point of water, in accord with Raoult's law. This behaviour could be responsible for the sub-zero

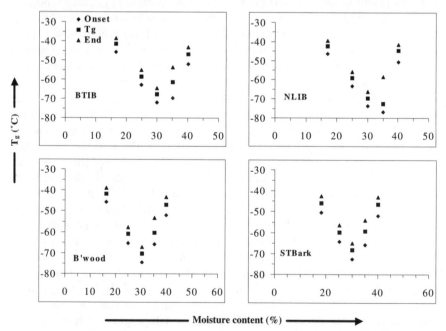

Figure 4 *Typical relationship between* T_g *and moisture content for honeys after annealing*

Table 3 *Parameters of the quadratic relationship between post − anneal T_g and moisture content*

Honey	Quadratic parameters					Freezable water (%)[a]	
	A	B	C	r	$M_{critical}$ (%)	35[b]	40[b]
BTIB	0.15	− 8.96	66.71	0.974	29.4	9.2	21.5
B'wood	0.16	− 9.43	69.71	0.975	29.0	9.7	20.8
Gumtop	0.16	− 9.56	73.89	0.982	29.3	8.9	21.0
Heath	0.16	− 9.47	71.11	0.979	29.2	9.8	21.6
L'wood	0.15	− 8.96	62.39	0.956	29.1	5.6	21.3
NLIB	0.18	− 10.91	91.85	0.903	29.6	2.6	22.2
STBark	0.16	− 9.38	71.54	0.974	29.2	10.4	21.8
Teatree	0.15	− 8.71	60.99	0.922	28.9	13.1	21.6
Yap	0.16	− 9.08	64.36	0.977	28.7	10.7	22.3
Y'Box	0.16	− 9.24	68.83	0.984	29.4	8.2	21.3

[a]Defined as the difference between initial moisture content of a sample and the moisture content corresponding to the pseudo-T_g, using the Gordon − Taylor model.
[b]Initial moisture content of the sample.

peak temperatures $(< − 10\,°C)$ obtained for endotherm **b** in honeys. We are currently investigating annealing honeys at other temperatures and for different times within the glassy and rubbery zones. This is essential for a better understanding of crystallisation and melting phenomena in honeys, which is necessary in order to construct phase diagrams for honeys. Nevertheless, the trend illustrated in Figure 4 was described by a quadratic relationship (Equation 11), and Table 3 summarises the relevant parameters.

$$\text{pseudo-}T_g = AM^2 + BM + C \qquad (11)$$

Differentiating Equation 11 with respect to M and equating to zero yields a critical moisture content $(M_{critical} = − B/2A)$, which was effectively 29% for all the honeys. It appeared that, in spite of differences in their composition, all the honeys showed an identical moisture level, above which ice crystallisation would occur, when they were held at $− 50\,°C$ for 30 min during a cooling-warming cycle. The temperature and duration of annealing would be expected to be influential, and this has been studied for simple sugar systems,[1] but the situation with honeys is relatively unknown. The role of annealing in defining maximum freeze-concentration for honeys demands detailed investigation, because honeys are naturally occurring mixtures of sugars, and the possibility of component interactions might differentiate honeys from mixtures of pure sugars. Annealing studies on honeys might also serve to reveal adulteration, which is a commercial concern.

Despite the preliminary nature of our annealing studies on these honeys, freezable water can be estimated for guidance. Rahman[3] stated that a rough estimation of freezable water could be obtained by dividing the enthalpy of a sample's peak by that of pure water, which is $334\ J\,°C^{-1}g^{-1}$. The faults with this approach are that the latent heat of fusion of ice depends on temperature, and the maximally concentrated state occurs at varying temperatures, due to the freezing point depression by solutes in a given system.[3] Also, the heat represented by the exo- or endothermic peak is a combination of the heat of fusion and the sensible heat taken up by freshly melted water. Perhaps a more accurate approach would be to calculate the moisture content corresponding to the pseudo-T_g, and subtract that from the total moisture in a honey. Freezable water values, calculated using the Gordon–Taylor model, are shown in Table 3. These results indicate that freezable water (at $-50\,^{\circ}\text{C}$ for 30 min) was essentially the same on diluting the honeys to 40% moisture, but for the 35% moisture level, the estimated freezable water varied with the honey. Hence, three different populations of non-freezable water in honeys might be inferred, and this would need to be further studied using more specialised techniques. Both nuclear magnetic resonance (NMR) and Fourier transform infrared (FTIR) have provided valuable information on the state of water in various foods,[30–32] and they would be useful for honeys as well. Additionally, the state of water is relevant to our understanding of the mechanism(s) of sugar crystallisation in honeys, and glass transition studies can contribute to that understanding.

4 Conclusions

Honey, after rapid cooling, can exist as a glass, but the glass transition depends on the type and moisture level of a given honey. Among the major constituents of honeys, water exerted the greatest influence (depression) on T_g. This dependence was better modelled when honeys were regarded as binary (water–anhydrous honey) systems, and the Gordon–Taylor model proved to be the best. At high moisture levels, freezable water was up to 50% of total water, but this would depend on the temperature and duration of annealing. Those aspects will be considered in future studies on honeys, which will also employ other techniques for evaluating the state of water in such a system, in order to explain crystallisation in honeys, and how the glass transition contributes to that phenomenon.

Acknowledgement

Financial and material support, respectively, from the Rural Industries Research Development Corporation and Capilano Honey Ltd, Brisbane, Australia, is gratefully acknowledged.

References

1. Y.H. Roos, 'Phase Transitions in Foods', Academic Press, New York, 1995.
2. J.M.V. Blanshard and P.J. Lillford, 'The Glassy State in Foods', Nottingham University Press, Loughborough, UK, 1993.
3. S. Rahman, 'Food Properties Handbook', CRC Press, Boca Raton, FL, 1995.
4. Y.I. Matveev, V.Y. Grinberg and V.B. Tolstoguzov, *Food Hydrocolloids*, 2000, **14**, 425.
5. L. Slade and H. Levine, *Crit. Rev. Food Sci. Nutr.*, 1991, **30**, 115.

6. B. Bhandari, B. D'Arcy and C. Kelly, *Int. J. Food Prop.*, 1999, **2**, 217.

7. P.D. Orford, R. Parker and S.G. Ring, *Carbohydr. Res.*, 1990, **196**, 11.

8. Y. Roos and M. Karel, *Int. J. Food Sci. Technol.*, 1991, **26**, 553.

9. Y. Roos and M. Karel, *J. Food Sci.*, 1991, **56**, 38.

10. T.R. Noel, S.G. Ring and M.A. Whittam, in 'The Glassy State in Foods', J.M.V. Blanshard and P.J. Lillford (eds.), Nottingham University Press, Loughborough, UK, 1993, 173.

11. C.A. Rubin, J.M. Wasylyk and K.G. Baust, *J. Agric. Food Chem.*, 1990, **38**, 1824.

12. Z. Kantor, G. Pitsi and J. Thoen, *J. Agric. Food Chem.*, 1999, **47**, 2327.

13. P.A. Sopade, B. Bhandari, P. Halley, B. D'Arcy and N. Caffin, *Food Australia*, 2001, **53**, 399.

14. Y.H. Roos, K. Roininen, K. Jouppila and H. Tuorila, *Int. J. Food Prop.*, 1998, **1**, 163.

15. J. Fan, J.R. Mitchell and J.M.V. Blanshard, *Int. J. Food Sci. Technol.*, 1996, **31**, 55.

16. C.-M. Chen and A.-I. Yeh, *J. Cereal Sci.*, 2000, **32**, 137.

17. D.C.P. Jardim, L.M.B. Candido and F.M. Netto, *Int. J. Food Prop.*, 1999, **2**, 227.

18. S. Khalloufi, Y. El-Maslouhi and C. Ratti, *J. Food Sci.*, 2000, **65**, 842.

19. J.W. White, in 'Honey: A Comprehensive Survey', E. Crane (ed.), Morrison and Gibbs, London, 1975, 207.

20. M.T. Kalichevsky, E.M. Jaroszkiewicz and J.M.V. Blanshard, *Polymer*, 1993, **34**, 346.

21. R.F. Tester, S.J.J. Debon and M.D. Sommerville, *Carbohydr. Polym.*, 2000, **42**, 287.

22. Y.P. Chang, P.B. Cheah and C.C. Seow, *J. Food Sci.*, 2000, **65**, 445.

23. R.W. Hartel and A.V. Shastry, *Crit. Rev. Food Sci. Nutr.*, 1991, **30**, 49.

24. C.E. Lupano, *Food Res. Int.*, 1997, **30**, 683.

25. C.J. Lomauro, A.S. Bakshi and T.P. Labuza, *Lebensm. Wiss. Technol.*, 1985, **18**, 111.

26. D.G. Mayer and D.G. Butler, *Ecological Modeling*, 1993, **68**, 21.

27. N. Wang and J.G. Brennan, *J. Food Eng.*, 1991, **14**, 269.

28. T.R. Noel, S.G. Ring and M.A. Whittam, *Carbohydr. Res.*, 1991, **212**, 109.

29. L.-M. Martinez and C.A. Angell, *Nature*, 2001, **410**, 663.

30. A.K. Horigane, H. Toyoshima, H. Hemmi, W.M.H.G. Engelaar, A. Okubo and T. Nagata, *J. Food Sci.*, 1999, **61**, 1.

31. A. Karim, M.H. Norziak and C.C. Seow, *Food Chem.*, 2000, **71**, 9.

32. H.-R. Tang, J. Godward and B. Hills, *Carbohydr. Polym.*, 2000, **43**, 375.

Rational Pharmaceutical Formulation of Amorphous Products

Rational Formulation Design – Can the Regulators be Educated?

Tony Auffret

PHARMACEUTICAL SCIENCES, PFIZER GLOBAL R&D,
SANDWICH, UK

1 The Development of Our Current Understanding of Amorphous State Science in the Pharmaceutical Arena

The rational design of formulations implies a knowledge-based choice of excipients calculated to perform to the desired specification. For pharmaceutical applications, this is almost exclusively concerned with the stabilisation of labile products, sufficient to provide a two-year shelf-life, preferably at ambient temperatures. The rational design of amorphous pharmaceutical formulations is historically based upon freeze-drying. Although other processes are available for the manufacture of amorphous materials, *e.g.* spray-drying, they have yet to gain prominence in this area. Process conditions may vary, but the physical principles underlying the stability of a spray-dried and a freeze-dried form of the same formulation are the same. This review will focus upon freeze-drying.

Originally coming to prominence as a method of preserving human blood plasma in the Second World War, freeze-drying has subsequently been applied to a diverse range of products, including new chemical entities, blood products and, more recently, biotechnology products. The process has gained a reputation for being difficult and unpredictable, although to a great extent, this may have arisen from attempts to use inappropriate thermodynamic (eutectic crystallisation) models to describe the behaviour of frozen solutions. Experience has subsequently taught us that the vast majority of freeze-dried products exist in an amorphous state and not as a crystalline phase.

As early as 1971, Ito recognised that the features of the collapse temperature (T_c) of supercooled solutions, which did not show eutectic crystallisation, 'correspond well to those of a glass transition'.[1] This insight has gone largely unrecognised and uncited. The concept that the behaviour of frozen aqueous solutions is governed by kinetic, non-equilibrium processes, and not by thermodynamic equilibria, gained general acceptance one to two decades later. A general and growing awareness of the glass transition, and its key importance in the storage of frozen foods and in the processing and stabilisation of freeze-dried products, can be traced back to between 1982[2] and 1988.[3] The publication of the glass transition temperature (T_g) of aqueous solutions of over 80 sugars, glycosides and polyhydric alcohols[3] left little doubt as to the generality and importance of glass formation in the freezing behaviour of small carbohydrate–water systems and introduced a rational basis for formulation design. The scientific principles governing successful freeze-drying were clearly laid out shortly thereafter.[4] That publication elucidated many key features that are still the subject of active research.[5]

- Cooling rate and ice morphology and distribution
- Freeze-concentration and rate enhancement as a cause of process loss
- pH shifts arising from crystallisation of buffer salts (particularly Na_2HPO_3)
- T_g' and product collapse
- Primary drying is ice sublimation
- Secondary drying is desorption of residual water
- Plasticising effects of water and salts
- The importance of T_g in the long term stability of the dried product
- Importance of water distribution for the stability of the 'dry' product

Shortly thereafter, the importance and industrial applicability of amorphous glasses for the stabilisation of labile products were recognised.[6,7] The importance of the glassy state, and not the process for producing it, was now clearly established. Although there may be differences in product morphology, drying, whether by freezing, evaporation, spray-drying or oven drying/baking, was established as a unified stabilisation technology for labile products.

It is curious that both of the advances above and two other early seminal papers in the field of practical process and formulation development appeared in the trade press[4,8,9] and not in the peer-reviewed scientific press. Certainly, one of the authors[4] made clear his view of some accepted principles (which may still be found today) in saying 'It is our strong belief, backed up by a growing body of evidence, that the concept of bound water, at least as applied to the process of freeze-concentration, has no valid scientific basis.' Given that twenty years had elapsed since Ito's first identification of the importance of the glass transition,[1] one may speculate on the reluctance of the peer-review process to accept new ideas that overturn established, if inappropriate, textbook explanations.

The application of the principles of stabilisation can be monitored by the number of related publications. A literature database was searched for publications, papers and patents, using the term 'protein stability', and the results are illustrated in Figure 1. The volume of publications can be seen to have risen dramatically post 1988, an approximate date by which it may be assumed that those skilled in the art would have been aware of the general principles.

Thus, the principles governing the rational formulation design of amorphous products were identified over a decade ago. In the intervening decade or so, these principles have been further elaborated. In particular, the concepts of fragility and structural relaxation have provided a theoretical basis for understanding the

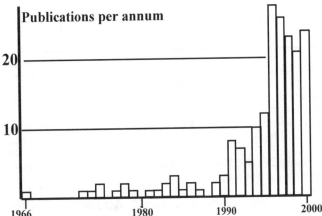

Figure 1 *The number of publications per annum on the subject of 'Protein Stability', as revealed by a literature search*

stability of amorphous products. Yet, despite the excitement generated by the rediscovery of the so-called Kauzmann anomaly,[10] there is still no general consensus regarding a measurable parameter to which amorphous state stability can be scaled.

Table 1 lists some papers that, for this author at least, mark memorable milestones in the development, over the last decade or so, of our understanding of the properties of the amorphous state. At the beginning of the decade, the glassy state itself was considered to be sufficient to ensure stability, and it should not be forgotten that this still remains true in many cases. In considering the stability of pharmaceuticals, biologics, food stuffs, cells and even organisms, we are considering a vast range of chemical reactions and interactions. We do not expect a chemical or kinetic uniformity in dilute solution, and neither should we in a more concentrated solution. The crystallisation of indomethacin from a glassy state[12] indicated clearly that there was mobility in the region of T_g, and thus that chemical reactions could not be ruled out in this region. Towards the end of the decade, Streefland,[18] by selecting a hydrolysis reaction with a rate-limiting step that was independent of water content, was able to measure reaction kinetics across a range of glasses. Within the errors of the measurement, it was not possible to identify a significant difference in reaction rates above and below T_g, but clearly not all glasses were equivalent. Some stabilisers were better than others, and there was a relationship with molecular weight. When scaled to T_g, reaction rates in solution were found to be of a similar magnitude to those in 'dried' glasses.

It also became clear during the decade that not only were all glasses not equivalent, but that the glass transition itself was not a simple transition. Rather, it represented a progressive transition of structural and chemical changes across temperature-related viscosity ranges. Moynihan's excellent review[14] of the glass transition in terms of relaxation processes redefines the glass transition as a strictly kinetic phenomenon. There can be no completely generic model, as 'each liquid will have its own unique structure and mode of structural rearrangement'.

Table 1 *Recent milestone publications in the development of amorphous state science*

Year	Concept	Author	Commentary
1988	Importance of non-equilibrium states in frozen foods	Levine and Slade[3]	
1990	From empiricism to predictability	Franks[11]	Basic formulation and process design principles of freeze-drying
1994	Indomethacin crystallises from solution below T_g	Zografi *et al.*[12]	The glassy state is not sufficient to ensure stability
1994	Uncoupled rotational mobilities of solute and solvent in concentrated solution	Girlich and Lüdemann[13]	Amorphous materials must be structured
1995	Structural relaxation and the nature of the glass transition	Moynihan[14]	Relaxation and glasses explained in plain English
1995	Fragility	Angell[15]	
1996	Why there are two glass transitions	Shalaev and Franks[16]	
1997	Frustration-limited domains	Kivelson[17]	A structural model of the glassy state
1998		Streefland and Franks[18]	Chemical kinetics measured in glasses

Whilst the pharmaceutical scientist was hoping for a uniform glassy-state solution to stability, the data that predestined this hope to failure already existed in other fields!

The concept that there is a single simple glass transition is erroneous. As the viscosity of a once-solid solution decreases, a number of phenomena are observable, and what is observed relates to the means of observation. The relationship between viscosity and observable phenomena, *e.g.* the appearance of two glass transitions in a DSC heating scan, as well as its practical implications, was discussed by Shalaev and Franks.[16] This concept was further extended by Sun,[19] who showed that such *physical* phenomena are predictable and primarily scaled to T_g, defined as a viscosity, across a range of materials.

It should be borne in mind that the events described by Sun are restricted to a temperature range of 40 to 50 °C. Considered over greater temperature ranges, this apparent physical equivalence does not hold. Paralleling the development of our understanding of chemical inequivalence of glasses was the demonstration of structural inequivalence, as exemplified by fragility.[15] Although the fragility of the glasses studied by Streefland[18] was not determined, the apparent relationship to molecular weight could equally be scaled to fragility, the better stabilisers being the more fragile.

A literal interpretation of the term amorphous would be 'shapeless', but when applied to uncrystallised, undercooled systems, a more precise definition is implicit. That is, without shape or form when *probed by methodologies that detect long-range order*. The uncoupling of the rotational mobilities of sucrose and water, as a glassy state is approached,[13] is most easily described in terms of clusters of sucrose molecules and of water molecules. Energetically transient clustering is compatible with a model of a 'landscape with megabasins',[20] or structurally as frustration-limited domains.[17] The latter theory also suggests that we cannot assume a seamless transition between glasses and dilute solutions, but, in the intervening milieu, exotic behaviour such as time-dependent phases may exist. This is not only of academic interest; although the physical chemistry of systems of say 30 to 50% water content has not been the focus of pharmaceutical attention, there is great economic importance, for example, in the process of wet granulation.

Thus, at the beginning of the twenty-first century, we can see the elements of a structural model of amorphous states that have high degrees of short-range order. Although the time scales of transitional motion may differ by orders of magnitude, such a model bears similarities to a structural theory of liquids proposed over 60 years ago.[21]

2 Progress in Rational Formulation Design

The commercial scientist, however, may ask the question, has our increasing understanding of the physical chemistry of amorphous states been translated into a practical guide for the stabilisation of pharmaceuticals and biologicals? Are formulations more rationally designed than they were 10 years ago? The evidence is not hopeful.

A consideration of the molecular mobility of amorphous pharmaceuticals below T_g leads to the general conclusion[22] that storage at $(T_g - 50)\,°C$ is necessary to ensure that there is no significant molecular mobility. It should not, however, be taken that, at a temperature above this, chemical degradation is unavoidable; much will depend upon the chemical reactions that are possible for a given formulation. Searching for a rational choice of stabilising excipient, one concludes that the choice appears to be sucrose or trehalose,[23] with a recommendation to avoid sodium or potassium phosphate buffers.

The advances of the last decade have given us a detailed view of *what* happens under certain circumstances, but have not provided the answer as to *why* it happens. That is the promise of the next decade.

3 Commercial Considerations

All pharmaceutical product formulations should deliver sufficient stability to ensure a two-year shelf-life, but chemical considerations are not the only important factors in developing a pharmaceutical product. Figure 2 summarises the other factors.

A formulation should not contravene a third party's intellectual property (IP), unless licensing is envisaged, and ideally should generate its own intellectual property. At the very least, freedom to operate must be guaranteed. The formula-

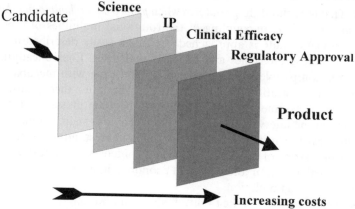

Figure 2 *The development of a pharmaceutical product*

tion must also deliver the product in a clinically efficacious manner. As many amorphous products are dissolved at the point of use, dissolution and solubility of all components are important. More imaginative formulation development can give rise to both IP and clinical efficacy, as exemplified by the Zydis technology of Scherer DDS and the Lyoc range of Laboratoires Lafon, both of which combine crystalline and amorphous phases to produce robust, rapidly dissolving dosage forms. Above all, a formulation must meet with regulatory approval. This absolute requirement, together with the exceptionally high costs of drug development, particularly in the phase III stage of clinical trials, and commercial pressure to shorten development times provide an almost irresistible pressure to use precedented excipients. This is, at the very least, a driving force in favour of an exquisitely detailed knowledge of the physical chemistry of such materials.

4 Practical Formulation Design

In looking for guidelines to assist in the rational design of amorphous formulations, one should look at the small specialist consultancy. Much used by 'big pharma', but often called in only when things have gone wrong, they have neither the time nor the opportunity to follow the modern trend of a multi-parallel array robot that can cover all possibilities. In reviewing the strategy of one such consultancy, Pafra Biopreservation, the following developmental steps can be identified, for a freeze-dried product:

- Buffer selection, if needed
- Freeze-thaw stability, to assess the need for a cryoprotectant
- Formulate, with excipient selection determined by freezing behaviour (DSC), tonicity and bulking requirements
- Selection of two formulations
- Process development, based on T_g'
- Produce samples of both formulations for evaluation

By using such a stepwise approach, the full development process could be completed within 12 weeks.

4.1 Buffer Selection

If appropriate, selection of a compatible buffer should be made as an initial step. As Figure 3 illustrates, mixtures of well-precedented phosphates and citrates cover a range up to about pH 8. Selection of buffers in the range pH 8–10 is more subjective; amino acids are used, but may present polymorph problems, and surprisingly to some, both diethanolamine and tris (masquerading under pharmacopoeial names of trometamol or tromethamine) are approved even for i.v. parenterals.

The propensity of phosphate buffers to crystallise on freezing[24,25] is often cited (*e.g.* ref. 23). In practice, however, the presence of the active material and/or any stabilising or bulking excipient is often sufficient to depress this potentially adverse effect.

4.2 Freeze–Thaw

A simple small-scale overnight freezing, of active in buffer, to $-20\,^{\circ}\text{C}$ is often sufficient to indicate undesirable chemical effects of freeze-concentration and the need for a cryoprotectant. If a product cannot be frozen in good yield, it cannot be freeze-dried. Equally so, if it can be frozen in good yield, it certainly can be freeze-dried successfully.

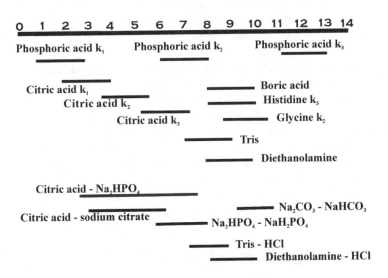

Figure 3 *Upper diagram: pK_a values ± 1 pH unit for some common buffer substances. Lower diagram: useful buffering ranges for some common buffer compositions*

4.3 Excipient Selection

Excipients may be needed for many reasons, *e.g.* tonicity, stability, bulking. The propensity of mannitol to crystallise from a frozen or dried (amorphous) solution, especially if the frozen solution is annealed, makes it ideal for use as a bulking agent. The same propensity means that an amorphous mannitol product is best avoided. Only materials that do not crystallise readily are suitable for use as stabilisers in amorphous products. Sucrose, lactose, PVP, maltose, dextran and serum albumin are precedented for i.v. parenterals, and at least one trehalose-containing injectable (Herceptin) has regulatory approval. For non-medical, *e.g.* diagnostic, products, Byco (a gelatin hydrolysate) and Ficoll (a cross-linked sucrose polymer) have high T_g' values and are useful for cost-effective processing.

4.4 Process Development

The key parameter needed for process development is the collapse temperature, which can be measured by a variety of methodologies, *e.g.* freeze-drying microscope. DSC analysis is particularly useful, as differing scanning rates and temperature ranges can be used to elucidate the freezing behaviour and indicate potential problems. At this stage, elimination of problematic formulations is more cost-effective than is resolution of the problem.

The state of the art does not yet allow us to predict the stability of a formulation, and thus potential reactions of excipients with actives (*e.g.* maltose with proteins) should be considered and tested on the dry product.

At this point, a lead and a back-up formulation can be selected for experimental drying. It is essential that the equipment used is capable of controlling shelf temperature. Primary drying at a shelf temperature below T_g' will ensure that melt-back does not occur, and controlled ramping of temperature during secondary drying will prevent collapse of the ice-free product. In the absence of algorithms to predict primary drying time, removal of samples from a laboratory drier is a simple and effective way to determine primary drying time. Extending primary drying beyond the end of ice sublimation is rarely harmful to the product; this author has only identified one such case in over 200 projects.

Characterisation of the product, typically by DSC behaviour and water content, is sufficient for post-processing quality control. Final excipient selection is verified by stability trials. It should be noted that the Arrhenius model of temperature dependence is not universally applicable. Heating a product with a T_g of 45 to 50 °C will inevitably lead to collapse. It does not follow that collapse at 20 °C will be predicted by the Arrhenius model.

5 Can the Regulators be Educated?

In a pharmaceutical context, the development of a pharmaceutical product must pass several hurdles, before it can be launched.

- Scientific
- Intellectual Property
- Clinical Efficacy
- Regulatory Approval

The scientific hurdle is to turn a new chemical entity into a stable dosage form. In order to underwrite an appropriate return on investment, the drug substance itself must generally be subject to patent protection, but further intellectual property can be generated by formulation and process design. The potential new product must be shown to be safe and clinically effective, with the final judgement of these issues being taken by the various national regulatory authorities.

As the development of a potential new drug proceeds, the associated costs increase markedly, ultimately reaching hundreds of millions of dollars. There is a great deal of pressure, therefore, to bring the product to market as early as possible, and thus the formulation chemist is subject to the twin drivers of 'speed of development' and 'ease of regulatory approval'. Thus, the use of precedented excipients and dosage forms tends towards common practice.

NF19/USP 24[26] lists just over 40 categories of precedented excipients, including buffering agents and bulking agents, for freeze-drying. The latter contains only two entries, creatinine and mannitol, of which the latter, mannitol, is by far the most precedented and popular. It should come as no surprise that there is not a category 'Stabilising Agent for Freeze-drying'. There is little consensus regarding suitable stabilisers, and the literature on the subject is contaminated by many publications that do not discern phase-separated crystalline forms from amorphous forms. An excipient that exists as a phase-separated crystal cannot contribute to the stabilisation of a residual amorphous phase. Furthermore, bearing in mind that some sugars, *e.g.* trehalose, are effective stabilisers as a result of partial crystallisation,[27] one needs to give careful consideration to any set of guidelines or approvals.

Although a patent office is not generally regarded as a regulatory body, the commercial importance of patent protection means that the national patent offices can wield significant regulatory authority. A failed patent application greatly affects the commercial value of both the product and the producer. The criteria by which patents are judged are wholly different from those which, for example, would be applied in the peer review of a submitted manuscript. Two recently approved patents,[28,29] selected randomly in this review of formulation design, exemplify long-term stability by 3 months at 50 °C in one case and 1 week at 37 °C in the other. Novelty, inventiveness and patentability of the formulations concerned are not disputed, and our knowledge of amorphous forms is increasing our arsenal of stabilising formulations. But on this evidence, one must also conclude that our ability to define stabilising formulations of universal applicability has not increased at a corresponding rate.

For pharmaceutical companies, the most important regulators are the various national drug-approving authorities, amongst whom the US FDA is particularly significant. The current FDA 'Guide to Inspections of Lyophilization of Parenterals' document[30] remains as it was in 1993.[31] Little reference is made to the amorphous state; the 'Lyophilization Cycle and Controls' section offers only advice for eutectic product: 'It is desirable after freezing and during primary drying to hold the drying temperature (in the product) at least 4–5 °C below the eutectic point. Obviously, the manufacturer should know the eutectic point...' Substitute 'glass transition temperature' for the 'eutectic point', and the declaration is equally applicable to an amorphous product. The term 'collapse temperature', defined as the temperature at which ice begins to melt, would not only cover amorphous and crystalline products but also encompass products containing both physical states.

It is disappointing that the only reference to the amorphous state exists in the 'Finished Product Inspection – Meltback' section: 'The amorphous form may exist in the 'meltback' portion of the cake, where there is incomplete sublimation'. The statement is undoubtedly true, but with the current state of the art, it cannot be considered to be a full description of the relevance of the amorphous state.

Rather than ask the question 'can the regulators be educated?', to which the answer is undoubtedly 'yes', one may more beneficially examine the behaviour of those whose role it is to educate – the academic and commercial scientific community. Is it still common to find descriptions, in the contemporary scientific and patent literature, of the type '... and the product was freeze-dried', without reference to collapse temperatures or process parameters. The maintenance of a product at 4–5 °C below its collapse temperature can most easily be achieved by maintaining the shelf at a temperature 4–5 °C below the collapse temperature, yet lyophilisation processes are still designed in which shelf temperature is held greatly above the collapse temperature, and the heat and mass transfer of sublimation are relied upon to maintain the product temperature at an appropriate level. There is good evidence[32] that the most efficient lyophilisation cycles employ a combination of low pressure and high shelf temperature. Such a combination is undoubtedly commercially attractive, and for many products, the risks are not serious. But for a product destined for parenteral administration, injection for example, into humans, is this the combination most likely to maintain product quality in the event of a process deviation? The answer is undoubtedly no!

It should be remembered that the duty of the regulatory body is to ensure that the manufacturers meet their obligation to provide safe medicines. Changes to guidelines and practices can only be effected after due consideration. The literature survey, illustrated in Figure 1, suggests that it takes about 10 years, before the scientific community, the vanguard of knowledge development, can absorb a new idea. This author would assert that it takes longer than that for a consensus opinion to emerge, and hence the time scales for embodying scientific change in regulatory documentation cannot be shorter than this.

6 Conclusions

Our knowledge of the amorphous state has increased dramatically in the past decade, but, from the point of view of rational formulation design, this has been descriptive rather than predictive. It is perhaps, therefore, not surprising that bodies regulating therapeutic products have not been hasty to incorporate developments into regulatory documentation.

A given formulation has a potential for stabilising a product, but that potential may not be realised unless the processing is appropriate.

$$\text{Product} = f(\text{formulation} \times \text{process}) \tag{1}$$

Equal weight should therefore be given to processing conditions, in research that seeks to advance our knowledge of formulation design of amorphous products.

Advances in the next decade will come from an understanding that the amorphous state is not formless – it simply has a structure that cannot be probed by methods suited to determining long-range order. A predictive model of the amorphous state is likely to come from advances in the following areas:

- A structural model of the amorphous state – importance of short-range order
- Structural-based prediction of dried product stability – scaling of stability to measurable parameters
- The amorphous–crystalline continuum and the existence of intermediate scales of order
- Design of stabilisers and stabilising mixtures – self-stabilising formulations

The concept of self-stabilising formulations exemplifies how our knowledge of materials could lead to a rational design of formulations. Table 2 lists a number of carbohydrates with the potential to form hydrated crystals.[27] The crystallisation of β,β-trehalose and raffinose could remove substantial amounts of plasticising water and thus increase the T_g of the residual amorphous phase. The complete crystallisation of lactose, in the example, would increase the water content of the residual product. The effect on the product would depend upon the glass curve of the active, although if the lactose was genuinely a stabiliser, this would be disastrous. An excipient that crystallised as a hydrate, below the glass transition, would be an ideal candidate for a self-stabilising formulation.

Table 2 *The potential for carbohydrate excipients to remove water from an amorphous product containing 5 g of excipient, 1 g of active and 5% w/w water (0.32 g), dried from 5% excipient, 10 mg ml^{-1} active*

	MW (anhydrous)	Hydrate (max)	Water capacity (g) of 5 g	Weight (g) to remove 0.32 g water
β,β-Trehalose	342	4	1.05	1.5
Raffinose	504	5	0.89	1.8
Stachyose	666	4	0.54	3
Trehalose	342	2	0.53	3
Melibiose	342	2	0.53	3
Mannotriose	504	3	0.53	3
Melezitose	504	2	0.35	4.6
Lactose	342	1	0.26	6

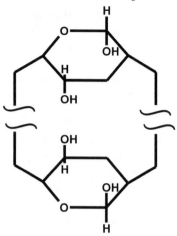

Figure 4 *Philosophose – a hypothetical molecule*

Crystalline *versus* amorphous behaviour could perhaps be better understood by a study of a rational, homologous series of compounds. Such a series is exemplified in Figure 4 and dubbed the 'Philosophose' series. Alterations to the hydroxyl substitution patterns of the rings can be used to moderate the potential for intra- or intermolecular hydrogen bonds. Alterations to the spacer arms can be used to modulate the strength of intramolecular hydrogen bonds, as well as to introduce other substituents. The present state of the art would not allow us to predict the crystalline or amorphous nature of such a compound, nor would it allow us to predict the shape of its glass curve, its fragility, nor the relaxation time of an amorphous form with a defined water content. Such things are essential to enable rational formulation design.

References

1. K. Ito, *Chem. Pharm. Bull.*, 1971, **19**(6) 1095.
2. F. Franks, 'Water: A Comprehensive Treatise', F. Franks (ed.), Plenum Press, New York, 1982, Vol 7, 99.
3. H. Levine and L. Slade, *J. Chem. Soc., Faraday Trans. 1*, 1988, **84**(8), 2619.
4. F. Franks, *Process Biochem.*, February 1989, Turret Group plc, Rickmansworth.
5. G. Gómez, M.J. Pikal, and N. Rodríguez-Horneda, *Pharm. Res.*, 2001, **18**(1), 90.
6. F. Franks, S. Mathias and R.H.M. Hatley, *Patent Application EP* 03835869 (1990), Priority date 1989.
7. F. Franks, R.H.M. Hatley and S.F. Mathias, *Pharm. Technol. Int.*, October 1991.
8. M.J. Pikal, *BioPharm*, 1990, **3**(8), 18.
9. M.J. Pikal, *BioPharm*, 1990, **3**(9), 26.
10. W. Kauzmann, *Chem. Rev.*, 1948, **43**, 219.
11. F. Franks, *CryoLetters*, 1990, **11**, 93.
12. M. Yoshioka, B. Hancock and G. Zografi, *J. Pharm. Sci.*, 1994, **83**, 1700.
13. D. Girlich and H.D. Lüdemann, *Z. Naturforsch., C: Biosci.*, 1994, **48**, 407.
14. C.T. Moynihan, *Rev. Mineral.*, 1995, **32**, 1.
15. C.A. Angell, *Science*, 1995, **267**, 1924.

16. E.Yu Shalaev and F. Franks, *Cryobiology*, 1996, **33**, 14.
17. G. Tarjus, D. Kivelson and S. Kivelson, *ACS Symp. Ser.*, 1997, **676**, 67.
18. E. Streefland, A.D. Auffret and F. Franks, *Pharm. Res.*, 1998, **15**, 845.
19. W.Q. Sun, *CryoLetters*, 1997, **18**, 99.
20. C.A. Angell, *Physica D*, 1997, **107**, 122.
21. N.F. Mott and R.W. Gurney, *Trans. Faraday Soc.*, 1939, **35**, 364.
22. B.C. Hancock, S.L. Shamblin and G. Zografi, *Pharm. Res.*, 1995, **12**, 799.
23. J.F. Carpenter, M.J. Pikal, B.S. Chang and T.W. Randolph, *Pharm. Res.*, 1997, **14**, 969.
24. N. Murase and F. Franks, *Biophys. Chem.*, 1989, **34**, 293.
25. F. Franks and N. Murase, *Pure & Appl. Chem.*, 1992, **64**, 1667.
26. 'National Formulary' **19**/'United States Pharmacopoeia' **24**. The US Pharmacopoeial Convention, Rockville, MD, 1999, 2404.
27. B.J. Aldous, A.D. Auffret and F. Franks, *CryoLetters*, 1995, **16**, 181.
28. T.A De Rosier, N.B. De La Cruz and R.K. Wilkosz, *Patent Application PCT/US97/11767*, 1998.
29. D.B. Volkin, C.J. Burke and S.-P. Sheu, *Patent Application PCT/US99/12026*, 1999.
30. 'Guide to Inspections of Lyophilization of Parenterals', http://www.fda.gov/ora/inspect_ref/lyophi.html (downloaded 28 February 2001).
31. 'Guide to Inspections of Lyophilization of Parenterals', Interpharm Press, Buffalo Grove, IL, 1993.
32. B.S. Chang and N.L. Fischer, *Pharm. Res.*, 1995, **12**, 831.

Solid–Liquid State Diagrams in Pharmaceutical Lyophilisation: Crystallisation of Solutes

Evgenyi Shalaev[1] and Felix Franks[2]

[1]GROTON LABORATORIES, MS-8156-004, GLOBAL R&D,
PFIZER INC., GROTON, CT 06340, USA
[2]BIOUPDATE FOUNDATION, 25 THE FOUNTAINS,
229 BALLARDS LANE, LONDON N3 1NL, UK

1 Introduction

Freeze-drying, as applied to pharmaceutical process technology, consists of several steps: freezing, primary drying (ice sublimation), and secondary drying (removal of non-frozen water, if any), all of which are combined in a single technological cycle. Solutions destined for lyophilisation are often of a complex composition, containing several solute components, such as the drug itself, pH buffers, other salts, bulking agents, and excipients added to protect drug substances against loss of biochemical activity and chemical degradation. The water content of such solutions usually exceeds 90%. During initial cooling of such solutions, water is normally the first component to crystallise, although not necessarily completely. At this stage, a biphasic system is formed, consisting of ice and residual freeze-concentrated solution.

As cooling proceeds, three general types of phase behaviour can be distinguished, according to the properties of the residual solution after initial (primary) ice crystallisation:[1]

(i) the freeze-concentrated solution forms a kinetically stable (but thermodynamically unstable) amorphous phase;

(ii) the freeze-concentrated solution forms a 'doubly unstable'[2] glass, in which partial ice/solute crystallisation may occur during annealing, thermal cycling, and/or drying; and

(iii) a binary ice/solute (eutectic) crystallisation may occur after primary ice crystallisation, with the remaining solution forming a kinetically stable glass.

Complete crystallisation of ice and *all* the solute(s) is rare. Setting of a

200

lyophilisation process, as well as the properties and quality of a final product, depends critically on the type of phase behaviour exhibited by a given formulation.

Phase changes that occur during freeze-drying of pharmaceutical formulations can be represented with the aid of phase diagrams. Phase diagrams make possible the determination of order of crystallisation and phase composition(s) of any particular system (*i.e.* the amounts and compositions of phases in equilibrium at any given temperature). Phase diagrams have been successfully used to describe complex phase relationships in different fields, starting from traditional applications in materials science, ceramics, geology, and metallurgy in the mid-twentieth century, and extended to cryobiology, beginning during the 1960s.[3] In freeze-drying technology, 'supplemented' binary phase diagrams were first introduced during the 1970s, usually with application to simple model systems (*e.g.* water–sucrose).[4] Such 'supplemented' phase diagrams (also known as solid–liquid state diagrams) describe both equilibrium and metastable (*e.g.* glassy) states. Details of such binary diagrams and their practical significance for freeze-drying have been described elsewhere.[5,6] Binary temperature–composition solid–liquid state diagrams can be used to trace physical changes during freeze-drying of systems in which only water crystallises. In practice, the situation is often more complex, *e.g.* when, in addition to ice, an excipient or drug substance also crystallises. In such cases, two-dimensional temperature–composition representations are not sufficient, and their uncritical use may result in misleading conclusions.

A gap exists between the practical use of complex freeze-drying systems and their correct phase description, despite the availability of a number of excellent textbooks on phase equilibria in multi-component systems (see refs. 7 and 8 for examples). In this paper, we attempt to remedy the situation. Our discussion consists of the following parts:

(i) a general description of routes of solute crystallisation in three-component systems, using phase diagrams; in particular, we concentrate on secondary crystallisation of solute, *i.e.* crystallisation that occurs after primary crystallisation of water. In a binary system, for example, eutectic crystallisation of water + solute represents secondary crystallisation;

(ii) a review of studies on the model systems water/NaCl/sucrose and water/glycine/sucrose, with emphasis on the effects of crystallisation of a bulking agent on collapse, rate of drying, and stability of biologically active preparations;

(iii) a review of existing literature on the impact of a co-solute on secondary solute crystallisation in frozen aqueous solutions.

2 General Description of Crystallisation of Solute During Freeze-drying

For purposes of freeze-drying, multi-component systems can, in the majority of cases, be considered as either pseudo-binary, when only water crystallises, or pseudo-ternary, when water and one solute crystallise. Crystallisation of two or more solutes is rare in practical freeze-drying processes and will not be considered here. When only ice crystallises, the process can be adequately described with the help of a two-dimensional, temperature–composition, solid–liquid state diagram.[4,9] It is useful to briefly describe the similarities and differences between equilibrium phase diagrams and solid–liquid state diagrams. For such a description, understanding of the phase rule is essential. The equilibrium phase rule is summarised as

$$P + F = C + 2 \tag{1}$$

or at constant pressure,

$$P + F = C + 1$$

where P is the number of phases (solid, liquid, gas), F is the number of degrees of freedom (temperature, pressure, composition), and C is the number of components (chemical species). For any solution, $C \geq 2$, but in almost all pharmaceutical preparations, $C \geq 3$. Neglecting the vapour phase, $P \geq 2$, but ordinarily, $P = 2$ during primary ice crystallisation. During the process of cooling, freeze-concentration and rewarming, solute crystallisation may occur, resulting in an increase in P to 3. Crystallisation of more than one solute is rare in pharmaceutical freeze-drying. It should be stressed that the phase rule can be applied to both (thermodynamic) equilibrium and metastable systems. It is essential that a system be in a state of thermal and mechanical equilibrium, *i.e.* temperature and pressure are uniform across the system, and composition gradients within each phase are absent. Hence, the phase rule may be applied to a system containing metastable supercooled liquid phases, as long as conditions of thermal and mechanical equilibrium are in effect.

Figure 1 illustrates an application of the phase rule to an equilibrium phase diagram and a solid–liquid state diagram. Elements of the equilibrium phase diagram are shown as solid lines, whereas broken lines describe elements of the solid–liquid state diagram. The water liquidus curve corresponds to $F = 1$ in both equilibrium (line ae) and metastable (line eW_g') regions, according to the phase rule. At the eutectic point (e), which is the point of intersection of the water liquidus with the solubility curve, three phases co-exist, and F becomes equal to zero. In other words, both the eutectic composition and eutectic temperature are independent of the initial concentration of the system. On the solid–liquid state diagram, the water liquidus intersects the glass transition curve at point W_g'. Point W_g' describes the composition and glass transition temperature (T_g) of the maximally freeze-concentrated solution.[5] According to a common assumption, water crystallisation stops beyond point W_g', at least in real time.[5] According to this assumption, the liquidus curve does not extend beyond point W_g'. Just as is the case for the eutectic point, the composition of the maximally freeze-concentrated solution is independent of the initial composition of a solution.[5] However, the difference is that the number of phases does not change at W_g', since this

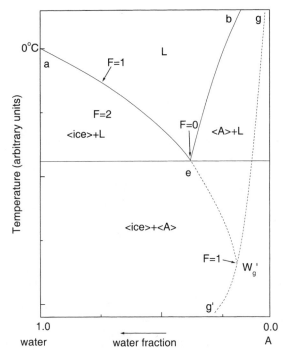

Figure 1 *Phase diagram for a hypothetical water–solute system. Solid lines describe an equilibrium phase diagram, and broken lines are part of a solid–liquid state diagram. The gg′ dashed line represents the glass transition curve, the aeW$_g$′ solid–dashed line represents the liquidus curve for water, and the eb solid line represents the solubility curve for solid. L represents liquid phase, and <A> represents crystalline solute. F is the number of degrees of freedom (Equation 1)*

point reflects purely kinetic rather than thermodynamic phenomena.

To describe crystallisation and phase relationships during freeze-drying in ternary or pseudo-ternary systems with crystallisation of a solute, a triangular representation is useful. Ternary aqueous systems containing one crystallisable and one non-crystallisable solute are convenient models to describe such complex phase behaviour during freeze-drying. Depending on the mass ratio of crystallisable/non-crystallisable solutes and the cooling rates, such systems can demonstrate all three types of behaviour described in the Introduction. A triangular phase diagram is the traditional way to describe the phase behaviour in such systems. The present section is based primarily on results obtained with model water/glycine/sucrose and water/NaCl/sucrose systems.[1,9–11] Figure 2 shows such a projection of a three-dimensional, temperature–composition, solid–liquid state diagram, for a hypothetical ternary or pseudo-ternary system, on a composition triangle.

Let us assume that solute 1 is crystallisable (*e.g.* glycine), whereas solute 2 is not crystallisable (*e.g.* sucrose). Note the term 'not crystallisable' means that crystallisation does not occur in real time, as under practical conditions of freeze-drying.

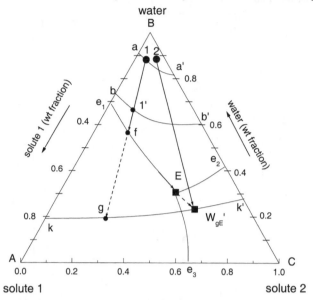

Figure 2 *Projection of a temperature–composition state diagram for a hypothetical water/
solute 1/solute 2 system on a composition triangle. The e_1, e_2, and e_3 points
correspond to the compositions of the binary eutectics for the solute 1/water, solute
2/water, and solute 1/solute 2 binary systems, respectively, and point E corresponds to the composition of the ternary eutectic mixture. Curves aa′ and bb′ are
isotherms for primary water crystallisation at temperatures T_1 and T_2, respectively. Curve kk′ describes the composition of the freeze-concentrated solution after
primary water crystallisation. The order of crystallisation (i.e. which of the three
components crystallises first) depends on the composition of the solution, with
respect to the ternary eutectic point. Compositions within field e_1Ee_3A correspond
to primary solute 1 crystallisation, e_1Ee_2B to primary ice crystallisation, and
e_3Ee_2C to primary solute 2 crystallisation*

Consider the freezing pattern for composition 1 in Figure 2. At temperature T_1
(*e.g.* −5 °C), ice starts to form (assuming that supercooling is avoided). The
composition of the freeze-concentrated solution depends on temperature and
corresponds to point 1′ at temperature T_2 (*e.g.*, −10 °C), for example. Further
cooling produces more ice, with a corresponding increase in concentration of the
unfrozen portion of the sample. The relative amounts of ice and unfrozen
solution can be determined using the lever rule (which is described elsewhere[7,8]).
However, the ratio of the solutes in the unfrozen portion remains constant. When
the composition of the freeze-concentrated solution reaches line e_1E (point f in
Figure 2), solute starts to precipitate, together with ice. In the pharmaceutical
literature, such a process is often called eutectic crystallisation. However, that is
not strictly correct for ternary or more complex systems, because the temperature of such secondary crystallisation is not invariant in a ternary system, but is a
function of the solute 1/solute 2 ratio. The composition of the freeze-concentrated solution changes along line e_1E, as ice + solute 1 continue to precipitate,
and both the water content and the ratio of solute 1/solute 2 in the freeze-

concentrated solution decrease as a result of secondary crystallisation. When the composition of the freeze-concentrated solution reaches the ternary eutectic point E, solute 2 should start to crystallise, and complete solidification of the system should occur. However, solute 2 usually does not crystallise; instead, solute 1 + ice continue to precipitate along line $e_1 W_{gE}'$ in Figure 2, until the viscosity of the freeze-concentrated solution becomes so high that crystallisation stops on the time scale of the experiment. It is commonly accepted that such kinetic retardation of crystallisation occurs when the viscosity approaches $\eta_g = 10^{12}$ Pa s, which corresponds to the viscosity at T_g.[5] The composition of such a maximally freeze-concentrated solution is shown as W_{gE}' in Figure 2, and corresponds to the intersection point of the secondary crystallisation line $e_1 W_{gE}'$ and the $T_g = f$(water, solute 1) surface. Depending on the conditions of cooling and the properties of any particular system, crystallisation may stop before the system reaches η_g. In that case, to maximise the amounts of ice and crystalline solute 1 formed, the system should be annealed at $T = T_{gE}'$, the T_g of the freeze-concentrated solution that is left after crystallisation of water + solute 1. Note also that this description is based on the assumption that crystallisation does not occur below T_g on a practical time scale. Exceptions may exist, *e.g.* as was demonstrated for indomethacin, where crystallisation below T_g was observed.[12] However, in the majority of practical situations, it is safe to assume that crystallisation does not occur below T_g on the time scale of a typical freeze-drying process (*i.e.* hours or days).

If the cooling rate is higher than a critical cooling rate, solute 1 fails to crystallise at point f in Figure 2, and ice alone continues to form, with a corresponding increase in concentration of the freeze-concentrated solution (line fg in Figure 2). In that case, *i.e.* when solute 1 does not crystallise, the final composition of the maximally freeze-concentrated solution corresponds to point g, and secondary crystallisation of solute 1 may be promoted by thermal treatment of the frozen solution, *via* heating to the devitrification temperature T_D. Both the critical cooling rate and T_D depend on the type of solutes and the solute 1/solute 2 ratio; T_D also depends on heating rate.[9] In addition, they may depend on a number of other experimental factors, *e.g.* sample size and the presence of heterophase surfaces. For example, it was observed for a water/glycine/sucrose system that T_D was decreased, when the solution was dispersed on a silica powder prior to cooling/heating cycles.[13]

It can be seen that, depending on the behaviour of solutions beyond point f (Figure 2), the compositions of maximally freeze-concentrated solutions may be different, even when the compositions of the initial solutions before cooling are identical. Note also that a composition of the freeze-concentrated solution, which corresponds to point W_{gE}', may be obtained, starting from a completely different initial solution (point 2 in Figure 2).

The phase behaviour described above was determined from detailed studies of two model systems, water/glycine/sucrose and water/NaCl/sucrose.[1,9–11] That phase behaviour can be summarised as follows:

(i) Crystallisation of a crystallisable solute 1 (glycine or NaCl) is always incomplete, *i.e.* the maximally freeze-concentrated solution after crystallisation of water and solute 1 contains all three components: water, sucrose and solute 1. The ratio of solute 1/sucrose in the maximally freeze-concentrated solution after secondary crystallisation of solute 1 corresponds to point W_{gE}' in Figure 2. In other words, the composition triangle is divided by line BW_{gE}' into two fields. Compositions located on the right-hand side of the BW_{gE}' line form kinetically stable freeze-concentrated glasses, in which crystallisation of solute 1 does not occur on a practical time scale, during either cooling, warming or annealing. For compositions located on the left side of the BW_{gE}' line, crystallisation of solute 1 may occur during either cooling, heating or annealing, depending on cooling/heating rates and annealing conditions.

(ii) The composition of the freeze-concentrated solution after primary water crystallisation depends on the solute 1/sucrose ratio and is independent of total solute content. However, the amount of ice formed varies with total solute concentration.

(iii) Secondary crystallisation of solute 1, *i.e.* crystallisation after primary water crystallisation, occurs as a simultaneous solute 1 + water crystallisation. This follows from the general features of ternary systems and the phase rule, and has been detected experimentally in frozen glycine/sucrose/water solutions, wherein a significant increase in ice X-ray diffraction peaks has been observed, together with glycine crystallisation.[14]

3 Practical Significance of Crystallisation of an Excipient on Collapse, Rate of Drying, and Stability of Biotechnological Products and Pharmaceuticals

In this section, we consider the practical significance of crystallisation of a solute to the development of pharmaceuticals and biopharmaceuticals, using sucrose/NaCl and glycine/sucrose mixtures as examples. As discussed below, crystallisation of a bulking agent in such systems may have a significant impact on freeze-drying conditions and biological activity of proteins. The significance of crystallisation of an active compound and buffer components to the chemical stability and biological activity of freeze-dried formulations has been described elsewhere[15,16] and is not considered here.

3.1 Collapse

The collapse temperature is the single most important parameter that determines conditions for primary drying. Collapse during freeze-drying is associated with loss of cake-like structure. Collapse results in a significant decrease in water removal rate, a poor pharmaceutical elegance, an increase in residual water content, and a decrease in reconstitution rate of a final product. All of these changes are highly undesirable, and thus collapse should be avoided. In order to avoid collapse, the product temperature during freeze-drying should be maintained below the collapse temperature. If the collapse temperature of a formulation is relatively low, it is more difficult and sometimes practically impossible to lyophilise such a formulation.

The collapse temperature of a mixture depends on the collapse temperatures of its individual components and their weight fractions. In particular, the collapse temperature of a mixture decreases as its salt content increases, and it is usually recommended that the salt concentration should be minimised for freeze-drying. It has been suggested, however, that in certain cases, an increase in the concentration of a salt (or another crystallising agent) may help to avoid collapse. Moreover, in certain cases, freeze-drying may be performed well above the collapse temperature, without visual macroscopic collapse. Such an approach has been considered, using the water/NaCl/sucrose system as a model.[1]

The collapse temperature as a function of sucrose/NaCl ratio is shown in Figure 3. Line *ab* represents the collapse temperature of the freeze-concentrated solution after water crystallisation, and line *cc'* represents the collapse temperature after primary water and secondary $<NaCl*H_2O>$ +water crystallisation. Given that freeze-drying of solutions with collapse temperatures below $-40\,°C$ is usually impractical, the sucrose/NaCl ratio should be $>10:1$, in order to have a formulation with a collapse temperature higher than $-40\,°C$ (as in Figure 3).

Consider a formulation with a sucrose/NaCl ratio of 6:1 (composition 1 in Figure 3) as an example. The collapse temperature of this formulation is lower than $-40\,°C$, and in order to avoid collapse during freeze-drying, the collapse temperature should be raised by increasing the sucrose/NaCl ratio, as indicated by the arrow from 1 to 1'. Another possible way to avoid collapse, by increasing the concentration of NaCl, is indicated by the arrow from 1 to 1'' in Figure 3. In that case, increasing the concentration of a crystallising solute (*e.g.* NaCl) promotes solute crystallisation. The solute crystals provide a physical support for the amorphous phase, and thus macroscopic collapse of the sample does not occur, even when the temperature is well above the collapse temperature of the amorphous phase. To illustrate this approach, four solutions with identical total solid concentrations and fill volumes, but with different $R=$ sucrose/NaCl ratios, were freeze-dried at a primary drying temperature of $-40\,°C$. Table 1 shows the phase compositions and collapse temperatures of those solutions. The solution with $R=5$ collapsed during freeze-drying, whereas the solutions with $R=1$ and 0.1 did not collapse.[1] Note that the T_g' for the R = 1 and 0.1 solutions was lower than both the primary drying temperature and the T_g' for the solution with $R=5$ (Table 1). These results demonstrate that a partially crystalline material can be freeze-dried above T_g', without collapse.

To avoid structural collapse in partially crystalline formulations, they should contain a significant amount of crystalline phase. There are no systematic data for critical fractions of crystalline phases, which are necessary to avoid collapse. However, according to Table 1, a weight ratio of amorphous/crystalline $=2.2$ (solution with $R=1$) was sufficient to prevent collapse, but it is possible that the necessary crystalline fraction may be significantly lower than that. It should be added that, while NaCl is an example of a crystallisable component, in practice,

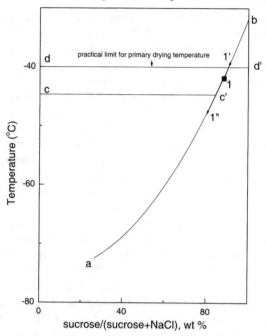

Figure 3 T_g' *of freeze-concentrated solutions as a function of sucrose/(sucrose + NaCl) ratio.*
T_g' *has been determined by DSC as the temperature of an endothermic step that*
precedes the melting endotherm in DSC heating curves for frozen solutions.[5] The
collapse temperature, as measured by freeze-drying microscope, is usually within
1–3 °C of the calorimetric T_g'.[36] Line ab represents T_g' of the maximally freeze-
concentrated solution after primary water crystallisation, and line cc' represents
T_g' *of the maximally freeze-concentrated solution after primary water and second-*
ary $(NaCl*H_2O + water)$ crystallisation

use of NaCl for such a purpose may be inadvisable, because of the high tonicity of
such a solution. However, this approach can be used with other crystallisable
solutes, *e.g.* glycine or mannitol.

3.2 Drying Rates

The impact of partial crystallisation of a bulking agent on the rate of dehydration
has been described previously for a water/glycine/sucrose system.[13] Three solutions with glycine/sucrose ratios (R) of 0.1, 0.77 and 1.8 were studied. In the first
two solutions, only water crystallised, resulting in a completely amorphous
solute matrix, whereas partial glycine crystallisation occurred in the third formulation, resulting in a partially crystalline solute matrix. The residual water
content was measured after 7, 18 and 25 hours of freeze-drying (data shown in
Table 2). The water content was consistently lower for the partially crystalline
sample than for the two completely amorphous samples, demonstrating that the
partially crystalline sample had a significantly higher drying rate. To understand
the observed differences in drying rates, phase compositions of the frozen and

Table 1 *Phase compositions (weight %) and collapse temperatures of frozen water/ sucrose/NaCl solutions containing 5 wt% total solids at different sucrose/ NaCl ratios*

	$R = sucrose/NaCl$ ratio			
Phase	10	5	1	0.1
Ice	93.9	93.8	93.9	94.9
$NaCl*H_2O$	0	0	1.9	4.6
Amorphous	6.1	6.2	4.2	0.5
Composition (wt%) of the amorphous phase				
Water	22.3	25.6	26.8	26.8
NaCl	7.1	12.4	26.8	26.8
Sucrose	70.6	62	55.6	55.6
T_g', °C	−41.2	−45.5	−45.9	−45.9

freeze-dried samples were determined (data shown in Table 3), based on the solid-liquid state diagram for the water/glycine/sucrose system.[13] It can be seen that the amount of amorphous phase in the sample with $R = 1.8$ was about 2.5 times lower than in the other two samples. This fact alone may have explained the higher drying rate for the partially crystalline sample. The residual water contents in the completely amorphous samples after different drying times were 1.5 to 2.4 times higher than that in the partially crystalline sample, which correlates with the relative amounts of the amorphous phase. It should be stressed that, for a partially amorphous sample, the local water content in the amorphous phase would be significantly higher than the overall residual water content, as measured by Karl Fischer titration or other appropriate methods. For example, the partially crystalline formulation with 6 wt% water contained 15 wt% water in the amorphous phase (Table 3), which might represent a major destabilising factor. In partially crystalline formulations, both overall water content and amount of crystalline phase should be measured, so that product quality can be ensured. This is an obvious suggestion, but one that is often ignored in the practice of pharmaceutical freeze-drying.

An opposite trend has been reported for a phenylalanine/sucrose system, wherein crystallisation of phenylalanine reduced the drying rate.[17] This was explained on the basis of the formation of lamellar-like crystals of phenylalanine, which created a diffusion barrier for water, resulting in a corresponding decrease

Table 2 *Residual water contents (wt%) in glycine/sucrose samples of different glycine/sucrose ratio (R), as a function of freeze-drying time*

R	7 hours	18 hours	25 hours
0.1	19.2 ± 1.0	6.0 ± 0.3	1.3 ± 0.2
0.77	17.0 ± 1.0	9.0 ± 1.5	1.6 ± 0.2
1.8	8.0 ± 1.0	4.0 ± 0.5	0.8 ± 0.4

Table 3 *Phase compositions (wt%) of frozen solutions and lyophilised samples, for water/glycine/sucrose systems with different glycine/sucrose ratios*

R = glycine/ sucrose	Frozen solution (10 wt% solute)			Freeze-dried material (6 wt% water)		
	ice	glycine	amorphous	glycine	amorphous	water content in amorphous phase
0.1	86.9	0	13.1	0	100	6
0.77	87.9	0	12.1	0	100	6
1.8	89.2	5.8	4.9	56.3	43.7	15

in drying rate. An alternative explanation may be suggested: phenylalanine forms a crystal hydrate with a lower vapour pressure than that of its amorphous state, thus reducing the thermodynamic driving force for dehydration.

3.3 Stability of Proteins

The significance of crystallisation of a bulking agent to the stability of freeze-dried biologicals has been considered in several publications. It was shown, for example, that crystallisation of a bulking agent in an anhydrous form during storage resulted in significant loss of activity of proteins, because of an increase in the water content of the remaining amorphous phase.[18] In another study, it was shown that crystallisation of the bulking agent mannitol during freeze-drying compromised the stability of protein formulations.[19] As a rule, crystallisation of a bulking agent is undesirable, because molecules of the bulking agent are thereby removed from any association with protein molecules, thus eliminating any cryo- and/or lyoprotective action of the excipient. In addition, such crystallisation increases the local concentration of water in the amorphous phase, which results in a major destabilising factor. It should be mentioned, however, that crystallisation of a bulking agent in the form of a crystal hydrate may be beneficial, as the water content of the remaining amorphous phase might thus be lowered.[20]

Crystallisation of a solute can occur at different stages of a freeze-drying process, and the properties and quality of the lyophilised product may depend on the conditions of such crystallisation, *i.e.* whether the crystallisation takes place during cooling, annealing, primary drying or secondary drying. A possible example of the significance of crystallisation conditions was given by Shalaev,[13] who reported on the stability of a freeze-dried conjugate of immunoglobulin G and horseradish peroxidase in a partially crystalline glycine/sucrose matrix. Two types of freeze-dried materials were prepared, which had identical chemical compositions but were frozen under different conditions. The solutions were cooled at two different rates: one formulation was cooled at a rate of 10 °C min^{-1}, which was higher than the critical cooling rate for glycine crystallisation (5 °C min^{-1}); the other was cooled at a rate of 0.5 °C min^{-1}, which was lower than the critical cooling rate. In the latter case, glycine crystallised during cooling, whereas in the former case, the crystallisation occurred during warming, when the product was heated above its devitrification temperature. Both materials

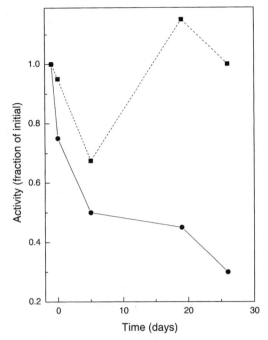

Figure 4 *Activity of a horseradish peroxidase/immunoglobulin G conjugate, in a freeze-dried formulation with sucrose and glycine, as a function of storage time at 37 °C. Squares: cooling rate was 0.5 °C min⁻¹; circles: cooling rate was 10 °C min⁻¹. Critical cooling rate for secondary crystallisation of glycine was 5 °C min⁻¹. Lines are shown as a visual aid*

were lyophilised simultaneously. Activity was measured, using an enzyme-linked immunosorption assay, before and after lyophilization, and during storage at 37 °C. Figure 4 shows the activity of the conjugate as a function of storage time. It can be seen that, for the material in which crystallisation occurred during cooling, the activity was significantly higher than in the material that crystallised during warming. This study indicated that the stage at which crystallisation of a bulking agent occurs may be an important parameter that impacts the stability of freeze-dried protein formulations. However, more detailed studies are needed to confirm those results and provide an understanding of the mechanism of the detrimental effect of devitrification on the activity of proteins.

4 Crystallisation of a Solute in Multi-component Frozen Solutions. Influence of a Second Solute on Solute 1 + Water Secondary Crystallisation

Depending on the properties of any particular drug molecule or dosage form, crystallisation of a solute may be either desirable or undesirable. For example, if a drug molecule is chemically unstable, a crystalline form is preferable, because of the generally higher stability of crystalline materials. On the other hand, if either the solubility or rate of dissolution of a crystalline drug is insufficient, then an amorphous form is the one of choice. As a rule, crystallisation of a buffer component is undesirable, whereas a bulking agent is usually preferred in a crystalline form, unless it also acts as a lyoprotector. The crystallisation behaviour of the solute of interest may be modified by adding another solute. From a general perspective, the addition of a second solute may either hinder, promote, or have little influence on the crystallisation of the first solute. Evidently, all three cases have been reported in the literature. In this section, we consider several examples of the influence of a second solute on the crystallisation of a first solute.

There are many cases of inhibition of crystallisation of a readily crystallisable solute (solute 1) by a second solute. For example, sucrose hinders the crystallisation of NaCl or glycine.[1,9–11] In both sucrose/NaCl/water and sucrose/glycine/water systems, neither glycine nor NaCl crystallised completely. The freeze-concentrated solution, after secondary water + solute 1 crystallisation, contained both crystallisable (*i.e.* either glycine or NaCl) and non-crystallisable (*i.e.* sucrose) solutes. The sucrose/(sucrose + solute 1) ratio in the maximally freeze-concentrated solution, after primary water and secondary water + solute 1 crystallisation, corresponded to composition W_{gE}' in Figure 2. This ratio was similar in both systems: 0.765 ± 0.075 in sucrose/NaCl/water and 0.785 ± 0.105 in sucrose/glycine/water. Note that solutions containing this or a higher sucrose/(sucrose + solute 1) ratio did not form crystalline solute 1, regardless of any thermal treatment of the frozen solutions. In other words, the freeze-concentrated solution, after primary water crystallisation, formed a kinetically stable (but thermodynamically unstable) glass. Sucrose at lower concentrations may suppress crystallisation of both glycine and NaCl, depending on cooling/warming rates and annealing treatments. However, if the sucrose content is less than that for the ratios shown above, then crystallisation of solute 1 may be expected under certain circumstances, *i.e.* a doubly unstable glass is formed during freezing of such solutions.

There have been other studies demonstrating that sucrose and other sugars hinder the crystallisation of salts (buffers) and other compounds that readily crystallise from binary mixtures with water. For example, sucrose inhibited the crystallisation of mannitol and lysine*HCl, at sucrose/solute ratios of 2:1 and 1:2, respectively.[21] Sorbitol inhibited the crystallisation of mannitol, when present at sorbitol:mannitol ratios $\geq 3:2$.[22] In a study of mannitol crystallisation in the presence of several non-crystallising co-solutes (sucrose, lactose, maltose, trehalose, dextran, lysozyme), it was found that the relative concentration threshold for mannitol was approximately mannitol/(mannitol + co-solute) = 0.3, independent of the nature of the second solute.[23] This was similar to the behaviour of sucrose/glycine/water and sucrose/NaCl/water mixtures, in which the critical threshold concentration of sucrose was approximately 0.2 for both systems (as discussed earlier).

Note that an absence of secondary crystallisation of a solute, which has been reported in the cited studies, does not necessarily mean that secondary crystallisation would not occur under other experimental conditions, *e.g.* with longer annealing. For example, a ternary solution of NaCl/glycine/water, which did not show any solute crystallisation during a cooling/warming scan, crystallised as a result of annealing.[24] A more rigorous determination of the boundary between kinetically stable and doubly unstable glasses formed after primary water crystallisation (*i.e.* point W_{gE}' in Figure 2) should be made, based on an analysis of the solid–liquid state diagram, as described in Section 2.

A very different result was observed for cefazolin/sugar mixtures, where sugars appeared to have little impact on the crystallisation of cefazolin.[25] Crystallisation of cefazolin in a binary water/cefazolin system occurred during heating of frozen solutions. In the presence of maltose and other sugars, the devitrification temperature increased slightly with increasing sugar concentration. However, the enthalpy of crystallisation was independent of sugar content, indicating that cefazolin crystallised completely in the presence of sugars. This complete crystallisation of cefazolin in ternary cefazolin/sugar/water solutions may suggest a liquid–liquid phase separation in the freeze-concentrated solution, after primary water crystallisation. Indeed, DSC scans for ternary frozen solutions (see Figure 1[25]) had a more complex appearance, possibly showing two T_g' steps, which might have been an indication of liquid–liquid demixing. Such demixing may produce two liquid phases, one mainly cefazolin-rich and the other sugar-rich, and crystallisation from the cefazolin-rich phase would be expected to be similar to that for the binary cefazolin/water system. It should also be noted that the study by Oguchi *et al.*[25] was performed with very concentrated solutions, containing 30% cefazolin (for which the eutectic composition is approximately 12%[26]), and that supersaturation of cefazolin is achieved before water crystallisation, with the possibility of partial crystallisation of cefazolin prior to ice formation. Such a possibility of primary crystallisation of cefazolin may introduce a further complication to the interpretation of experimental data. The impact of liquid–liquid demixing in freeze-concentrated solutions on the inhibition of crystallisation of polyethylene glycol (PEG) by different sugars and polymers has been reported.[27] In particular, it was demonstrated that mono- and disaccharides, which were relatively less miscible with PEG, inhibited crystallisation of PEG to a lesser degree.

Promotion of solute crystallisation in frozen solutions by a second solute has been reported for several drug molecules. For example, Korey and Schwartz[28] reported that 100% crystalline lyophilised cakes were obtained, when atropine sulfate, cefoxitin sodium or doxycycline hydrate was co-lyophilised with glycine, and the glycine content was above a certain critical value. For glycine contents below the critical level, partially crystalline materials were formed. In the absence of glycine, these freeze-dried drugs were completely amorphous. Similar results were observed in the presence of several other amino acids, urea and niacinamide.[28] In another study, the crystallisation of a cephalosporin compound was enhanced by the use of a relatively small amount of a co-solvent, isopropyl alcohol.[29]

Secondary crystallisation of a solute may produce either a pure crystalline solute or its crystal hydrate. There are many examples of crystal hydrates formed by both excipients and drug molecules during lyophilization, *e.g.* mannitol,[30] NaCl,[31] phosphate buffer[32] and cephalosporins.[33] In the majority of cases, crystal hydrates that were formed during a freezing step lost their water of hydration during subsequent drying under vacuum. Removal of water of hydration may produce an anhydrous crystal, *e.g.* in the case of mannitol or NaCl. On the other hand, there are cases where removal of water of hydration produces an amorphous material. Amorphisation as a result of loss of water of hydration has been reported for sugars,[34,35] cephalosporins[33] and phosphate buffer.[32] Such formation of an amorphous state from a crystal hydrate during drying is usually undesirable, because it may compromise the chemical or physical stability of a lyophilised formulation.

5 Conclusion

One may suggest a number of different ways to explain experimental results on solute crystallisation in frozen solutions. For example, inhibition of NaCl or glycine crystallisation by sucrose may be explained in terms of an increased viscosity of the freeze-concentrated solution in the presence of sucrose. Another possible explanation is that, in the presence of sucrose, the freeze-concentrated solution becomes dilute with regard to solute 1, which results in increasing diffusion distances between solute 1 molecules, and hence a slowing down of crystal growth. To explain facilitation of crystallisation of certain drug molecules by a second solute, one may consider the epitaxial growth of crystals of a drug, together with crystals of an excipient. When a second solute does not crystallise (*e.g.* crystallisation of cephalosporin in the presence of isopropyl alcohol), the enhanced crystallisation of the cephalosporin may be explained on a thermodynamic basis, if we assume that the solubility of cephalosporin in a freeze-concentrated solution is reduced, due to the presence of the co-solvent. However, all of these suggestions are purely qualitative and have no quantitative predictive power. In order to be able to predict the crystallisation behaviour of a solute during freeze-drying, one needs data from detailed systematic studies of model systems. A mechanistic understanding of crystallisation in frozen solutions should derive from the analysis of a combination of thermodynamic and kinetic factors, including the thermodynamic driving force for crystallisation (*i.e.* extent of supersaturation), and features of nucleation and crystal growth. A knowledge of phase diagrams would be essential in such studies, because the thermodynamic driving force for crystallisation depends on the extent of supersaturation, and the only way to express that parameter is to use phase diagrams that describe the composition of phases in equilibrium, as a function of temperature. It should also be noted that another important problem, concerning the stability of crystal hydrates under conditions of freeze-drying, can be considered with the aid of pressure–temperature–composition phase diagrams. Such a diagram has been applied[1] to express the thermodynamic driving force for water removal from amorphous phases and crystal hydrates. We believe that significant progress in the understanding of the scientific basis of freeze-drying and the design of freeze-dried dosage forms will be achieved, when such state diagrams are applied to describe and predict the behaviour of model and real systems.

References

1. E.Y. Shalaev and F. Franks, *Cryobiology*, 1996, **33**, 14.
2. C.A. Angell, E.J. Sare, J. Donnella and D.R. MacFarlane, *J. Phys. Chem.*, 1981, **85**, 1461.
3. D. Rasmussen and B. Luyet, *Biodynamica*, 1969, **10**, 319.
4. A.P. MacKenzie, *Dev. Biol. Stand.*, 1977, **36**, 167.
5. H. Levine and L. Slade, *J. Chem. Soc. Faraday Trans.* 1, 1988, **84**, 2619.
6. F. Franks, *Cryo-Letters*, 1990, **11**, 93.
7. V.J. Anosov, M.I. Ozerova and Y.A. Fialkov, 'The Basis of Physicochemical Analysis', Nauka, Moscow, 1976 (in Russian).
8. M. Hillert, 'Phase Equilibria, Phase Diagrams and Phase Transformations', Cambridge University Press, Cambridge, 1998.
9. E.Y. Shalaev and A.N. Kanev, *Cryobiology*, 1994, **31**, 374.
10. E.Y. Shalaev, F. Franks and P. Echlin, *J. Phys. Chem.*, 1996, **100**, 1144.
11. T. Suzuki and F. Franks, *J. Chem. Soc. Faraday Trans.*, 1993, **89**, 3283.
12. V. Andronis and G. Zografi, *J. Non-Cryst. Solids*, 2000, **271**, 236.
13. E. Shalaev, PhD Thesis, Koltsovo, Russia, 1991.
14. E.Y. Shalaev and A.I. Vavilin, unpublished results, 1990.
15. F. Franks, 'Biophysics and Biochemistry at Low Temperatures', Cambridge University Press, Cambridge, 1985.
16. M.J. Pikal, A.L. Lukes, J.E. Lang and K. Gaines, *J. Pharm. Sci.*, 1978, **67**, 767.
17. C. Roth, G. Winter and G. Lee, *J. Pharm. Sci.*, 2001, **90**, 1345.
18. K. Izutsu, S. Yoshioka and S. Kojima, *Pharm. Res.*, 1994, **11**, 995.
19. K. Izutsu, S. Yoshioka and T. Terao, *Pharm. Res.*, 1993, **10**, 1232.
20. B.J. Aldous, A.D. Auffret and F. Franks, *Cryo-Letters*, 1995, **16**, 181.
21. B. Lueckel, D. Bodmer, B. Helk and H. Leuenberger, *Pharm. Dev. Technol.*, 1998, **3**, 325.
22. K. Ito, *Chem. Pharm. Bull.*, 1971, **19**, 1095.
23. A.I. Kim, M.J. Akers and S.L. Nail, *J. Pharm. Sci.*, 1998, **87**, 931.
24. M.J. Akers, N. Milton, S.R. Byrn and S.L. Nail, *Pharm. Res.*, 1995, **12**, 1457.
25. T. Oguchi, E. Yonemochi and K. Yamamoto, *Pharm. Acta Helv.*, 1995, **70**, 113.
26. L. Gatlin and P.P. DeLuca, *J. Parent. Drug Assoc.*, 1980, **34**, 398.
27. K. Izutsu, S. Yoshioka, S. Kojima, T.W. Randolph and J.F. Carpenter, *Pharm. Res.*, 1996, **13**, 1393.
28. D.J. Korey and J.B. Schwartz, *J. Parent. Sci. Technol.*, 1989, **43**, 80.
29. Y. Koyama, M. Kamat, R. DeAngelis, R. Srinivasan and P.P. DeLuca, *J. Parent. Sci. Technol.*, 1988, **42**, 47.
30. L. Yu, N. Milton, *e.g.* Groleau, D.S. Mishra and R.E. Vansickle, *J. Pharm. Sci.*, 1999, **88**, 196.
31. E.Y. Shalaev and F. Franks, *Thermochim. Acta*, 1995, **255**, 49.
32. A. Pyne, K. Chatterjee and R. Suryanarayanan, *AAPS Pharm. Sci.*, 2001, **3**, No. 3.
33. R.K. Cavatur and R. Suryanarayanan, *Pharm. Dev. Technol.*, 1998, **3**, 579.
34. K. Kajiwara, F. Franks, P. Echlin and A.L. Greer, *Pharm. Res.*, 1999, **16**, 1441.
35. A. Saleki-Gerhardt, J.G. Stowell, S.R. Byrn and G. Zografi, *J. Pharm. Sci.*, 1995, **84**, 318.
36. M.J. Pikal and S. Shah, *Int. J. Pharm.*, 1990, **62**, 165.

Miscibility of Components in Frozen Solutions and Amorphous Freeze-dried Protein Formulations

Ken-ichi Izutsu and Shigeo Kojima

NATIONAL INSTITUTE OF HEALTH SCIENCES, TOKYO, JAPAN

1 Introduction

Sugars and polyols have been widely used to protect proteins from structural change and inactivation during freezing and freeze-drying.[1,2] Many freeze-dried protein formulations are multi-component systems that contain a protein, stabilizers, salts and other excipients. Freezing of an aqueous solution concentrates the solutes into an unfrozen, supercooled solution that exists among ice crystals. Recent studies have shown that various solutes in aqueous solutions are freeze-concentrated into a single amorphous phase or into multiple phases, depending on solute combinations and concentration ratios.[3–9] Solute miscibility in frozen solutions and freeze-dried solids is getting increasing attention as an important factor for protein stability during formulation processes and subsequent storage. This report introduces analytical methods, examples, and possible effects of multiple solute phases in frozen solutions.

2 Separation of Solutes in Frozen Solutions

Ice formation causes various solutes to concentrate to as high as 70–80% (w/w) in supercooled solutions, regardless of their concentrations in initial solutions. Solute miscibility in a concentrated aqueous solution depends on a variable thermodynamic balance of molecular interactions between components and entropic energy gain on mixing.[10] Many solute combinations (e.g. polymers, polymer and salt) separate in a freeze-concentrate, basically by the same thermodynamic mechanism that causes aqueous two-phase separation at room temperature.[5] Solute molecules with weak repulsive interactions are miscible in dilute initial solutions, whereas the same combinations separate into different phases in frozen solutions, because the increased concentration crosses a bimodal curve into a two-phase region of a system's phase diagram. Other solute combinations are concentrated as compatible mixtures, because the mixing is

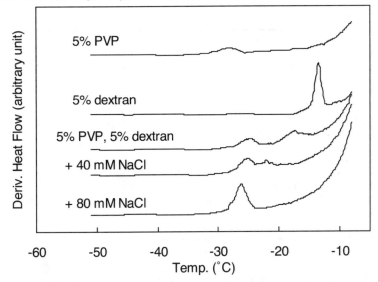

Figure 1 *DSC derivative thermograms for frozen polymer solutions. Aliquots (10 μl) of frozen solutions in aluminum cells were scanned from −100 °C at 5 °C min⁻¹.*

thermodynamically favorable and/or increased viscosity kinetically prevents phase separation. Freeze-concentration induces solute immiscibility, even from initial solutions much lower in concentration than that required for aqueous two-phase separation. In contrast to the separation of some crystallizing solutes (*e.g.* salts, mannitol) in multi-solute frozen solutions and freeze-dried solids, the significance of amorphous solute miscibility has not been recognized until recently.

Thermal analysis of frozen sugar or polymer solutions by differential scanning calorimetry (DSC) shows a characteristic transition that shifts the baseline of a thermogram (T_g': glass transition temperature of maximally freeze-concentrated solution).[11] Because each single-solute frozen solution has an intrinsic T_g' that does not depend on initial concentration, T_g' transitions for multi-solute systems provide valuable information on solute miscibility in frozen solutions. Figure 1 shows DSC derivative thermograms for frozen solutions containing a single polymer or polymer combinations. The T_g' transition is observed as a sharp peak in derivative thermograms.[11] Polyvinylpyrrolidone (PVP, 10 k), dextran (11 k) and DEAE-dextran (500 k) showed T_g' peaks at approximately −28, −13 and −26 °C, respectively. A combination of dextran and DEAE-dextran (5% w/w each) shows a single T_g' peak in between the two component T_g's (*i.e.* at −21 °C), indicating freeze-concentration into a single mixed-solute phase. Two T_g' peaks, for a frozen solution of dextran and PVP combined, indicated separation into individual freeze-concentrated phases.[12] Miscibility of polymer combinations in frozen solutions, determined by thermal analysis, is summarized in Table 1.

Small molecule and/or ionic co-solutes greatly affect polymer miscibility in frozen solutions. Many salts decrease the miscibility of polyelectrolytes and

Table 1 *Miscibility of polymers in frozen solutions*

Miscible (single mixture phase)
   DEAE-dextran/dextran
   dextran sulfate/dextran
   *PAANa/dextran
   PVP/ovalbumin
   Ficoll/ovalbumin
   Ficoll/ovalbumin + NaCl
Phase separation (two amorphous phases)
   PVP/dextran
   Ficoll/dextran
   PAANa/PVP
   PVP/ovalbumin + NaCl
Phase separation (crystal and amorphous phases)
   PEG/PVP
   PEG/dextran

*PAANa: poly(acrylic acid, sodium salt)

non-ionic polymers at relatively low concentrations, by shielding attractive electrostatic interactions. Relatively higher concentrations of salts alter polymer miscibility by changing a polymer's hydration state. For example, some salting-in salts (*e.g.* NaSCN) improve the miscibility of non-ionic polymers in a freeze-concentrate.[7]

3   Effects of Component Miscibility on Formulation Quality

Many polymer excipients are included in protein formulations in which they are intended to stabilize protein conformation and freeze-dried cake structure. Miscibility of proteins and polymer excipients in frozen solutions varies, depending on specific polymer combinations and co-solute compositions. Some spherical proteins (lysozyme, ovalbumin, BSA) and PVP can be freeze-concentrated into a single mixed-solute phase in the absence of salts, whereas addition of salts causes separation of the polymers into different amorphous phases, as illustrated in Figure 2. Many protein solutions do not show an obvious T_g' peak (located at approximately $-10\,^{\circ}\mathrm{C}$ by other analytical methods) in DSC derivative thermograms. Such proteins are generally more miscible with Ficoll than with PVP. Other factors, such as pH and freezing rate, would also affect solute miscibility. A given degree of solute miscibility in frozen solutions is likely to be maintained during a freeze-drying process. Increased viscosity should provide only a small opportunity for further immiscibility, but not for mixing, during freeze-drying.

Solute miscibility in frozen solutions and freeze-dried solids would affect the quality of a protein formulation, mainly in two ways: (i) by altering protein/excipient molecular interactions required to maintain protein conformation during freeze-drying processes; and (ii) by altering the molecular mobility of an amorphous freeze-dried solid, which determines the chemical stability of a protein

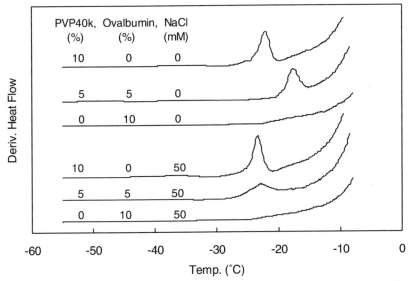

Figure 2 *Effect of NaCl on DSC derivative thermograms for frozen PVP and ovalbumin solutions*

during storage. Development of various recombinant protein pharmaceuticals requires the rational design of stable formulations. Monitoring and controlling component miscibility, in both frozen solutions and freeze-dried solids, should be important for stable freeze-dried formulations. Further study, using other analytical methods (*e.g.* SEM), would help our understanding of the physical properties (*e.g.* phase size) of multiple amorphous phases in frozen solutions.

References

1. F. Franks, *Eur. J. Pharm. Biopharm.*, 1998, **45**, 221.
2. M.J. Pikal, 'Freeze-drying/Lyophilization of Pharmaceutical and Biological Products', L. Rey and J.C. May (eds.), Marcel Dekker, New York, 1999, Chapter 6, 161.
3. L.M. Her, M. Deras and S.L. Nail, *Pharm. Res.*, 1995, **12**, 768.
4. M.C. Heller, J.F. Carpenter and T.W. Randolph, *J. Pharm. Sci.*, 1996, **85**, 1358.
5. T.W. Randolph, *J. Pharm. Sci.*, 1997, **86**, 1198.
6. K. Izutsu, S. Yoshioka, S. Kojima, T.W. Randolph and J.F. Carpenter, *Pharm. Res.*, 1996, **13**, 1393.
7. K. Izutsu, M.C. Heller, T.W. Randolph and J.F. Carpenter, *J. Chem. Soc., Faraday Trans.*, 1998, **94**, 411.
8. K. Izutsu and S. Kojima, *Phys. Chem. Chem. Phys.*, 2000, **2**, 123.
9. K. Izutsu and S. Kojima, *Pharm. Res.*, 2000, **17**, 1316.
10. S.L. Shamblin and G. Zografi, *Pharm. Res.*, 1998, **15**, 1828.
11. H. Levine and L. Slade, *J. Chem. Soc., Faraday Trans. 1*, 1988, **84**, 2619.
12. L. Slade and H. Levine, *Crit. Rev. Food Sci. Nutr.*, 1991, **30**, 115.

Investigations into the Amorphous and Crystalline Forms of a Development Compound

D. O'Sullivan, G. Steele* and T.K. Austin

PREFORMULATION TEAM, PHARMACEUTICAL AND
ANALYTICAL R&D, ASTRAZENECA R&D CHARNWOOD,
BAKEWELL ROAD, LOUGHBOROUGH, LEICESTERSHIRE LE11 5RH,
ENGLAND

1 Introduction

During initial development of the compound under investigation, a sodium salt, the final stage of production involved freeze-drying to produce bulk supplies of the drug. However, this process produced an amorphous solid that was chemically unstable at ambient temperature. As a result of this instability, the bulk drug had to be stored at $-20\,°C$ to prevent its decomposition. Since this was not practical on a large scale, investigations were conducted in an attempt to produce a crystalline form of the compound, which it was hoped would be more stable. Unfortunately, conventional crystallisation approaches failed to produce suitable material; however, it was found that exposure of the amorphous solid to high relative humidities ($>80\%$ RH) induced crystallisation.

The amorphous and crystalline phases of the compound were characterised using a variety of techniques, *e.g.* X-ray powder diffraction (XRPD), dynamic vapour sorption (DVS), and isothermal microcalorimetry (IM) using a thermal activity monitor (TAM). XRPD, IM and solution calorimetry (SOLCAL) were then used in the construction of calibration curves of crystalline and amorphous mixtures of the drug. The assays were developed in order to quantify amounts of each component in binary-phase mixtures.

2 Results

2.1 Characterisation

XRPD confirmed that crystallisation occurred on exposure of the compound to high levels of moisture, and Figure 1 shows the diffractograms of the amorphous and crystalline forms before and after exposure to 84% RH.

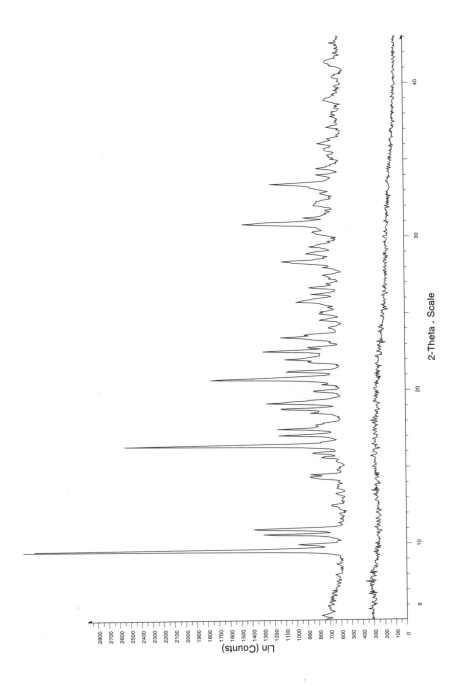

Figure 1 *XRPD patterns of the amorphous and crystalline forms of the development compound*

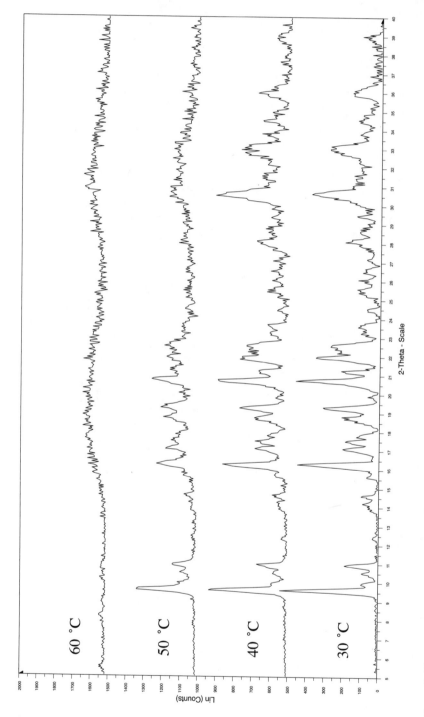

Figure 2 *Temperature-XRPD patterns obtained by heating the crystalline hydrate*

Thermogravimetric analysis (TGA) showed that the amorphous and crystalline phases contained ~ 8 and $\sim 16\%$ w/w water, respectively. In both forms of the compound, the water was thermally labile, and variable temperature-XRPD experiments showed that the crystalline hydrate could be converted back to the amorphous state by heating it to approximately 60 °C. Figure 2 shows the XRPD pattern of the crystalline hydrate obtained by *in situ* heating experiments.

Figure 3 shows the DVS plot for the compound. In this experiment, the amorphous solid sorbed approximately 12% w/w moisture at a constant rate from 0 to 60% RH. Between 60 and 70% RH, a short plateau region was observed, which preceded a further 6% w/w increase in moisture uptake (to 19% w/w total moisture content) between 70 and 80% RH. The final change was a surprising decrease in mass to 12% w/w moisture between 80 and 95% RH. On reduction of the RH, no change in mass was observed, until $\sim 12\%$ RH was reached. Below this RH, the hydrate water in the sample was rapidly lost, to give a value less than that measured in the starting material. XRPD analysis showed that this material was amorphous.

A TAM ramp-RH experiment (illustrated in Figure 4) was carried out in a manner analogous to the DVS experiment, whereby a sample was subjected to an atmosphere of increasing RH. After an equilibration period, increasing the RH resulted in an increase in the heat flow. This continued up to $\sim 40\%$ RH, after which the heat flow decreased slightly from 40–70% RH. At this RH, which corresponded to the plateau region observed in the DVS experiment, a definite decrease in the heat flow was recorded. It is probable that this corresponded to structural collapse of the sample. As the RH was increased further, the heat output then increased to a second peak, decreased slightly, and then reached a sharp maximum, corresponding to crystallisation of the sample. A small secondary peak was observed, as the heat output returned to baseline.

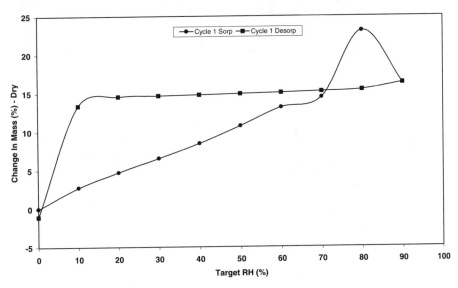

Figure 3 *DVS isotherm plot*

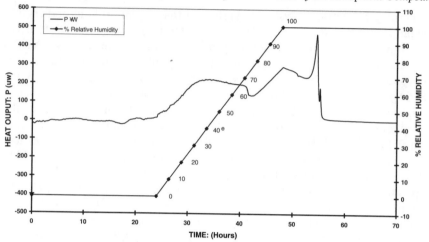

Figure 4 *TAM ramp-RH experiment*

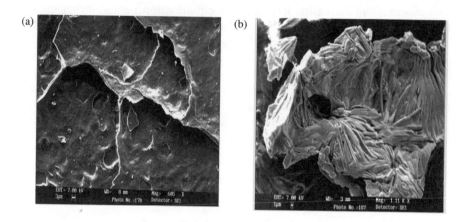

Figure 5 *SEM images of the development compound – (a) amorphous, (b) crystalline*

Figure 5(a) and (b) shows the scanning electron microscopy (SEM) images of the compound before and after exposure to 84% RH. In Figure 5(a), the appearance of the material is typical of that seen for freeze-dried materials, *i.e.* it shows pores from which the water had sublimed during the primary freeze-drying stage. It can also be seen that the amorphous form had a glassy appearance, compared to the more structured crystalline form (Figure 5(b)).

2.2 Quantitative Analysis

In order to be able to carry out a quantitative analysis of mixtures of amorphous and crystallisation phases of the compound, samples ranging from 0 to 100%

w/w amorphous content were prepared by mixing known proportions of the components by weight. The mixtures were then analysed by XRPD, SOLCAL and IM.

2.2.1 XRPD. Figure 6(a) shows the diffractograms of the amorphous/crystalline mixtures, and Figure 6(b) shows the area of the largest peak at 9° 2θ *versus* the % crystallinity of the sample. As can been seen from Figure 6(a), the peak at 9° 2θ was still detectable in mixtures containing 2% w/w crystalline component in 98% amorphous. The plot of the area of this peak *versus* the % crystallinity of the samples was reasonably linear up to 60% w/w crystalline component in amorphous, with a correlation coefficient of 0.977 being measured.

2.2.2 SOLCAL. The SOLCAL is used in conjunction with the TAM apparatus. In this technique, approximately 3 mg of sample is dissolved in water, and

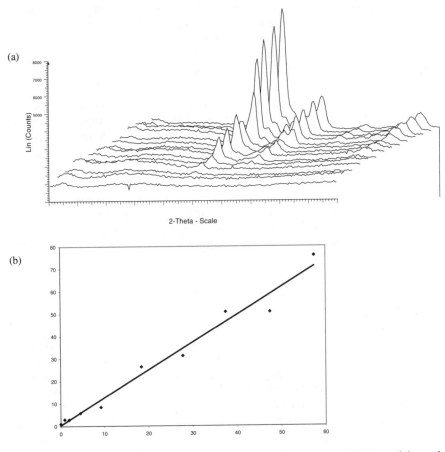

Figure 6 (*a*) *XRPD diffractograms of amorphous/crystalline mixtures*; (*b*) *Area of the peak* 9° 2θ versus % crystallinity

heat conduction is measured. Figure 7 (a) shows the results of heat output *versus* % crystallinity. Two types of behaviour were observed. In the case of the amorphous sample, dissolution was exothermic, whilst that of the crystalline sample was endothermic. The calibration curve was constructed using the cumulative heat flow. Figure 7 (b) shows the enthalpy of solution versus % crystallinity of the mixtures. The plot obtained was linear, with a correlation coefficient of 0.921 being measured.

2.2.3 Isothermal Microcalorimetry.   IM analysis of the moisture sorption and crystallisation of the compound was carried out using the ampoule method described by Briggnar *et al.*[1] In this technique, the compound was exposed to 75% RH (saturated sodium chloride solution), 84% RH (saturated potassium chloride solution) or 100% RH (water) in sealed ampoules, and the amount of heat liberated on crystallisation was measured.

Figure 8(a) shows the profiles obtained, when the pure amorphous phase was exposed to these conditions. In all cases, the initial thermal events corresponded to a cumulative of instrumental artefacts and surface sorption of moisture. The profile at 75% RH showed no crystallisation event, even after 30 hours. However, in the case of the 84 and 100% RH samples, crystallisation exotherms were detected after ~ 15 hours and 5 hours, respectively. Therefore, from these data, it is clear that there is a critical RH between 75 and 84%, where crystallisation will be initiated. Figure 8(b) shows the heat output of the three samples *versus* % crystallinity. The occurrence of two peaks during the crystallisation of amorphous materials is a relatively common phenomenon representing the crystallisation and water desorption events seen in the DVS experiments.[2]

2.2.4 Stability.   A short stability study at 40 °C (in sealed ampoules) showed that the crystalline form was indeed more stable than the amorphous form of the compound. For example, after 6 months' storage, the crystalline and amorphous forms showed 0.1 and 1.6% increases in impurities, respectively.

3 Discussion

3.1 Characterisation

Although the compound could not be crystallised using conventional techniques, XRPD studies showed that exposure of the amorphous form to humid environments of > 80% RH resulted in the formation of a crystalline hydrate.

Variable-temperature XRPD showed that dehydration led to the formation of the amorphous phase above 60 °C. The ease with which the water was lost on heating suggests that it was not strongly bound within the crystal structure. Mechanistically, it is proposed that dehydration proceeds by the formation of uncoordinated water in the parent lattice, followed by elimination of the water through channels. This causes the formation of molecular vacancies, which

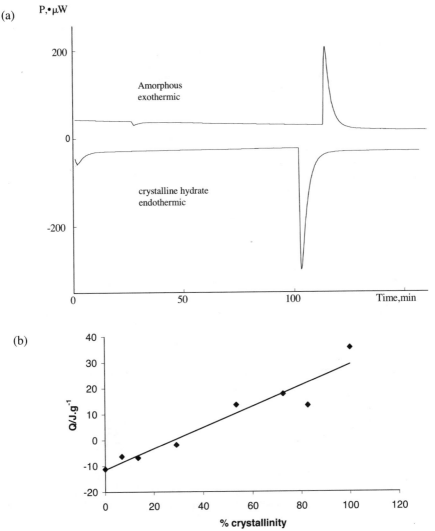

Figure 7 (*a*) *Heat flow for 100% amorphous and crystalline samples*; (*b*) *Enthalpy of solution versus % crystallinity*

results in disruption *via* structural collapse, giving rise to the observed amorphicity.

Crystallisation of the compound occurred in several stages, and DVS and IM experiments provided some insight into the mechanism. Results showed that during the initial sorption phase (0–60% RH), the mass increase was approximately 12% w/w, and was exothermic in nature. This represented sorption and condensation of moisture onto the surface of the powder. The net effect would be to increase molecular mobility and hence lower the glass transition temperature (T_g) of the amorphous phase, *i.e.* the sample would be plasticised.[3] Between 60

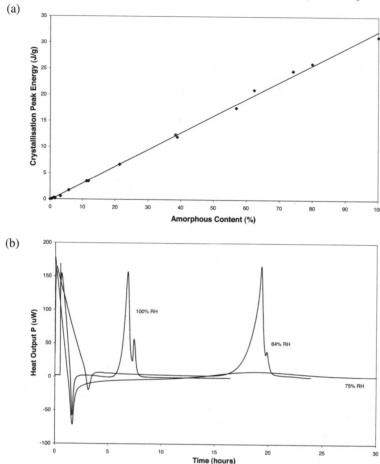

Figure 8 (*a*) *Heat output obtained by crystallisation of a 100% amorphous sample exposed to 84% RH;* (*b*) *Crystallisation peak energy versus % crystallinity*

and 70% RH, the DVS showed a plateau in moisture sorption, corresponding to an endothermic event, probably representing structural collapse of the sample. From IM measurements, made while increasing the humidity beyond 75% RH, crystallisation was observed to be initiated. The decrease in mass between 80 and 95% RH in the DVS experiment was accompanied by a decrease in heat flow, which is thought to be due to the sample expelling 'excess' water (in terms of its stoichiometric requirements). Such water expulsion has been observed for partially amorphous, micronised substances.[4] The hydrate formed was stable with respect to changes in RH, until an environment of less than $\sim 12\%$ RH was achieved, at which point dehydration was initiated.

3.2 Quantitative Analysis

Mixtures of pure amorphous and crystalline phases were used to construct calibration curves. In all instances, a good linear relationship was obtained between the parameter measured and crystalline content of the mixtures. However, limitations to these methods include, for example, processing effects such as sample homogeneity, sample anisotropic effects (in the case of XRPD), and variation in dissolution/wetting behaviour.

The XRPD calibration plot was linear up to 60% w/w crystalline, and the limit of detection was approximately 2% w/w. It may be possible to extend the linear region and improve the linearity by using an internal standard such as lithium fluoride.[5] Alternatively, the slurry technique, as described by Bugay *et al.*,[6] could be used. This method makes the mixture more homogeneous by slurrying the components in a solvent in which they are not soluble. However, it assumes that there are no polymorphic interconversions or crystallisation of the amorphous phase. Nonetheless, the data are encouraging; the relative lack of sensitivity of XRPD derives from the fact that X-ray diffraction measures an average degree of order, and as such, it is an indirect measure of amorphous content.[7]

Dissolution of a solute involves several thermal events associated with wetting, breakage of lattice bonds, and solvation energy. Bearing this in mind, we obtained a good linear plot, when the heats of solution of mixtures of amorphous and crystalline samples were measured. By use of this technique, 5% w/w crystalline content in amorphous was detectable, and it may be possible to detect lower levels of the crystalline material, as shown by other workers.[8] Potential advantages of this technique include rapid data acquisition and consumption of small amounts of material.

IM gave the best results, whereby a plot with excellent linearity and a limit of detection of 0.5% w/w amorphous content in crystalline was obtained. This was in line with other workers who have used IM in this manner.[2] IM results are less affected by mixing efficiency, since the moisture will move through the powder bed as a whole and crystallise all amorphous regions of the sample, provided the RH is above the critical level. A potential disadvantage of IM measurements, when used to generate a calibration curve, is the relatively lengthy acquisition time, which may be reduced by using smaller sample sizes. However, this may decrease the sensitivity of the measurements and raise the limit of detection. It should be noted that other tests, such as accuracy and precision, would have to be measured to fully validate the IM and the other techniques used in this study. Finally, all of these measurements assumed that the two-state model used in these experiments is valid for measurements where amorphous and crystalline samples exist in the one-state model.[9]

It is well known that amorphous forms of compounds are less stable in the solid state, compared to their corresponding crystalline phases.[10–12] This was also the case for the development compound described here, which showed superior thermal stability in the crystalline state.

4 Conclusions

Two phases of a development compound were identified. An amorphous form, when exposed to high RH (in excess of 75%), crystallised to form a hydrate. The crystalline form was shown to have superior physicochemical properties, when compared to the metastable amorphous form, and consequently represented a more attractive form of the compound from a production and storage point of view.

Binary mixtures of the two pure phases enabled calibration plots to be constructed. From these studies, the detection limit of amorphous content in a crystalline matrix was shown to be less than or equal to 5% w/w. On the basis of the data presented, IM represents the best method of determining small quantities, approximately 0.5 w/w, of amorphous material in an otherwise crystalline matrix. However, both SOLCAL and XRPD, with more method development, could be improved in terms of their linearity and limits of detection.

References

1. L.E. Briggner, G. Buckton, K. Bystrom and P. Darcy, *Int. J. Pharm.*, 1994, **105**, 125.
2. G. Buckton, P. Darcy, D. Greenleaf and P. Holbrook, *Int. J. Pharm.*, 1995, **116**, 113.
3. B. Hancock, S.L. Shamblin and G. Zografi, *Pharm. Res.*, 1995, **12**, 799.
4. G.H. Ward and R.K. Schultz, *Pharm. Res.*, 1995, **12**, 773.
5. S.D. Clas, R. Fraizer, R.E. O'Connor and E.B. Vadas, *Int. J. Pharm.*, 1995, **121**, 73.
6. D.E. Bugay, A.W. Newman and W.P. Findlay, *J. Pharm. Biomed. Anal.*, 1996, **15**, 49.
7. H. Imaizumi, N. Nambu and T. Nagai, *Chem. Pharm. Bull.*, 1980, **28**, 2565.
8. A. Saleki-Gerhardt and G. Zografi, *Pharm. Res.*, 1994, **11**, 1166.
9. R. Suryanarayanan and A.G. Mitchell, *Int. J. Pharm.*, 1985, **24**, 1.
10. E.R. Oberholtzer and G.S. Brenner, *J. Pharm. Sci.*, 1979, **68**, 863.
11. M.J. Pikal, A.L. Lukes and J.E. Jang, *J. Pharm. Sci.*, 1977, **66**, 1312.
12. J.T. Carstensen and T. Morris, *J. Pharm. Sci.*, 1993, **82**, 657.

Thermophysical Properties of Amorphous Dehydrated and Frozen Sugar Systems, as Affected by Salts

M.F. Mazzobre,[+] M.P. Longinotti,* H.R. Corti*# and M.P. Buera[+]

+DPTO. DE INDUSTRIAS, #INQUIMAE, FCEyN, UBA,
1428 BUENOS AIRES, ARGENTINA
*UNIDAD DE ACTIVIDAD QUÍMICA, COMISIÓN NACIONAL DE
ENERGÍA ATÓMICA, PCIA. BUENOS AIRES, ARGENTINA

1 Introduction

Crystallization of sugars or ice formation from the amorphous state can lead to a loss of the protective effects of amorphous matrices on biomolecules and biological structures.[1-3] Inorganic salts have a universal presence in biological systems, and they are often added to therapeutic and diagnostic formulations, in order to maintain the pH or isotonicity of a rehydrated product.[4] The effect of electrolytes on the thermophysical properties of sugar systems is of special interest, due to their major influence on water structure and their possible interactions with biomolecules. However, little attention is generally paid to the effect of salts on the physical properties of frozen or dehydrated materials. The synergistic effects of sugars and divalent cations on protein stabilization has been reported,[5] and several physical methods have provided evidence for the existence of sugar complexes with inorganic salts in solution and in solid form.[6-9] However, there have been few studies on the effect of salts on the glass transition temperature (T_g) of aqueous solutions.[10-12] Sugar–salt–water interactions may play an important role in biomolecule stabilization.[13] The objective of the present work was to explore the effects of salts on water sorption, sugar crystallization or ice formation characteristics, and conductivity behaviour in sugar–water systems involved in biomolecule stabilization.

2 Materials and Methods

Figure 1 shows phase/state diagrams for trehalose–water and sucrose–water systems, on which are illustrated the temperature/composition regions where the effects of salts on sugar or water crystallization and transport properties were studied. Amorphous systems were obtained by freeze-drying solutions containing 20% (w/v) trehalose (T) (Pfanstiehl, Waukegan, IL, USA) or sucrose (S) (Mallinckrodt, St. Louis, MO, USA), or mixtures of T or S with $MgCl_2$, $CaCl_2$, NaCl or KCl (all of which were p.a. grade), at sugar:salt molar ratios (R) of 5, 2 or 1. $R = \infty$ represents a pure sugar system. A Heto-Holten A/S, cooling-trap model CT 110 freeze-dryer (Heto Lab Equipment, Denmark) was operated at a conden-

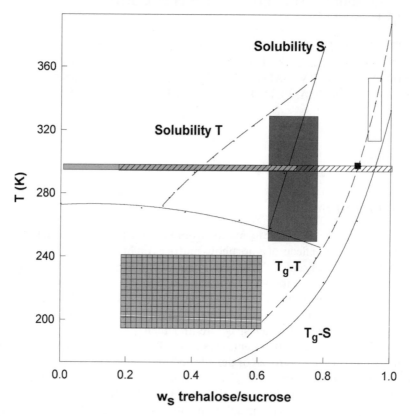

Figure 1 *Phase/state diagrams for sucrose–water and trehalose–water showing the regions where the effects of salts were investigated. Solubility and T_g curves for trehalose (T) and sucrose (S) are plotted as a function of the mass fraction of solids (w_s):* ■ *Crystallization kinetics for trehalose and trehalose–salt;* □ *Crystallization kinetics for sucrose and sucrose–salt;* ▨ *Conductivity of $MgCl_2$ or NaCl in sucrose or trehalose solutions of varying concentration, under isothermal conditions (298 K);* ■ *Conductivity of $MgCl_2$ or NaCl in sucrose or trehalose solutions of constant sugar concentration, under conditions of varying temperature;* ▨ *Water sorption isotherms for sucrose, sucrose–salt, trehalose and trehalose–salt;* ▦ *Water crystallization*

ser plate temperature of 163 K and a chamber pressure of 4.10^{-4} mbar and temperature of 298 K. After freeze-drying, samples were held over saturated salt solutions in the range of 11–85% relative humidity (RH) for at least 2 weeks under isopiestic conditions at 298 K, and the mass fraction of water (w) was determined after drying at 343 K under vacuum.

A differential scanning calorimeter (DSC) system (Mettler TA 4000 with a TC11 TA processor and Graph Ware TA72 thermal analysis software) was used to determine T_g, crystallization temperature (T_c), and heat of crystallization (ΔH_c) and melting (ΔH_m) for sugars or water. The instrument was calibrated with indium and zinc. $40\mu l$, hermetically sealed aluminium pans (Mettler) containing 10–15 mg samples were utilized. An empty pan was used as a reference. Two replicates of each sample were analysed at a scanning rate of 10 K min^{-1}, and averaged values were reported.

A thermostatted glass cell (of 30 cm^3 volume), with plated platinum electrodes and an AC bridge, was used for conductivity measurements on supercooled trehalose–salt or sucrose–salt solutions. Viscosity changes were achieved by varying the concentration of sugar under isothermal conditions (298 K), or by varying the experimental temperature (from 248 to 323 K) at constant sugar concentration. The conductivity cell constant was determined at temperatures above and below 273 K, employing previously reported data,[14,15] and corrections were made for the contributions of ionic impurities or CO_2.

3 Results and Discussion

3.1 Water Sorption

Water sorption isotherms for amorphous, freeze-dried samples of sucrose or trehalose, and their mixtures with several chlorides (sugar:salt molar ratio $R = 5$), have been reported recently.[16] Figures 2(a) and 2(b) show results plotted as moles of sorbed water per mole of solute (open symbols), q, as a function of relative vapour pressure (RVP or 'water activity', a_w), for sucrose–MgCl$_2$ and trehalose–MgCl$_2$ mixtures at 298 K. Water sorption data for mgCl$_2$ alone, calculated from reported osmotic coefficients up to saturation and for less-than-saturated solutions,[17,18] and for each disaccharide alone were also plotted. Sucrose adsorbs less than one water molecule at $a_w < 0.2$, and then q decreases, reflecting the fact that amorphous sucrose can crystallize to an anhydrate above that a_w. At $a_w > 0.80$, q increases due to sucrose dissolution. Sorption by trehalose increases monotonically with increasing a_w up to $a_w = 0.4$, then levels off at higher a_ws, corresponding to the formation of a stable crystalline dihydrate ($q = 2$). The water sorption isotherms shown in Figures 2(a) and 2(b) for the two disaccharides are in close agreement with previously reported results.[19-21] Experimental water sorption data were also plotted as moles of sorbed water per mole of salt (closed symbols); this was useful for comparing experimental results with those expected if we assume that crystalline sugars do not adsorb water. Figure 2 shows that the amount of water sorbed by sucrose–MgCl$_2$ (a) and trehalose–MgCl$_2$ (b), in both cases, the values fell between the amounts

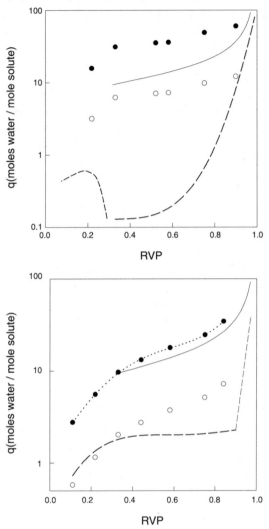

Figure 2   *Salt effects on the water sorption properties of (a) sucrose–MgCl$_2$ (top) and (b) trehalose–MgCl$_2$ (bottom) at 298 K. Solid lines correspond to water sorption data for mgCl$_2$ alone and dashed lines to sorption data for sucrose or trehalose alone. Water sorption is reported per mole of salt (•) and per mole of salt–sugar mixture (○)*

corresponding to the isotherms for the binary systems, sugar–water (dashed lines) and salt–water (solid lines). For the sucrose system, the water sorption was greater than that expected for the salt alone, suggesting that the presence of MgCl$_2$ inhibited the formation of anhydrous sucrose crystals, and that a mixture of amorphous sucrose and salt could adsorb water. This behaviour was also found for NaCl–sucrose mixtures, but for mixtures of sucrose with KCl or CaCl$_2$, the salts were responsible for most of the water adsorbed. It was observed[16] that water sorption at a given RVP increased with increasing charge/size ratio of the cation ($Mg^{2+} > Ca^{2+} > Na^+ > K^+$). For the $MgCl_2$ — trehalose system, the effect of salt on water sorption, illustrated in Figure 2(b), was masked by the fact that trehalose alone does not crystallize at RVP ≤ 0.44, and is thus able to sorb a higher amount of water than sucrose does prior to crystallization. However, the salt–trehalose isotherm reflected an increase in water uptake above that RVP, suggesting a possible effect on trehalose crystallization (discussed later in Section 3.3).

3.2 Ice Formation

Water crystallization was analysed by dynamic DSC runs on samples containing $w = 0.4$, 0.6 and 0.8. Samples with $w = 0.4$ were subjected to rapid cooling by placing the DSC sample pans at 123 K prior to DSC runs. This rapid cooling and the low mass fraction of water resulted in prevention of sugar or ice crystallization during cooling in these samples. Samples with $w = 0.6$ or 0.8 were cooled slowly by placing them at 247 K for 24 h, conditions under which ice formation was expected to occur readily in those samples.[22,23] Results shown in Table 1 suggest that the presence of salts, in samples with $w = 0.4$, did not significantly affect T_g values, which were close to ones reported previously.[12,22,24] In the presence of salts, the water crystallization temperature (T_{wc}) increased for both sugar systems, except for the sucrose–KCl sample. The addition of NaCl to trehalose–water systems has been reported to delay ice formation without affecting T_g.[12] The apparent amount of ice formed (directly related to the area of the exothermic peak during cooling in a DSC run) was in the order $mgCl_2 < NaCl < KCl < sugar$ alone. The shift in T_{wc} values toward higher temperatures, and the lower amounts of ice formed, could be attributable to water–ion interactions, which became critical under the conditions studied, where kinetic restrictions to ice formation occurred.

Table 1 *Glass transition temperature* (T_g), *onset temperature for water crystallization* (T_{wc}), *and heat of water crystallization* (ΔH_{wc}), *obtained for samples with sugar:salt ratios of 5:1 (R = 5) and w = 0.4, which were previously quenched at $-150\,^\circ C$*

Sugar	Salt	T_g (°C)	T_{wc} (°C)	ΔH_{wc} (J g^{-1})
Trehalose	None	−72.0	−49.6	42.7
	KCl	−71.5	−39	28.8
	NaCl	−70.	−35	11
	$MgCl_2$	−70	–*	–*
Sucrose	None	−75.5	−63	39.8
	KCl	−74.2	−63	36.7
	NaCl	−73	−44.7	24.3
	$MgCl_2$	−73.2	−34.2	6

* No exothermic peaks were observed.

Results in Table 2 show the effect of $MgCl_2$ concentration ($R = 5$ to 1) on ice formation in trehalose or sucrose solutions with $w = 0.6$, after storage at 247 K for 24 h, and the w_g' and T_g' values determined[23] from the point of intersection of the T_g and calculated liquidus curves for each system.[16] The amount of water in the amorphous phase (w_A) after storage was estimated from T_g values. It should be noted that, after storage at 247 K for 24 h, samples with $w = 0.8$ exhibited T_g values close ($\pm 2\,°C$) to those shown in Table 1 for samples with $w = 0.6$, confirming that the composition of the freeze-concentrated matrix obtained under those conditions was independent of the initial water content of the samples.[23] Samples with $w = 0.6$ and 0.8, subjected to slow cooling and then stored at a temperature at which little diffusional restrictions on ice crystal formation would be expected, formed freeze-concentrated samples having w_A values close to the theoretical extrapolated w_g' for each system (for samples with $R = 1$, the greater differences between w_A and theoretical w_g' values were expected). Several reported w'_g values for both sugar–water systems are close to $w = 0.2$.[12,22,25] Experimental T_g' values between -35 and $-30\,°C$ for trehalose[12] and between -32 and $-45\,°C$ for sucrose have been reported.[22–25] Maximal freeze-concentration for a sugar–water system is defined by T_g' and w_g';[23] when salts are added, T_g' values decrease, while w_g' values increase. The amount of unfrozen water remaining in the supercooled matrix (excluded from the ice) was higher in salt-containing systems than in sugar–water systems, and this explained the former's lower T_g values. For $w = 0.6$ or 0.8 at $-26\,°C$, we suggest that water–ion interactions were not sufficiently strong to prevent significant ice crystallization. An abrupt change in molecular mobility in sucrose systems at $w < 0.6$ has been reported,[26,27] which could further support the notion that water–sugar interactions may become critical for samples with $w = 0.4$, in which effects of salts were evident.

3.3 Sugar Crystallization

Trehalose crystallization was evaluated over time, during storage under condi-

Table 2 *Values of experimental T_g and mass fraction of water in the amorphous phase (w_A), for freeze-concentrated sugar–$MgCl_2$ samples with $w = 0.6$, after storage at $-26\,°C$ for 24 h. w_g' and T_g' values, in each case, were calculated from the point of intersection of the liquidus and T_g curves*

Sugar	R	$T_g(°C)$	w_A	$T_g'\,(°C)$	w_g'
Trehalose	∞	-44	0.238	-31	0.215
	5	-52.5	0.256	-42	0.225
	2	-57	0.270	-58	0.310
	1	-65	0.305	-80	0.42
Sucrose	∞	-55	0.238	-39	0.180
	5	-54.2	0.232	-50	0.220
	2	-63	0.250	-67	0.280
	1	-90.2	0.414	-89	0.400

tions that ensured the minimum mass fraction of water (0.1) needed to form the dihydrate crystal (0.44 RVP at 298 K, 15 K above T_g), and crystallization was expected to be completed within the experimental time frame. The degree of crystallization, ϕ_m, of trehalose dihydrate was calculated from the ratio of the area of the endothermic melting peak (ΔH_m) for a sample and the calorimetric enthalpy of melting of pure trehalose dihydrate (139 J g^{-1}). Table 3 shows the T_g and crystalline fraction of trehalose, after 22 days under the specified conditions. While pure trehalose systems were totally crystalline after that time, in systems containing mgCl$_2$ or CaCl$_2$, ϕ_m was 50%, and no further crystallization was observed after 90 days.

Crystalline fractions (ϕ), for sucrose and sucrose:salt ($R = 5$) systems having a T_g value of 303 ± 2 K, were obtained by integrating the exothermal crystallization peaks[28] measured under isothermal conditions. Typical sigmoid curves were obtained in all cases, when ϕ was plotted *versus* time. Data were analysed using the Johnson–Mehl–Avrami–Kolmogorov (JMAK) equation:[29–31]

$$1 - \phi = \exp(-k\theta^n) \tag{1}$$

where k is the crystallization rate constant [(time)$^{-n}$], θ is the elapsed time from the onset of crystallization (induction time was subtracted), and n is the so-called Avrami index. The parameters of Equation 1 were calculated by non-linear regression analysis. The n values obtained were in the ranges of 3–4, 3–3.5, 2–2.5, 1.5–2, and 1.5–1.7 for sucrose alone, and its mixtures with KCl, NaCl, CaCl$_2$ and mgCl$_2$, respectively. The value of the exponent n increases with increasing 'dimensionality' of growth.[30–34] A value of n close to 4 might indicate heterogeneous nucleation and development of spherical crystals from sporadic nuclei. Values of $n = 3$ also indicate spherulitic growth, where the majority of nuclei form near the beginning of crystallization. Values of n close to 2 indicate plate-like growth (primarily in two directions), with a very high initial nucleation rate. Samples containing mgCl$_2$ showed n values < 2. Analysis using the JMAK equation allowed us to hypothetize that, in the presence of salts, nucleation occurs very rapidly at the beginning of heating and that salts constrain the dimensionality of crystal growth. Coefficient k has different units (min^{-n}) in each case, and a direct comparison between cases is not possible. In order to compare all the analysed systems, crystallization rate was expressed as:

$$\frac{d\phi}{dt} = K\phi^N \tag{2}$$

Table 3 T_g *and degree of crystallization (ϕ_m) for trehalose systems with w $= 0.1$, after 22 days of storage at 43% RH at 25°C*

Salt	$T_g(°C)$	ϕ_m
None	10.1	100
KCl	9.4	95
NaCl	9.1	85
MgCl$_2$	8.2	50

where N is the reaction order, and K is a rate coefficient. N was between 0 and 1, and the following kinetic model has been proposed:[35]

$$\frac{1}{d\phi/dt} = \frac{1}{K_o} + \frac{1}{K_1\phi}$$

(3)

where K_o and K_1 are zero-order and first-order kinetic coefficients, respectively. When ϕ is small (for short reaction times), the second term prevails, and the reaction order is close to 1; when ϕ is large, the second term can be neglected, and an expression of zero-order is obtained. K_o and K_1 were calculated by non-linear correlation, allowing us to fit experimental data up to a ϕ value of 90%. Since T_g reflects the mechanical or dynamic relaxation behaviour of a solid in the amorphous state, the crystallization rate at a particular temperature, T, is often directly linked to the difference, $T - T_g$.[23] Figure 3 shows calculated K_o and K_1 values as a function of $T - T_g$. Sucrose systems clearly showed a different kinetic behaviour from salt-containing systems. The rate coefficient, K_o (related to the slope of the linear part of crystallization curves), had a smaller temperature dependence than did the rate coefficient, K_1 (which accounted for non-linearity at short times), thus suggesting that temperature effects were more pronounced in the initial stages of crystal growth. In most of the salt-containing systems, delayed crystallization was observed at very large $T - T_g$ values, which would have corresponded to instant crystallization in a salt-free system. The presence of salts clearly delayed sucrose crystallization, without affecting the T_g value of the system.

3.4 Electrical Conductivity

Results for water sorption and water/sugar crystallization behaviour suggested that the effect of salts on sugar properties was directly related to the charge/mass ratio of the cations present ($Mg > Ca > Na > K$). The molar conductivities of NaCl and $mgCl_2$ in dilute aqueous solutions of trehalose and sucrose have been measured,[36,37] over a wide range of viscosities, to determine if sugar–salt–water interactions are reflected in the transport properties of the electrolytes. The results are shown in Figure 4. Measurements covered the one-phase (liquid) and supercooled regions of the phase diagrams for the sugar–water systems. By means of cooling concentrated solutions of trehalose or sucrose, the supercooled region was approached, covering viscosities between 1 and 10^6 mPa s^{-1}. The salt concentration used for these measurements ensured that the molar conductivities were close to the values at infinite dilution.

As found in previous studies of sucrose[38] and maltose,[39] the molar conductivity of salts did not decrease in proportion to the inverse of solution viscosity, as predicted by Walden's law, but rather followed a relationship of the type:[40]

$$\Lambda\eta^\alpha = constant$$

(4)

where $\alpha < 1$ is a measure of the uncoupling of conductivity and viscosity. The deviation from Walden's law (which applies for $\alpha = 1$) could be related to the

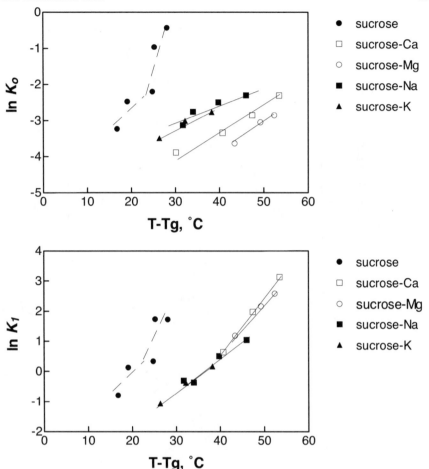

Figure 3 *Zero-order* (K_0) *and first-order* (K_1) *rate coefficients for sucrose crystallization in sucrose-only and sucrose–salt systems, as a function of* $T - T_g$

presence of structural heterogeneities induced by the preferential solvation of ions in aqueous systems. Previous results from molecular dynamics simulations[36] demonstrated that the residence time for water in the solvation layer of sodium ions in aqueous trehalose solutions is greater than the corresponding residence time for the disaccharide. It has been proposed that the decoupling of viscosity and diffusion takes place at a so-called crossover temperature, T_{cr}, which is close to 1.18 T_g for most glass-forming liquids.[41] Therefore, for homogeneous solvents, Walden's law would be valid ($\alpha = 1$) for $T > T_c$, while a decoupling ($\alpha < 1$) at $T < T_{cr}$, related to a dynamical phase transition, could be observed at $T = T_c$. The systems shown in Figure 4, however, were far above the crossover temperature. Therefore, the deviation from predictions of the viscous model could be related to preferential solvation, which resulted in structural

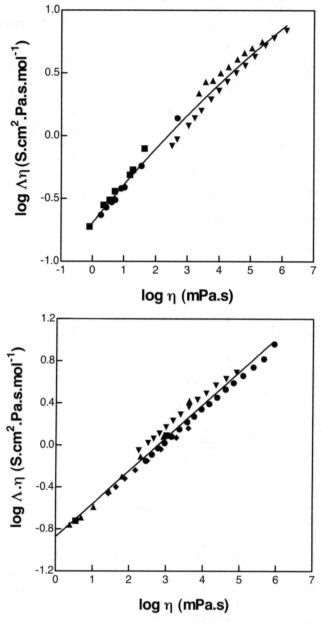

Figure 4 *Plots of Walden product for* $MgCl_2$ *and NaCl in aqueous sucrose and trehalose systems as a function of viscosity. Top* $(MgCl_2)$ *– viscosity changed by varying sugar concentration at 298 K:* ● *sucrose,* ■ *trehalose; viscosity changed by varying temperature:* ▲ *trehalose, 70.5% w/w;* ▼ *sucrose, 77% w/w. Bottom (NaCl) – viscosity changed by varying sugar concentration at 298 K:* ▲ *trehalose,* ■ *sucrose; viscosity changing by varying temperature;* ● *sucrose, 76.2% w/w;* ▼ *trehalose, 68.1% w/w;* ♦ *sucrose, 65% w/w*

inhomogeneities. Equation 4, with α between 0.64 and 0.75, described quite well the experimental results. The reason for enhanced ion mobility in aqueous sugar solutions could be the existence of structural microheterogeneities in the fluid. The liquid environment around the ions would be richer in water, and consequently, the local viscosity would be lower than the bulk viscosity. A lower slope was obtained, in the case of trehalose solutions with $MgCl_2$, by changing temperature (high viscosity points) than by changing concentration. The higher hydration of Mg^{2+} ion, compared to Na^+, caused the water-rich region to be exhausted at lower viscosity. The conductivity of $MgCl_2$ was lower than that of NaCl in aqueous supercooled sucrose. Ion association to form $MgCl^+$ could also have contributed to the observed behaviour. The results obtained for transport properties in sugar–salt–water systems revealed a high degree of local inhomogeneities, which in turn were induced by preferential solvation as a consequence of greater ion–water interactions, in comparison to ion–disaccharide interactions. Therefore, while ion mobility was enhanced by a low-viscosity local environment, it would be expected that sugar molecule mobility would be inhibited by a high local viscosity. The situation could also be described in terms of spatially heterogeneous dynamics,[42] as was found, by other techniques, for other supercooled liquids. The dynamics in regions separated by just a few nanometers could be different by several orders of magnitude. Dramatic changes in the short-range dynamics of sugar–water systems could occur, without modifying the T_g of the system, which would be a result of suprastructural relaxations. Thus, the presence of salts could retard the crystallization of sugars, even when T_g remained unchanged, and the effect should be more pronounced for ions with stronger interactions with water.

4 Conclusions

Water–ion interactions were observed in systems of reduced water concentration ($w \leq 0.4$), but they did not affect the T_g values of the systems. Thus, the interaction of water with ions takes place at a molecular level, without affecting supramolecular or macroscopic properties. On the basis of experimentally observed behaviour in terms of the conductivity of sugar–salt systems, and that reported from molecular dynamics simulations, we conclude that salts interact preferentially with water, increasing the extent of supersaturation and consequently the nucleation rate in sugar-rich regions, thus affecting the kinetics of sugar crystal growth, by way of a site-saturation effect.

Sugars and salts are important components of biomaterials subjected to drying/freezing/freeze-drying. Our experiments showed that water sorption properties and ice and sugar crystallization characteristics are modified by the presence of salts commonly found in biological materials. It must be emphasized that care should be taken, when extrapolations from data for pure sugars are employed to predict the characteristics of complex materials containing salts.

Acknowledgements

The authors acknowledge financial support from Secretaría de Ciencia y Técnica, Universidad de Buenos Aires (Project EX 107), Agencia Nacional de Promoción Científica y Tecnológica (Project 06-5066), and from the International Foundation for Science (Sweden). MPL acknowledges student fellowships from Universidad de Buenos Aires and Comisión Nacional de Energía Atómica.

References

1. T. Suzuki, K. Imamura, K. Yamamoto, T. Satoh and M. Okazaki, *J. Chem. Eng. Japan*, 1997, **30**, 609.
2. W.Q. Sun and A.C. Leopold, *Physiol. Plant.*, 1993, **89**, 767.
3. S. Cardona, C. Schebor, M.P. Buera and J. Chirife, *J. Food Sci.*, 1997, **62**, 105.
4. L.M. Her, M. Deras and S.L. Nail, *Pharm. Res.*, 1995, **12**, 768.
5. J.F. Carpenter and J.H. Crowe, *Cryobiology*, 1988, **25**, 459.
6. S.J. Angyal, *Pure Appl. Chem.*, 1973, **35**, 131.
7. W.J. Cook and C.E. Bugg, *Carbohydr. Res.*, 1973, **31**, 265.
8. N. Morel-Desrosiers, C. Lhermet and J.P. Morel, *J. Chem. Soc. Faraday Trans.*, 1991, **87**, 2173.
9. P. Rongere, N. Morel-Desrosiers and J.P. Morel, *J. Chem. Soc. Faraday Trans.*, 1995, **91**, 2771.
10. W. Goetze, 'Liquids, Freezing and Glass Transition', Elsevier, 1991.
11. C.A. Angell, R.D. Bressel, J.L. Green, H. Kanno, M. Oguni and E.J. Sare, *J. Food Eng.*, 1994, **22**, 115.
12. D.P. Miller, J.J. de Pablo and H.R. Corti, *Pharm. Res.*, 1997, **14**, 578.
13. M.F. Mazzobre and M.P. Buera, *Biochim. Biophys. Acta*, 1999, **1473**, 337.
14. J. Barthel, F. Feuerlein, R. Neueder and R. Wachter, *J. Solution Chem.*, 1980, **9**, 209.
15. J. Barthel, R. Neueder, F. Feuerlein, F. Strasser and L. Iberl, *J. Solution Chem.*, 1983, **12**, 449.
16. M.F. Mazzobre, M.P. Longinotti, H.R. Corti and M.P. Buera, *Cryobiol.*, 2001, **43**, 199.
17. R.N. Goldberg and R.L. Nuttall, *J. Phys. Chem. Ref. Data*, 1978, **7**, 263.
18. R.A. Robinson and R.H. Stokes, 'Electrolyte Solutions', Butterworths, London, 1955.
19. M. Karel, 'Water Relations of Foods', R.B. Duckworth, (ed.), Academic Press, London, 1981, 639.
20. H.A. Iglesias, J. Chirife and M.P. Buera, *J. Sci. Food Agric.*, 1997, **75**, 183.
21. Y. Roos and M. Karel, *Food Technol.*, 1991, **45**, 66.
22. Y. Roos and M. Karel, *Cryo-Lett.*, 1991, **12**, 367.
23. L. Slade and H. Levine, *Crit. Rev. Food Sci. Nutr.*, 1991, **30**, 115.
24. Y. Roos and M. Karel, *J. Food Sci. Technol.*, 1991, **26**, 553.
25. D. Simatos and G. Blond, 'Water Relationships in Foods', H. Levine and L. Slade, (eds.), Plenum Press, New York, 1991, 139.
26. J.M. Flink, 'Physical Properties of Foods', M. Peleg and E.B. Bagley, (eds.), AVI, Westport, 1983, 473.
27. M.A. Hemminga, M.J. Roozen and P. Walstra, 'The Glassy State in Foods', J.M.V. Blanshard and P.J. Lillford, (eds.), Nottingham University Press, Loughborough, 1993, 157.
28. C.J. Kedward, W. MacNaughtan and J.R. Mitchell, *J. Food Sci.*, 2000, **2**, 324.

29. W.A. Johnson and R.F. Mehl, *Trans. AIME*, 1939, **135**, 416.
30. M. Avrami, *J. Chem. Phys.*, 1939, **7**, 1103.
31. A.N. Kolmogorov, *Bull. Acad. Sci. USSR (Sci. Mater. Nat.)*, 1937, **3**, 3551.
32. R.H. Doremus, 'Rates of Phase Transformations', Academic Press, Orlando, 1985, 24.
33. J.W. Graydon, S.J. Thorpe and D.W. Kirk, *J. Non-Crystal. Solids*, 1994, **175**, 31.
34. Z. Chvoj, J. Sestak and A. Triska, 'Kinetic Phase Diagrams', Elsevier, Amsterdam, 1991, 169.
35. C. Petriella, S. Resnik, R.D. Lozano and J. Chirife, *J. Food Sci.*, 1985, **50**, 622.
36. D.P. Miller, P.B. Conrad, S. Fucito, H.R. Corti and J.J. de Pablo, *J. Phys. Chem. B*, 2000, **104**, 10419.
37. M.P. Longinotti, M.F. Mazzobre, M.P. Buera and H.R. Corti, *Phys. Chem. Chem. Phys.*, 2001, **4**, 533.
38. R.H. Stokes, 'The Structure of Electrolytic Solutions', W.J. Hamer, (ed.), Wiley, New York, 1959, 298.
39. T.R. Noel, R. Parker and S.G. Ring, *J. Chem. Soc. Faraday Trans.*, 1996, **92**, 1921.
40. M. Spiro, 'Organic Solvent Systems', A.K. Covington and T. Dickinson, (eds.), Plenum Press, New York, 1973, Chapter 6.
41. E. Rössler, *Phys. Rev. Lett.*, 1990, **65**, 1595.
42. M.D. Ediger, *Ann. Rev. Phys. Chem.*, 2000, **51**, 99.

Glass-forming Ability of Polyphosphate Compounds and Their Stability

Kiyoshi Kawai, Toru Suzuki, Tomoaki Hagiwara and Rikuo Takai

DEPARTMENT OF FOOD SCIENCE AND TECHNOLOGY, TOKYO UNIVERSITY OF FISHERIES, 4-5-7 KONAN, MINATO-KU, TOKYO 108-8477, JAPAN

1 Introduction

Many kinds of phosphate compounds, *e.g.* DNA, ATP, phospholipids, *etc.*, exist widely in biological systems. Phosphate salts are often used as pH buffers in many kinds of protein solution systems. An inorganic polyphosphate, polymerized from 100 phosphate repeats, has even been found in microbial and mammalian cells.[1] Recently, it was reported that inorganic phosphates play an important role in the tolerance of *E. coli* to starvation.[2] Synthetic inorganic polyphosphates, such as di- or tripolyphosphates, have often been used as food additives, to prevent water release after freeze-thawing, to improve adhesion between meat materials, or to control pH.[3-5] Thus, phosphate compounds exist in many situations; however, their physicochemical properties, especially at low temperatures, are not well known. Murase *et al.*[6,7] reported that some monophosphate salt solutions, which are often used as buffers, can exhibit a eutectic point, leading to drastic pH changes, but they may instead transform into a glassy solid state. There is little other information available about the physical properties of polyphosphates at low temperatures.

In contrast, much knowledge about the glass transition behaviour for many kinds of carbohydrate and protein systems has been accumulated in the past decade.[9] One of the mechanisms by which carbohydrates can stabilize proteins during freezing or freeze-drying has come to be understood through state diagrams that illustrate the relationship between glass transition temperature (T_g) and moisture content.[8,9] Such a viewpoint on glass transitions has enabled many successful and practical applications.[10,11] Exploratory studies on improved glass-formers have also been conducted, not only involving new materials but also mixtures.[12,13]

Again, however, there are virtually no reports about polyphosphates in the above context, even though, in the field of inorganic glass science, it is well known that highly polymerized polyphosphates have been categorized as glass-formers as good as silica.[14,15] So, as a first step in the present study, we investigated – using differential scanning calorimetry (DSC) thermal analysis – whether or not frozen solutions and freeze-dried powders of polyphosphates, such as di- and tripolyphosphates and ATP-related compounds, would form glassy states. T_g values, for di- and tripolyphosphates and ATP-related compounds with different moisture contents, were determined over a wide temperature range.

Various recent studies on glassy materials have focused on an evaluation of molecular mobility in the glassy state around or below T_g.[16–18] The following opinion has been discussed: just because a material is in a glassy state does not always imply that it is an effective stabilizer; its molecular mobility in the glassy state should be assessed. Therefore, in order to obtain such information, ATP and ADP were chosen, from among the polyphosphates examined in this study, and their molecular mobility below T_g was investigated through measurements of enthalpic relaxation during aging.[19,20]

2 Materials and Methods

In this study, the following phosphate compounds were used: tetrasodium diphosphate decahydrate, pentasodium tripolyphosphate anhydrous (Wako pure chem., Japan), adenosine 5′-triphosphate (ATP) disodium salt (Sigma, ultra, minimum 99%), adenosine 5′-diphosphate (ADP) sodium salt (Sigma, 95–99%), and adenosine 5′-monophosphate (AMP) sodium salt (Sigma, 99 + %). Aqueous solutions of 5–15% (w/w) concentration were prepared in vials, and about 10 mg of sample was put into an aluminum DSC pan and sealed. Thermal analysis was performed with a Shimadzu DSC-50 calorimeter, which was calibrated for temperature and heat capacity with indium. Samples were heated to 20 °C at 2 °C min^{-1}, after cooling to -60 °C at -6 °C min^{-1}. Solutions of tripolyphosphate, ATP and ADP were freeze-dried to prepare low-moisture samples. Those samples were also sealed in aluminum DSC pans and scanned from 0 °C to 120 °C at 2 °C min^{-1}. After DSC measurements, those pans were pin-holed on top and then dried in an oven at 120 °C, in order to determine moisture content from final weight.

For study of enthalpy relaxation, glassy samples of ATP, ADP and D(+)-trehalose dihydrate, prepared by freeze-drying, were sealed in aluminum DSC pans and scanned according to the following sequence. They were heated once to $T_g + 30$ °C to erase their thermal history, then cooled to $T_g - 50$ °C. Samples were then aged in the DSC for 2–48 h at some temperature in the range of 10–45 °C below T_g. Aged samples were cooled again to $T_g - 50$ °C, and then reheated to $T_g + 30$ °C. Each scan was run at 5 °C min^{-1}. Enthalpy relaxation (ΔH) was determined from the area of the endothermic peak around T_g, which appeared on the DSC heating curve.

3 Results and Discussion

3.1 Thermal Properties of Polyphosphate Compounds

Figure 1 shows DSC heating curves for frozen aqueous phosphate solutions, each of which indicated a baseline shift below the ice melting peak. That shift was taken to correspond to the T_g of a maximally freeze-concentrated solution, referred to as T_g'.[9] The T_g' midpoint values – for di- and tripolyphosphate, ATP, ADP and AMP – were estimated to be -15, -39, -24, -27 and $-22°C$, respectively. Only the AMP sample showed an exothermic recrystallization peak above T_g'. All freeze-dried samples were also found to be in the glassy state, as indicated by their DSC heating curves in Figure 2, which exhibited typical glass transitions. Furthermore, the T_g values for freeze-dried samples decreased with increasing moisture content, due to the plasticizing effect of water,[9] as illustrated by the T_g–moisture content glass curves plotted in Figure 3. The solid lines in Figure 3 represent curves fitted by the following Couchman equation:[21,22]

$$T_g = \frac{X_s \cdot \Delta C_{ps} \cdot T_{gs} + X_w \cdot \Delta C_{pw} \cdot T_{gw}}{X_s \cdot \Delta C_{ps} + X_w \cdot \Delta C_{pw}} \tag{1}$$

where X_s and X_w are the mole fractions of solute and pure water, respectively. T_{gw} and ΔC_{pw} are the T_g and heat capacity increment for pure water, which were taken as $-138°C$ and $35\ J\ mole^{-1}\ °C^{-1}$ from the literature.[23] T_{gs} and ΔC_{ps} for

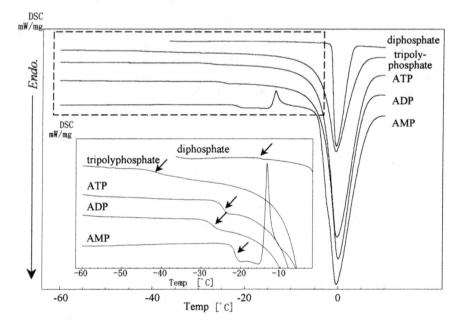

Figure 1 *DSC heating curves for $\sim 10\%$ polyphosphate solutions. Each scan was run at $2°C$ min^{-1}, except for the diphosphate sample, which was run at $0.5°C\ min^{-1}$. Arrows indicate T_g'*

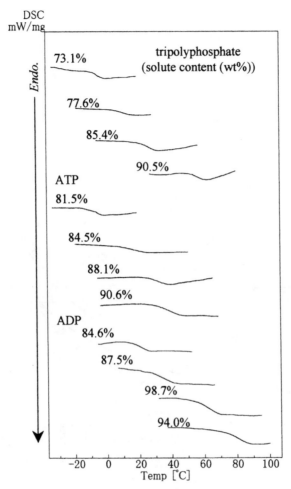

Figure 2 *Typical DSC heating curves for lyophilized tripolyphosphate, ATP and ADP, adjusted to each indicated solute content. Each scan was run at 2°C min^{-1}*

pure solute were adjustable parameters in fitting calculations. For comparison, the T_g curves for trehalose and sucrose[24] are also shown in Figure 3. Interestingly, it was found that tripolyphosphate, ATP and ADP had the same or higher T_g' and T_g values as carbohydrates of similar molecular weight. On the other hand, corresponding values for C_g', the solute concentration at T_g',[9] were found to be lower than those for trehalose and sucrose. This result suggested that polyphosphates have higher water-holding capacities. Those data are summarized in Table 1. From those results, it was suggested that the polyphosphates examined in this study may be effective stabilizers for freezing or freeze-drying, since they have high T_g and T_g' values, but do not show eutectic points.

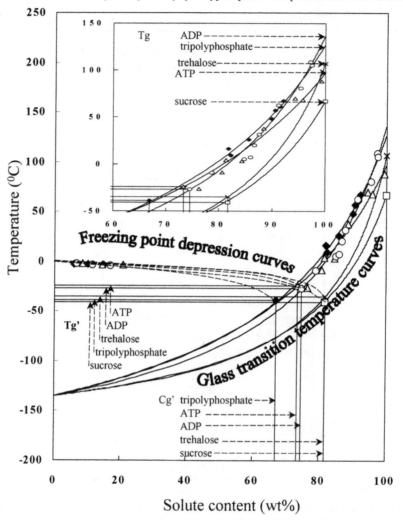

Figure 3 *State diagrams for solute–water binary systems. Data for sucrose and trehalose were taken from Roos[24]*

3.2 Stability of Glassy States

Figure 4 shows typical DSC heating curves for fresh and aged glassy samples. Whereas the fresh glassy samples showed a normal-looking glass transition, the aged samples also showed an endothermic peak around T_g, which represented recovery of relaxation enthalpy for the glassy samples. In each case, ΔH was estimated from the peak area. The samples shown in Figure 4 were all aged under the same conditions, *i.e.* 48 h at $T_g - 15\,°C$. ΔH values for ATP and ADP were significantly lower than that for trehalose. Results for aging under other conditions showed similar trends. Figure 5 illustrates the relationships between en-

Table 1 *Thermal properties of polyphosphate compounds*

Compound	T_g' (°C)	T_g (°C)	ΔC_p ($J\ mol^{-1}\ {}°C^{-1}$)	C_g' (%)	Molecular weight ($g\ mol^{-1}$)
Diphosphate	−15	–	–	–	266.06
Tripolyphosphate	−39	127	200	66.9	367.86
ATP	−24	97	355	73.4	551.1
ADP	−27	137	185	74.6	427.2
AMP	−22	–	–	–	347.2
Sucrose[24]	−41	67	195.38	81.7	342.3
Trehalose[24]	−35	107	187.27	81.6	342.3

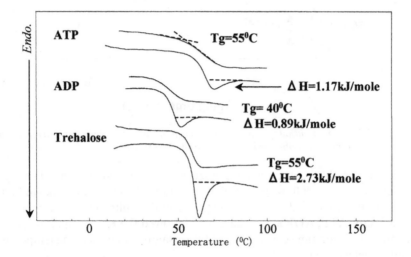

Figure 4 *Typical DSC heating curves for fresh and aged (48 h at T_g −15°C) samples of glassy ATP, ADP and trehalose*

thalpy relaxation and aging time for different aging temperatures. From Figure 5, the enthalpy relaxations for glassy ATP and ADP were found to be considerably smaller than that for trehalose. To obtain the mean relaxation time constant (τ), the enthalpy relaxation–aging time data were fitted to the Kohlrausch–Williams–Watts (KWW) equation (2):[19,25,26]

$$\Phi_t = \exp - (t/\tau)^\beta \qquad (2)$$

where t is aging time, β is a relaxation time distribution parameter, and Φ_t is a function defined by Equation 3:

$$\Phi = 1 - (\Delta H_t/\Delta H_\infty) \qquad (3)$$

where ΔH_t is the relaxation enthalpy after aging time t, and ΔH is the maximum relaxation enthalpy, calculated from Equation 4:

$$\Delta H_\infty = (T_g - T) \cdot \Delta C_p \qquad (4)$$

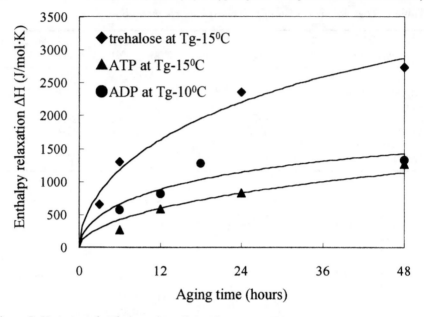

Figure 5 *Variation of enthalpy relaxation with aging time at various aging temperatures. Solid lines were obtained by fitting to the KWW equation*

where ΔC_p is the heat capacity change at a measured T_g, and T is aging temperature. The best fit with the KWW equation, shown by solid lines in Figure 5, was found to well describe the experimental results. Thus, the relaxation processes for ATP, ADP and trehalose all followed the KWW equation. Therefore, the relaxation time constants τ and β for samples at each aging temperature could be determined.

Furthermore, it was found that the relaxation rate decreased significantly with decreasing aging temperature (data not shown). At $T_g - 30°C$, enthalpy relaxation for ATP and ADP could not be detected after 48 h aging; the same was true for trehalose at 45°C below T_g. That finding may suggest that the molecular mobility of glassy ATP and ADP decreases rapidly with a small reduction in temperature.

Hancock *et al.*[19] applied the Vogel–Tammann–Fulcher (VTF) equation (Equation 5 below) for analysis of relaxation processes in glass materials below T_g. In accord with those authors, we also attempted to use the VTF equation:

$$\tau = \tau_0 \cdot \exp(T_0 \cdot D/(T - T_0)) \tag{5}$$

where τ_0, D and T_0 are defined as a hypothetical relaxation time constant for an unrestricted molecule, a parameter correlated with fragility, and an ideal temperature at which molecular mobility is completely prevented, respectively.[27–29] A plot of log τ vs T_g/T, with best-fit curves based on Equation 5, is shown in Figure 6. From those results, it was found that below T_g, τ did not follow the Arrhenius equation, in that the semi-log plot showed no linearity, but rather followed the VTF equation, increasing rapidly with decreasing temperature. The

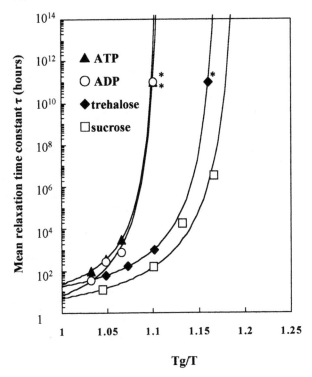

Figure 6 *Relationship between mean relaxation time, τ, and T_g/T. Solid lines were calculated from the VTF equation. Points denoted by * signify that no enthalpy recovery was observed, at least within 48 h. Data for sucrose were taken from Hancock et al.,[19] and the fitted curve was recalculated in this study*

best-fit values for parameters τ_0, D and T_0, and β obtained from Equation 2, are listed in Table 2.

In the case of amorphous polymers below T_g, parameter β is thought to represent the distribution of domain sizes that induces cooperative mobility within the glassy state.[30] In general, $\beta \approx 0.5$ for most polymers at T_g.[19,26] Furthermore, it is known that β increases as molecular weight decreases.[19,30] β values, for each material listed in Table 2, showed somewhat wide ranges that depended on aging temperature. Additionally, β values showed small differences that depended on the type of material. Interestingly, β for ATP, of relatively high molecular weight, was smaller than the other values. Indeed, all the β values obtained were considered to be within a reasonable range. It has been proposed that the lower the $T_g - T_0$ value, the more fragile and effective a material is as a stabilizer.[31] The $T_g - T_0$ values obtained for ATP, ADP and trehalose, even though they were determined from experiments below T_g, were 34, 33 and 51 °C, respectively, and ATP and ADP not only have high $T_g{'}$ and T_g, but also high T_0. Although not a definitive conclusion, speculation suggests that amorphous ATP and ADP may be more effective stabilizers than are disaccharides such as trehalose.

Table 2 *Parameters of the KWW and VFT equations*

Compound	β	τ_0 (h)	T_0 (K)	T_g-T_0 (K)	D	Moisture content (wt%)
ATP	0.53–0.55 (T_g −10 to −20 K)	0.81	294.2	34	0.39	6.7
ADP	0.54–0.62 (T_g −10 to −20 K)	0.12	278.2	33	0.48	11.2
Trehalose	0.56–0.87 (T_g −15 to −30 K)	1.10	274.5	51	0.52	8.2
Sucrose[19]	0.40–0.80 (T_g −15 to −45 K)	0.25	290	60	0.62	–

τ_0 and D values listed in Table 2 were both far from literature values for several other materials, which had been determined from experiments above T_g, *i.e.* our τ_0 values were larger and our D values were smaller.[27] These inconsistencies may be due to differences in measurement conditions, *e.g.* above *vs* below T_g. Results from this study may suggest that parameters obtained from measurements above T_g do not always coincide with those measured below T_g. Judging from our parameters, it can be said that amorphous ATP and ADP are relatively stable. However, difficulty remains with the theoretical interpretation of such parameters, as Hancock *et al.* have suggested.[19] At this stage, we think it worthwhile that the parameters were determined and could be compared, as listed in Table 2.

4 Conclusion

It was found that polyphosphates, including ATP-related compounds, show glass-forming abilities as good as those of small carbohydrates, and they have relatively high T_g values. In particular, ATP and ADP were good glass-formers with high T_g' and T_g. Moreover, from results of stability studies below T_g, it was found that ATP and ADP may form more stable, fragile glasses, even in comparison to trehalose and sucrose. Therefore, these polyphosphate materials may be expected to be protein stabilizers as effective as carbohydrates. However, the mechanism by which ATP and ADP form such stable glasses could not be determined in this study. It should be noted that ATP and ADP have many charged phosphate ester groups. Such a characteristic may affect the glass transition behaviour and stability of amorphous polyphosphates.

References

1. A. Kornberg, *J. Bacteriol.*, 1995, **177**, 491.
2. A. Kuroda, K. Nomura, R. Ohtomo, J. Kato, T. Ikeda, N. Takiguchi, H. Ohtake and A. Kornberg, *Science*, 2001, **293**, 705.
3. T. Ohsima, T. Suzuki and C. Koizumi, *Trends Food Sci. Technol.*, 1993, **4**, 157.
4. M.E. Hoke, M.L. Jahncke, J.L. Silva, J.O. Hearnsberger, R.S. Chamul and O. Suriyaphan, *J. Food Sci.*, 2000, **65**, 1083.
5. X.L. Xiong, X. Lou, C. Wang, W.G. Moody and R.J. Harmon, *J. Food Sci.*, 2000, **65**, 96.

6. N. Murase and F. Franks, *Biophys. Chem.*, 1989, **34**, 293.

7. N. Murase, P. Echlin and F. Franks, *Cryobiology*, 1991, **28**, 364.

8. F. Franks, *CryoLetters*, 1990, **11**, 93.

9. L. Slade and H. Levine, 'Water Relationships in Foods', H. Levine and L. Slade (eds.), Plenum Press, New York, 1991, 29.

10. H. Levine and L. Slade, *BioPharm*, 1992, **5**, 36.

11. L.M. Crowe, D.S. Reid and J.H. Crowe, *Biophys. J.*, 1996, **71**, 2087.

12. B.S. Chang and C.S. Randall, *Cryobiology*, 1992, **29**, 632.

13. Y.H. Roos, 'Phase Transitions in Foods', Academic Press, London, 1995, 109.

14. M.T. Averbuch-Pouchot and A. Duriff, 'Topics in Phosphate Chemistry', World Scientific Publishing, Singapore, 1996, 196.

15. H.D. Robert, 'Glass Science', 2nd ed., John Wiley & Sons, New York, 1994, 25.

16. R.H.M. Hatley and J.A. Blair, *J. Mol. Catal. B-Enzym.*, 1999, **7**, 11.

17. S.P. Duddu, G. Zhang and P.R. DalMonte, *Pharm. Res.*, 1997, **14**, 596.

18. V. Andronis and G. Zografi, *Pharm. Res.*, 1998, **15**, 835.

19. B.C. Hancock, S.L. Shamblin and G. Zografi, *Pharm. Res.*, 1995, **12**, 799.

20. P. Tong and G. Zografi, *Pharm. Res.*, 1999, **16**, 1186.

21. P.R. Couchman, *Polym. Eng. Sci.*, 1987, **27**, 618.

22. P.R. Couchman, *Macromolecules*, 1987, **20**, 1712.

23. M. Sugisaki, H. Suga and S. Seki, *Bull. Chem. Soc. Jpn.*, 1968, **41**, 2591.

24. Y.H. Roos, 'Phase Transitions in Foods', Academic Press, London, 1995, 115.

25. R.N. Haward and R.J. Young, 'The Physics of Glassy Polymers', 2nd ed., J.M. Hutchinson (ed.), Chapman & Hall, London, 1997, 85.

26. D. Champion, M. Le Meste and D. Simatos, *Trends Food Sci. Technol.*, 2000, **11**, 41.

27. C.A. Angell, R.D. Bressel, J.L. Green, H. Kanno, M. Oguni and E.J. Sare, *J. Food Eng.*, 1994, **22**, 115.

28. C.A. Angell, *J. Non-Cryst. Solids*, 1991, **131–133**, 13.

29. C.A. Angell, *Science*, 1995, **267**, 1942.

30. S. Matsuoka, 'Relaxation Phenomena in Polymers', Oxford University Press, New York, 1992, 42.

31. R.H.M. Hatley, *Pharm. Dev. Tech.*, 1997, **2**, 257.

Chemistry in Solid Amorphous Matrices

Chemistry in Solid Amorphous Matrices: Implication for Biostabilization

Michael J. Pikal

SCHOOL OF PHARMACY, UNIVERSITY OF CONNECTICUT, STORRS, CT

Introduction

This presentation is a summary of much of the recent unpublished data from our laboratories, with supporting information from the literature. The focus is on the relationship between chemical and physical stability in glassy systems, and the dynamics of a glass as measured by structural relaxation time.

The Meaning of Pharmaceutical Stability

- Many pharmaceuticals are 'unstable'
  - Degrade chemically and physically
- Instability during processing
- Instability during distribution and storage
  - Need stability for – 2 years (25 °C, if possible)
- Order of Stability:
  - Crystalline ≫ amorphous solid ≫ solution
- Many products cannot be crystallized

Stability Considerations

- How can we stabilize?
  - mechanisms of stabilization?
- Can we quickly (by a physical test) predict stability?
  - isothermal calorimetry
- direct measure of heat from degradation reaction
- determination of relaxation kinetics → predict degradation kinetics?

We wish to better understand mechanisms of stabilization, so that formulation development may proceed with greater efficiency. Secondly, a quick physical test predictive of stability would be of great value. Calorimetry could provide such a test, by measurement of relaxation time, if stability and relaxation are correlated.

Key Factors in Stabilization

- *Structure and Dynamics*
- Mobility is Important: The Formulation Should be in the Glassy 'Solid State'
  - *i.e.* a protein should be diluted in an inert *glassy* matrix to minimize mobility! (F. Franks,[1] H. Levine & L. Slade,[2] ...)
- Structure is Important for Proteins: The Formulation Should Provide a 'Non-Reactive' Solid State Conformation After Freeze-Drying
  - *i.e.*, presumably the 'native' conformation (J. Carpenter, S. Prestrelski,[3] ...)

However, glassy does not mean perfect stability, and a 'native' conformation is not always stable during storage (*i.e* otherwise, why add stabilizers & freeze-dry?)

It is commonly held that stability is much superior below the glass transition temperature, and for a protein, it is acknowledged that protein structure could be critical. That is, stability is better in the 'solid state', and with proteins, better in the native conformation.

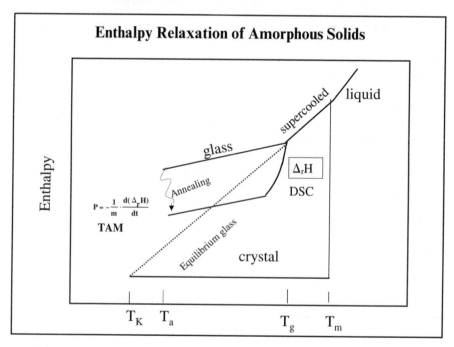

As a liquid cools and passes its melting point, T_m, it becomes more and more viscous and eventually falls out of equilibrium at the glass transition temperature, T_g. Below T_g, the H vs T curve decreases in slope, reflecting the loss of much or most of the configurational heat capacity. If the sample at temperature T_a is allowed to 'anneal' for a time, the energy decreases toward the equilibrium curve, drawn as a dashed line in the above figure. The intersection of the equilibrium curve with the crystal energy marks the 'Kauzmann temperature', T_K, at which the configurational heat capacity in the *equilibrium glass* becomes zero, and configurational mobility also becomes zero. If the sample temperature is then increased, as in a DSC experiment, the enthalpy lost during annealing is recovered at the glass transition as $\Delta_r H$. The rate of enthalpy loss, or relaxation, may be directly measured with a Thermal Activity Monitor (TAM), as the rate of heat evolved from a sample as it anneals.

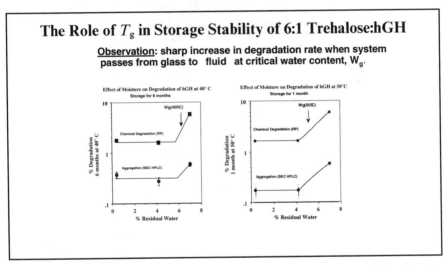

The above slide illustrates the impact of the glass transition on stability. Chemical stability here is degradation by deamidation and oxidation, while aggregation is non-covalent aggregation, mostly to dimer. Both classes of degradation processes accelerate by about a factor of five, as the water content increases to the point where the glass transition temperature has been lowered below the storage temperature, marked in the plots by the arrow. Thus, to the left of the arrow, the material is a glass with much better stability than to the right of the arrow, where the system is above the glass transition temperature.

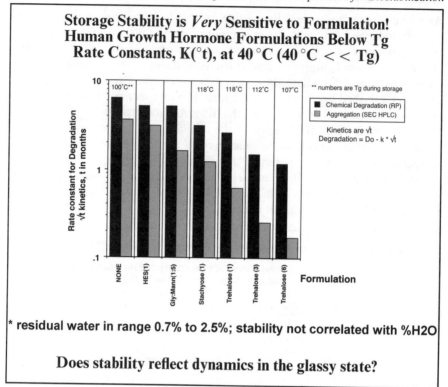

The above data show that stability of a protein, human growth hormone (hGH), is extremely sensitive to formulation, even when the storage temperature, 40 °C, is well below the T_g of all formulations. Note that the trend is qualitatively the same for the distinctly different pathways of chemical degradation and aggregation; it appears that physics, not chemistry, is dominating the stabilization.

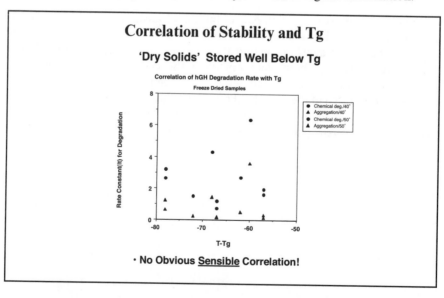

One might guess that the further the T_g is above the storage temperature, the more 'solid' is the formulation, and therefore, the more stable. The above data, taken from the bar graph on the previous page, show that there is no sensible correlation between stability and $T_g - T$. Note that the comparisons here are for samples well below the T_g.

Mobility (Motion) in Glasses

- "Low Mobility" = "Low Reactivity"
  - Motion is "fast" above T_g
- Below T_g, mobility slows greatly, but is not zero!
  - Therefore, stability is not perfect even below T_g
- Mobility in a Glass Depends on More than "$T - T_g$"
  τ = structural relaxation time, mobility = $1/\tau$
  Modified VTF Equation: $\rightarrow \tau$ depends on:
  1. $T - T_g$
  2. Strength of Glass (D)
  3. Zero mobility temperature, T_0
  4. Thermal history: fictive temperature, T_f

While mobility is much lower in the glassy state than above the glass transition temperature, mobility is not zero in a glass, and it is well known that mobility, or structural relaxation time, τ, depends not only on the temperature relative to T_g, but also depends on the strength of the glass, the Kauzmann temperature (or zero-mobility temperature), and on thermal history through the fictive temperature.

Extension of the VTF Equation to Glasses

Fictive temperature: temperature at which equilibrium glass has same configurational entropy as real glass.

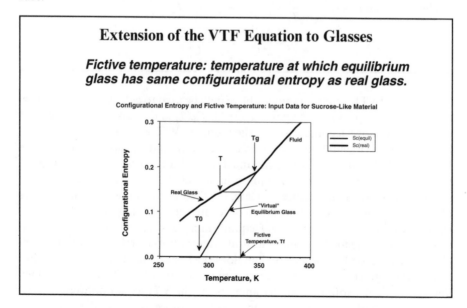

Configurational Entropy and Fictive Temperature: Input Data for Sucrose-Like Material

The extension of the VTF equation to glasses proceeds through introduction of the concept of 'fictive temperature'. As illustrated in the above figure, fictive temperature is the temperature at which the 'equilibrium glass' has the same configurational entropy as does the real glass at the temperature of interest, T. The 'equilibrium glass' is a virtual state that is an extrapolation of the 'fluid' curve above T_g into the glassy state, shown in the figure as a dashed line.

Relaxation Time in Glasses: Theory

| statistical mechanics of equilibrium fluid | + | thermodynamics of difference between glass and equil. fluid | → |

$$\tau(T,T_f) = \tau_0 \cdot \exp\left(\frac{D \cdot T_0}{T - (T/T_f) \cdot T_0}\right)$$

D = 'strength parameter', from experiment (*i.e.*, viscosity of fluid)

T_0 = 'zero mobility' temperature (*i.e*, where zero free volume), from experiment

T_f = Fictive Temperature = temperature where equilibrium fluid has same configurational entropy (*i.e.*, free volume) as the glass at T

Thermodynamics →

$$\frac{1}{T_f} = \frac{(1-\gamma_c)}{T} + \frac{\gamma_c}{T_g}; \ T \leq T_g$$

$$\gamma_c \equiv \frac{\Delta C_p^{1,g}}{\Delta C_p} = \frac{C_p^1 - C_p^g}{C_p^1 - C_p^{xstal}}$$

The modified VTF equation shown above may be used to calculate the structural relaxation time, provided the parameters D, T_0, T_f and τ_0 are all available. The above figure shows a comparison of the modified VTF equation (*i.e.* the dashed line) with experimental structural relaxation time data measured using both DSC enthalpy recovery experiments and TAM relaxation rate data. At temperatures above T_g, the system is in equilibrium, and the fictive temperature is the same as the actual temperature, T. Here, the theoretical curve has a high degree of curvature, and the extension of this curve into the glassy region (*i.e.* $T_g/T > 1$) is the relaxation curve for the 'virtual' equilibrium glass. The modified VTF equation, which is intended to be an accurate description of relaxation in the real glass, shows much less curvature than does the relaxation curve for the equilibrium state. Agreement between experiment (*i.e.* symbols) and theory (*i.e.* dashed line) is excellent! Note that the parameters for the VTF equation were derived from experiments completely independent of those used to determine the

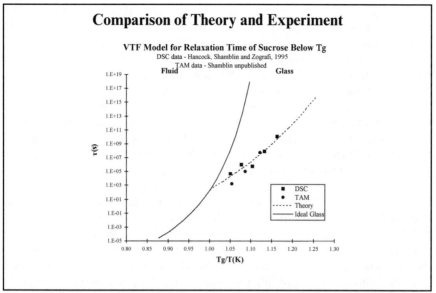

Comparison of Theory and Experiment

structural relaxation times, and the theoretical curve is not a 'fit' to the experimental τ data. The excellent agreement between theory and experiment, with no 'adjustable' parameters, suggests that we do indeed understand the major factors that influence structural relaxation in the glassy state. Clearly, although temperature relative to T_g is admittedly important in determining glass dynamics, other factors (*i.e.* the parameters in the modified VTF equation) are critical.

Is Structural Relaxation Time Relevant to Stability?

…maybe, after all, both require mobility!

Kohlraush–Williams–Watts (KWW) Eqn

Relaxation Function, $\Phi = \exp\left(-\left(\dfrac{t}{\tau}\right)^{\beta}\right)$

Relaxation Rate, relative to 'stretched time', t^{β}

$$\frac{\partial \ln \Phi}{\partial t^{\beta}} = -1/\tau^{\beta}$$

$$stability \propto \left[\frac{1}{|(\partial \ln \Phi / \partial t^{\beta})|}\right]^{c} = (\tau^{\beta})^{c} \;????$$

$c = $ 'coupling coefficient'

Note: If reaction is diffusion controlled *and* Stokes–Einstein equation is valid, $c = 1.00$

Note that the relaxation function is not a simple exponential; rather, it is a stretched exponential, where $0 > \beta < 1$. Likewise, we commonly find that degradation in glassy systems does not follow 'first-order' kinetics, with the degradation function being a simple exponential; rather, it is a stretched exponential. The glassy system is not in configurational equilibrium, and structural relaxation (as well as degradation) proceeds in parallel fashion, from a distribution of states with differing kinetics of relaxation (or degradation). It is easily shown that a number of parallel exponential decays may be described by the stretched exponential function. Since both relaxation and instability demand molecular motion, or the ability to change molecular configurations in an experimental time frame, one does expect, in general, a correlation or 'coupling' between structural relaxation and pharmaceutical stability. However, because the free volume requirement for the degradation process of interest may not be exactly the same as the free volume requirement for the configurational change involved in structural relaxation, we do not expect the relationship between stability and structural relaxation time to be one of direct proportion. Rather, we expect that stability, or reciprocal of relevant rate constant, might be proportional to some power, c, of the structural relaxation time on the stretched time scale, denoted τ^{β}. Perfect coupling, as with a diffusion-controlled reaction, where the diffusion constant is inversely proportional to viscosity (*i.e.* Stokes–Einstein equation is valid), would demand that the coupling coefficient, c, is unity. In general, c would be expected to be less than unity.

Determination of Enthalpy Relaxation by Isothermal Calorimetry (TAM) and DSC

TAM	*DSC*
Directly, the relaxation rate during aging experiment (better for kinetic studies)	Indirectly, enthalpy recovery of aged material
'Continuous' trace of relaxation of isothermal aging processes (more data points, better for comparison with theoretical models)	Measuring the 'ΔH' of samples aged for various times → limited number of data points
Experimental time: short (2–4 days)	Experimental time: long (week–months)
Higher sensitivity, in principle, no limits	Lower sensitivity: materials that relax slowly or release not very much heat cannot be studied by DSC
Sample: > 200 mg	5 × 5 mg
Sample reusable	Non-reusable
Less labor	Much more labor

The kinetics of relaxation are commonly studied by measurement of enthalpy recovery during a DSC scan after 'annealing' the sample; a direct measurement of the rate of enthalpy relaxation with a Thermal Activity Monitor (TAM) offers some advantages. Here, we fit the rate of energy release to the derivative version of the relaxation function, Φ. However, when the KWW expression is used for Φ, the derivative of the energy becomes infinite, as time approaches zero. Clearly, the KWW expression is inexact at small times, and we find that the alternative

'Modified Stretched Exponential' expression (MSE) is a better representation of the relaxation function. The MSE contains one more parameter than the KWW expression, but reduces to the KWW expression at longer times, and the derivative remains finite at zero time.

$$\Phi(t) = \exp\left[-\left(\frac{t}{\tau_0}\right) \cdot \left(1 + \frac{t}{\tau_1}\right)^{\beta-1}\right] \; (0 < \beta < 1)$$

$$\text{with } t/\tau_1 \gg 1, \; \Phi \rightarrow \exp\left(-\left(\frac{t}{\tau_D}\right)^{\beta}\right), \; \tau_D = \tau_0^{1 \cdot \beta} \cdot \tau_1^{\left(\frac{\beta-1}{\beta}\right)}$$

$$t \rightarrow 0, \; (t \ll \tau_1), \; \Phi(t) \rightarrow \exp\left(-\left(\frac{t}{\tau_0}\right)\right) \rightarrow 1 - t/\tau_0; \; \frac{d\Phi}{dt} \rightarrow -1/\tau_0$$

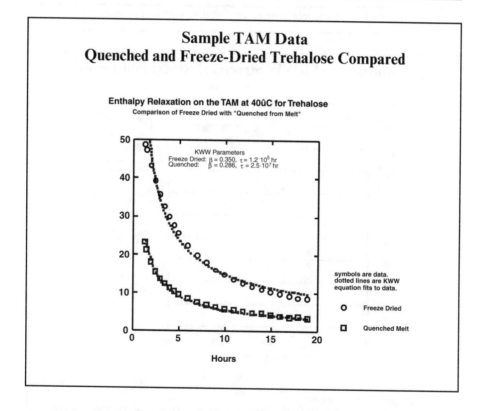

Sample TAM Data
Quenched and Freeze-Dried Trehalose Compared

Enthalpy Relaxation on the TAM at 40ûC for Trehalose
Comparison of Freeze Dried with "Quenched from Melt"

KWW Parameters
Freeze Dried: $\beta = 0.350$, $\tau = 1.2 \cdot 10^5$ hr
Quenched: $\beta = 0.286$, $\tau = 2.5 \cdot 10^7$ hr

symbols are data.
dotted lines are KWW
equation fits to data.

O   Freeze Dried

□   Quenched Melt

Hours

Sample TAM relaxation data are shown above. Note that the method of sample preparation has a profound influence on the relaxation behavior. As shown below, some evidence exists for a small amount of additional relaxation heat measured with the TAM, but as a first approximation, the heat evolved during the relaxation experiment in the TAM, as measured by integration of the TAM power signal, is equal to the enthalpy recovery measured by DSC.

Comparison of Enthalpy Relaxation:

TAM during annealing *vs* DSC enthalpy recovery

Material	$t°C$	Anneal, hr	TAM, MSE J/g	DSC J/g
Sucrose Quenched	40	90	4.1	4.4
Sucrose Freeze-dried	50	118	7.2	5.4
Sucrose Freeze-dried	50	116	7.2	6.3
Trehalose Quenched	40	90	0.95	<1

Maybe TAM measures more relaxation than enthalpy recovery???

Comparison of KWW Structural Relaxation Times

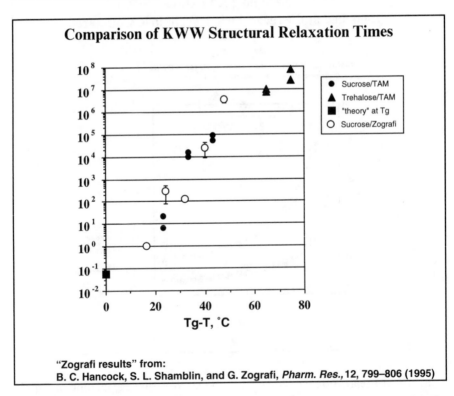

- ● Sucrose/TAM
- ▲ Trehalose/TAM
- ■ "theory" at Tg
- ○ Sucrose/Zografi

Tg-T, °C

"Zografi results" from:
B. C. Hancock, S. L. Shamblin, and G. Zografi, *Pharm. Res.,* **12, 799–806 (1995)**

Above, we compared structural relaxation times for sucrose, measured with the TAM, with those determined by DSC enthalpy recovery. Within experimental error, DSC and TAM relaxation times are the same. Below, the effect of selected variables on relaxation time is illustrated. As expected, τ becomes smaller as temperature and moisture content increase. Also, trehalose, with its much higher

T_g than sucrose, has a much longer τ at a given temperature. Finally, τ depends on method of sample preparation.

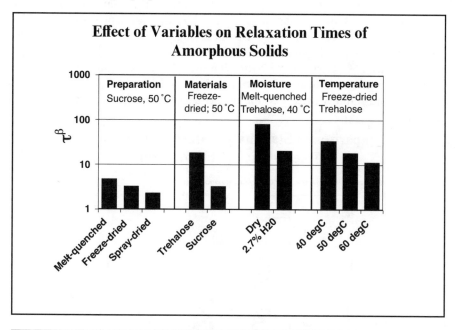

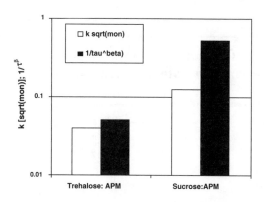

> The degradation of aspartame in sucrose and trehalose glasses shows *"square root time kinetics"* (*i.e., stretched exponential kinetics*)

> The stability of the small peptide (aspartame), with no secondary structure, *does* show correlation with the structural relaxation time.

Previous work (Bell and Hageman[5]) suggested no correlation between τ and stability in aspartame:polyvinylpyrrolidone systems. Our data suggest a good correlation in *small molecule* glassy systems. Below, the degradation rate for cefamandole Na is about a factor of 20 less than that for cephalothin Na. The rate of heat generation from the chemical degradation process, denoted as limiting power, is also about a factor of 20 less, as expected if the heats of reaction are the same. The reciprocal of τ for cefamandole Na is slightly more than a factor of 20 smaller than that for cephalothin Na, indicating good correlation between τ and stability, with a coupling coefficient somewhat less than unity.

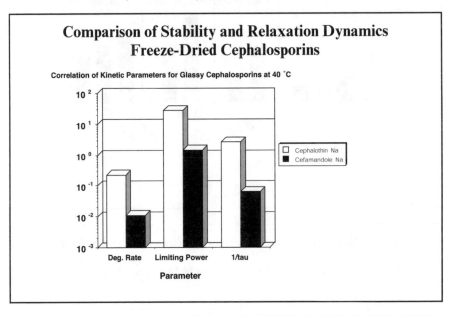

Comparison of Stability and Relaxation Dynamics Freeze-Dried Cephalosporins

Correlation of Kinetic Parameters for Glassy Cephalosporins at 40 °C

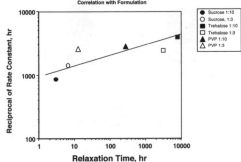

Stability of Sodium Ethacrynate: Correlation of Dimerization Rate with Relaxation Time

Correlation of Degradation Rate with Relaxation Time for Freeze-dried NaEC at 60°C

Correlation with Formulation

• **Rough correlation, but clearly other important factors.**

Sodium ethacrynate is a 'small molecule' that undergoes dimerization in amorphous systems. As noted above, much *but not all* of the stability variation with formulation correlates with structural relaxation time.

Literature data (S. Duddu and co-workers[6]), comparing aggregation rate for an antibody in a series of sucrose and trehalose formulations, suggest that stability during isothermal storage scales with the 'reduced' time, where 'reduced' time is the ratio of storage time to relaxation time. Thus, the concept of a correlation between stability and relaxation time is supported. However, it is also obvious that, at constant reduced time, stability is better in sucrose systems. One might speculate that better retention of structure is obtained in sucrose formulations.

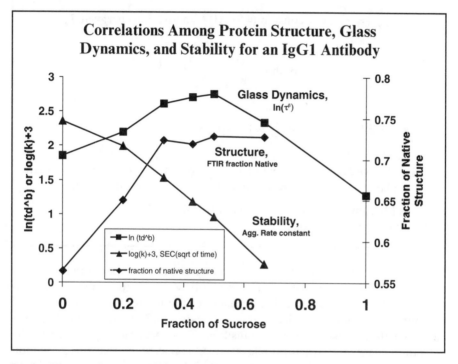

In the Figure above, stability data (rate constant for dimerization) for a high molecular weight protein are compared with structural relaxation time and retention of 'native structure' upon freeze-drying. Retention of native structure is measured by FTIR spectroscopy on the solid, compared to 100% native in the aqueous state. As the fraction of sucrose in the formulation increases in protein-rich systems, the degree of retention of native structure, denoted 'structure' above, increases, the relaxation time constant, τ^β, increases, and the rate constant for aggregation shows a steady decrease. Here, we have the expected correlations between stability and structure and between stability and relaxation time. Above about 0.4 weight fraction sucrose, the 'structure' is invariant to further increase in sucrose content, and above 0.5 weight fraction sucrose, further increases in sucrose content actually decrease the relaxation time constant, τ^β. However, stability continues to improve monotonically, as the level of sucrose increases. Clearly, in sucrose-rich systems, stability does not correlate in sensible fashion with either glass dynamics or protein structure, at least structure as measured by FTIR. It would appear that either we have structural variations that are not observed by FTIR, or another effect dominates stability behavior.

Can We Estimate τ (or D, T_0, T_f) from Width of Glass Transition?

- Moynihan's Correlation: high T_g inorganic glasses

$$\frac{\Delta H^*}{RT_g^2} = \frac{C}{\Delta T_g}; \quad C = \text{constant}(4.8); \quad \Delta H^* = \text{activation energy at } T_g$$

$$\text{may show: } \frac{\Delta H^*}{RT_g} = \frac{(T_g/T_0)}{(T_g/T_0 - 1)^2}$$

- Angell's relationship $\dfrac{T_g}{T_0} = 1 + 0.0255 \cdot D$
- Observation: for quenched fresh glass, $T_f - T_g$
- *Bottom Line*: measure width of glass transition,[2] T_g and may evaluate glass parameters, D & T_0, and then estimate relaxation time.

Is Moynihan's Correlation Generally Valid??
Answer: no! (not exactly)

Since the width of the glass transition region marks the temperature range over which the system regains (or loses) the mobility necessary to support conformational equilibrium, the width is a measure of the temperature dependence of structural relaxation. Indeed, a number of years ago, Moynihan found empirically that the width of the glass transition region was directly related to the activation energy for structural relaxation. One may show that this activation energy is also related to the parameters in the VTF equation (*i.e.* D, T_0), and with the use of the Angell relationship between T_g, T_0, and D, we can evaluate the parameters in the VTF equation. Therefore, with an estimate or evaluation of fictive temperature, one may estimate the relaxation time at any given temperature. This procedure, if it works, is of great practical significance, since one may then evaluate relaxation time as a function of temperature, simply by the usual routine DSC characterization of a solid formulation. The real question is, 'Is Moynihan's correlation of general validity?' From a combination of theoretical modeling of relaxation during a DSC scan and direct experimental studies, we find that Moynihan's correlation is only valid for a comparison of systems with essentially the same KWW stretched exponential constant, β, a condition that Moynihan alluded to himself.

Modification of Moynihan's Correlation

- Theoretical rationale of Moynihan's Correlation – assumes: τ(onset)/τ(offset) = constant
- not valid, according to our simulations
- We find (theory): τ^β(onset)/τ^β(offset) = constant

$$\frac{\tau^\beta(\text{onset})}{\tau^\beta(\text{offset})} = \exp\left(\beta \frac{\Delta H^*}{RT_g}\left(\frac{\Delta T_g}{T_g}\right)\cdot k\right) = \text{constant (7.1)}; \, k \approx 0.667$$

Therefore,

$$\beta\frac{\Delta H^*}{RT_g} = \frac{C}{\Delta T_g/T_g}; \, C \approx 3.0$$

And, with $T_f - T_g$ in fresh glass,

$$\ln(\tau^\beta) = \beta\ln \tau_{T_g} + \beta\frac{\Delta H^*}{RT_g}(T_g/T - 1) \approx \text{Constant} + \beta\frac{\Delta H^*}{RT_g}(T_g/T - 1)$$

The Moynihan correlation essentially assumes that the ratio of relaxation time at the end of the glass transition region to that at the start is a constant, regardless of material. Our modeling studies have shown that this ratio of relaxation times, raised to the β power, is roughly a constant, and therefore the relationship between activation energy and glass transition width also involves the stretched exponential power, β. With this relationship, and assuming the simplest approximation for fictive temperature, i.e. $T_f = T_g$, we may integrate the defining derivative expression for activation energy and find an expression for τ^β in terms of the activation energy for structural relaxation (i.e. in terms of the width of the glass transition region). Comparison of τ^β estimated in this way with direct evaluation of τ^β from calorimetry suggests that the width of the glass transition region may be a useful approximation (data not shown).

The following figure compares stability of human growth hormone formulations, presented in an earlier bar graph, with values of τ^β estimated from the widths of the glass transition regions. It is obvious that there are two correlation lines, one for rate constants for chemical degradation and one for rate constants for aggregation, with the slope of the aggregation correlation being greater. Recognizing that the slope of the correlation line is directly proportional to the coupling coefficient, we maintain that a steeper slope for aggregation is reasonable, since one might anticipate greater coupling between structural relaxation and aggregation than between structural relaxation and chemical decomposition. Thus, we conclude that structural relaxation and stability of hGH formulations are correlated.

Stability of hGH and Estimated τ^β

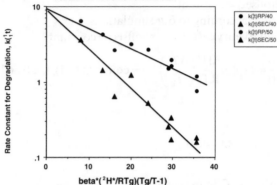

Correlation of hGH Stability with Properties of the Quenched Glass
Calculations from Glass Transition Width

Acceptable correlation; stronger coupling with aggregation, as expected.
* annealing during storage not considered here; will increase tau^beta

Conclusions

- Relaxation Time (or τ^β) does correlate with stability *sometimes*!
- Causes of poor correlation
  - With proteins, structure is important
    - Evidence from FTIR (not discussed)
  - We assume no phase separation
    - Not always a valid assumption; phase separation *may* occur in a non-crystalline system.

References

1. F. Franks, R.H.M. Hatley and S.F. Mathias, *BioPharm*, 1991, **4(9)**, 38.
2. H. Levine and L. Slade, *BioPharm*, 1992, **5(4)**, 36.
3. J.F. Carpenter, S.J. Prestrelski and T. Arakawa, *Arch. Biochem. Biophys.*, 1993, **303**, 456.
4. B.C. Hancock, S.L. Shamblin and G. Zografi, *Pharm. Res.*, 1995, **12**, 799.
5. L.N. Bell and M.J. Hageman, *J. Agric. Food Chem.*, 1994, **42**, 2398.
6. S.P. Duddu, G. Zhang and P.R. Dal Monte, *Pharm. Res.*, 1997, **14**, 596.

Residual Water, its Measurement, and its Effects on Product Stability

Residual Water in Amorphous Solids: Measurement and Effects on Stability

David Lechuga-Ballesteros, Danforth P. Miller and Jiang Zhang

INHALE THERAPEUTIC SYSTEMS, INC., SAN CARLOS, CALIFORNIA, USA

1 Introduction

Pharmaceutical and food technologies rely on drying processes as a means of improving the shelf-lives of commercial products, which are usually complex mixtures. For example, the main components of low-moisture foods are carbohydrates, lipids, proteins and water. Similarly, freeze- or spray-dried therapeutic products that contain low molecular weight drug molecules, peptides or proteins are usually prepared with saccharides, polyols, amino acids, buffer salts, polymers and water.

To increase product stability, liquid water is typically removed by sublimation or evaporation. In many cases, these processes begin with a dilute solution and end with the formation of a solid phase, which exists in a molecularly disordered amorphous state, as it undergoes a glass transition. Long-range molecular mobility of the solid decreases dramatically below the temperature range of the glass transition (T_g), and the rates of physical and chemical processes are significantly reduced as well.[1]

In practice, glassy solids ordinarily contain a finite amount of water, which is a function of the final drying conditions and overall composition. During storage and use, water vapor may permeate packaging materials and further hydrate the glassy phase. As a means to ensure the stability of amorphous foodstuffs and pharmaceutical products, the importance of controlling product water content and temperature during manufacturing and storage has been well-established during the last decade.[1] One practical, albeit simplistic, conclusion is that a thorough understanding of the effects of water on the stability of glassy solids is necessary to design any successful commercial product.[1-3]

2 Hydration of Glassy Solids

2.1 Residual Water

The term 'residual water' is often arbitrarily used in the literature to refer to the total amount of water present in a solid. Water content in amorphous solids is commonly measured by chemical, gravimetric and spectroscopic methods. Karl Fischer methods use coulometric or volumetric titrations in non-aqueous solvents to stoichiometrically determine the amount of water that participates in a chemical reaction. Thermogravimetric techniques are used to infer residual water content from weight loss upon heating. More recently, spectroscopic means (IR, near IR, Raman) have been used to non-destructively measure residual water. Hybrid techniques have also been used to add specificity. For example, TGA-IR uses infrared spectroscopy to identify the solvent lost during heating. The classic 'weight loss on drying' technique to quantify water content has been made specific by using molecular sieves to selectively quantify organic and water vapor streams liberated from a sample during heating. Although these methods provide a measure of the total amount of water in a solid product, not all of them provide information on the 'state' or distribution of water in a glassy solid. Determination of the distribution of water of hydration is useful to establish appropriate controls for water content during manufacturing, storage and use.[3,4]

The state of water within a glassy solid is determined by the degree of interaction, primarily through hydrogen bonding, between water molecules and the glassy solid. The distribution of water within an amorphous solid is greatly influenced by the chemical nature (*i.e.* hydrogen-bonding ability) and physical state (*i.e.* glassy, partially crystalline, crystalline hydrate, *etc.*) of the solid. In addition, the physicochemical properties of the glassy state (*i.e.* molecular mobility, stickiness, viscosity changes, structural collapse, crystallization, *etc.*) are functions of hydration.[1]

2.2 Water Vapor Sorption Isotherms

Gravimetric determination of water sorption isotherms has become one of the most common experimental approaches to investigate interactions between water vapor and amorphous or partially amorphous solids. A water sorption isotherm shows the functional relationship between the *equilibrium* water content of an amorphous solid and the relative humidity (RH) at a constant temperature (see Figure 1). The term *sorption* is used to denote both *adsorption* and *absorption*. Automated commercial instruments equipped with sensitive microbalances allow for rapid measurement of an isotherm with minimal sample requirements.

The distinction between thermodynamic equilibrium and steady state deserves special attention, since by definition, glassy solids represent a non-equilibrium state, and technically, water vapor and a glassy solid never reach thermodynamic equilibrium. However, it is not uncommon to denote an equilibrium condition, when changes are not observed during the time course of an experiment. This is a reasonable assumption for poly(vinyl pyrrolidone) [PVP] K90 and lyophilized trypsin, as shown in Figure 1. However, in the case of sucrose, state changes during the experimental determination of a sorption isotherm need to be considered. As shown in Figure 1, the water-induced crystallization of sucrose is observed as a loss of water content.

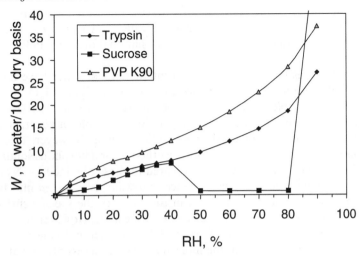

Figure 1 *Water sorption isotherms for lyophilized trypsin, sucrose and PVP K90 at 25 °C. Isotherms were determined gravimetrically using an automated water sorption system*

2.2.1 Water Vapor Sorption Isotherm Models. The use of adsorption models to fit a water sorption isotherm, in combination with measurements of T_g as a function of water content, has served as a practical basis for the development of stable glassy commercial products.[3-5] Typically, isotherm data are measured at discrete RH values, and the isotherm is constructed by fitting data to an equation, which may or may not have physical meaning. Some understanding of the mechanism of water sorption can be gained through use of the BET and GAB adsorption models.[4,6] The GAB[7-9] model can often describe a water sorption isotherm over the entire range of relative humidity, using three fitting parameters. In contrast, the BET[10] model requires only two fitting parameters, but is often unable to fit an isotherm over the entire RH range. The concept of monolayer coverage of adsorbate is common to both models.

The classical description of the BET model refers to the adsorption of gases onto a molecularly impervious solid surface. In the context of BET theory, the change in curvature in a Type II adsorption isotherm is related to the saturation of energetically equivalent adsorption sites, also known as monolayer coverage or monolayer capacity solids.[10] While there is some debate about the true meaning of monolayer coverage for water sorption into amorphous solids, the concept provides a useful framework for describing many important effects of water on the molecular structure, mobility and stability of glassy solids.

2.2.2 Analogy to Adsorption Isotherms. Organic glassy solids can be classified into several families, among them mono-, oligo- and polysaccharides; amorphous polymers; and peptides and proteins. In general, these glassy solids interact with water vapor in a unique way in comparison to crystalline or inorganic glassy solids (such as window glass). In the latter cases, water molecules are only *adsorbed* to solid surfaces. In contrast, for organic glassy solids, water molecules are not confined to surfaces, but rather they freely diffuse throughout the solid phase. Thus, at a given water activity, the water content of an amorphous organic solid is much greater than that of its crystalline counterpart.

Such differences between the types of interactions of water have been illustrated by calculations of specific surface areas based on N_2 and H_2O 'adsorption' isotherms. In principle, both adsorbates should reflect the same surface area, as dictated by the amount of adsorbate at the 'BET' monolayer coverage and the adsorbate's respective molecular cross-sectional area. However, calculations based on BET analysis of water sorption isotherms for amorphous solids greatly overestimate specific surface areas, compared to those obtained for N_2 adsorption.[11-13] This has been used as evidence that water is *absorbed* into the bulk structure of amorphous solids. Thus, in this context, the specific surface area calculated from BET analysis has no geometric meaning. However, it does raise a relevant question: what is the 'surface' available to water molecules? The answer would involve a description of the molecular regions accessible to water vapor within the structure of an organic glass.

The term 'hydrated glassy solid' is often used, when organic glassy solids incorporate water molecules into their solid structure. When the solvent is water, the term 'hydration' is preferred to the more general term 'solvation'. Usually, the component in excess is regarded as the solvent, though the definition of what constitutes a solvent or a solute is arbitrary. Thus, it can be said that, when water vapor is incorporated into a glassy solid, the glassy solid solvates the water molecules. Solvation denotes interaction between solvent and solute. For solvation to occur, solvent–solvent and solute–solute bonds are broken, and non-covalent solvent–solute bonds are formed. Therefore, hydration of a glassy solid involves hydrogen bonding between water molecules and the molecules that constitute the glassy solid.

Given that absorption is not only a surface phenomenon, and it actually resembles a solvation process, the term 'hydration limit' would seem more appropriate than 'BET monolayer' or 'monolayer hydration' in the context of water vapor absorption. For consistency and accuracy in the description of this parameter, we have chosen to replace, where appropriate, reference to the 'BET monolayer' with the term 'hydration limit'.

2.2.3 Plasticization (Water-induced T_g Depression).   The hydration of an organic glassy solid affects its T_g; this effect is known as plasticization.[1] Plasticization is not unique for water; any miscible solvent or low molecular weight species (buffers, salts, additives) that is incorporated into an amorphous phase will depress its T_g. For hydrated glasses, the degree of plasticization is a function of water content, and it is material-dependent, as shown in Figure 2. Plasticization of pharmaceutical materials[14] and model food systems[1,15] has been extensively studied.

A non-linear fit of experimental $T_g(W)$ data is often performed to describe

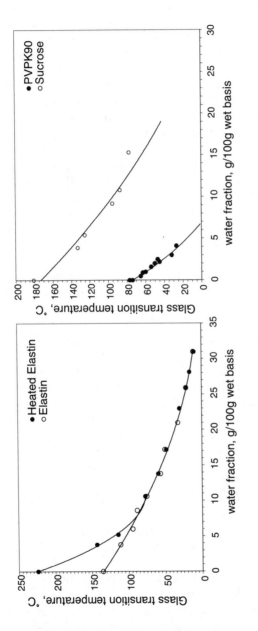

Figure 2 *The effect of residual water on the calorimetric* T_g *of (left) lyophilized elastin in globular and denatured (after heat treatment) states and (right) amorphous PVP K90 and sucrose. The extent of* T_g *depression is material-dependent and reflects changes in the free volume of the hydrated glass.* T_g *data were measured by differential scanning calorimetry. Lines denote trends in the data and do not represent curve fits to any equation*

plasticization as a function of other operational variables such as RH.[3] In general, the T_g of a mixture can be predicted based on either the free volume or thermodynamic properties of the individual components. Many models have been used to predict the water sorption-induced plasticization effect. For a comprehensive review, the reader is referred to the book on the fundamentals of glassy systems by Roos.[6]

The Gordon–Taylor (GT) model[16] has been used to predict the T_g of mixtures based on T_g of the pure components and their volume fractions. For a binary mixture, the single adjustable parameter of the GT model takes into account the change in free volume upon ideal mixing. For ideal mixing of water and a glassy phase, there is no excess free volume, and the GT equation should accurately predict T_g. In practice, this equation has been shown to be reasonably accurate for many types of pharmaceutical mixtures. Furthermore, the GT equation is sometimes used to estimate composition based on a single T_g measurement. Differences between the actual T_g and that predicted using the GT model have been interpreted as deviations from ideal mixing of free volume. Such deviations can be positive or negative, depending on the strength of the interactions between water and the solid phase.[14]

The GT equation can also be used to predict the T_g of multi-component mixtures. However, estimation of the equation's parameters becomes difficult, even for ternary mixtures such as protein/excipient/water. In that case, it is often difficult to measure the calorimetric T_g of a pure globular protein.

Another commonly used model, known as the Couchman–Karasz (CK) model, treats the glass transition as a thermodynamic second-order phase transition. The CK model has a functional form that is similar to the GT equation. For a binary mixture, the CK model has a single adjustable parameter that is a ratio of quantities typically measured using differential scanning calorimetry, *i.e.* the ΔC_p at T_g for each of the pure components. This model uses a linear mixing rule for heat capacity changes at T_g. The CK approach has been successfully applied to food and pharmaceutical systems.[6] However, experimental determination of ΔC_p changes at T_g for strong glasses, such as those of globular proteins, is often difficult.

Like the GT model, the CK model has been successfully extended to multi-component systems. Also like the GT model, the CK model is often difficult to apply to simple protein/excipient/water mixtures. Typically, the difficulty in measuring T_g of a pure protein precludes measurement of its heat capacity change at T_g.

Plasticization of many amorphous solids is characterized by a much steeper decrease in T_g at lower water contents than at higher water contents (see Figure 2). Although this effect depends on the specific chemical nature of a glassy solid, it is thought that it is due to the larger free volume change per mole of sorbed water molecules induced by the first water molecules. As the water content increases, the fraction of solid–water interactions decreases, and the plasticizing efficiency of water is reduced. Eventually, the affinity of water for the solid decreases so much that water preferentially associates with itself. It has also been suggested that, for amorphous proteins, plasticization is greatly reduced when the hydrogen-bonding capacity of the solid is exceeded.[17,18]

2.2.4 Water Sorption Isotherms and Solution Models. Solution theories provide an alternative approach to binding site-based (*i.e.* BET) interpretations of water vapor sorption. In this approach, water vapor absorption into amorphous solids is treated as a dissolution process. Classical solution models such as Flory–Huggins[19] describe the enthalpic (solid–solvent interactions) and entropic (volumetric changes) contributions to mixing free energy. This model has been shown to successfully predict water sorption isotherms for several amorphous solids in the rubbery state (*i.e.* $T_{storage} > T_g$).[18,20]

Two variations of the Flory–Huggins model have been independently developed to incorporate the free energy change due to non-ideal volumetric changes induced by water absorption in an amorphous solid. These refinements have been shown to predict the entire water sorption isotherm of an amorphous solid, in both the glassy and rubbery states.

In one approach, measurements of water sorption-induced compressibility changes in keratin were used to estimate free energy changes. These free energy changes were then used to predict the water sorption isotherm for keratin.[21]

In the second approach, water sorption-induced free energy changes have been described based on plasticization of glassy polymers, such as poly(methyl acrylate).[22,23] The Vrentas approach has been successfully applied to predict the sorption isotherms for other more hydrophilic glassy solids such as PVP,[18,20] glucose, trehalose and dextran.[20] The Vrentas model has also been used to predict the sorption properties of binary amorphous dispersions of sucrose–PVP and trehalose–PVP.[24]

The isotherms of co-lyophilized sugar–PVP mixtures have been reasonably described with an extended three-component Flory–Huggins solution model and Vrentas structural relaxation model. The solute–water interaction parameters for a binary solute system are similar to those for the corresponding single solute systems, suggesting that solute–water interactions are not significantly affected by sugar–PVP interactions.[24] This is consistent with results from spectroscopic analysis of sucrose–PVP dispersions, which suggested that water and sucrose 'bind' to different sites on a PVP molecule.[25-27] The nature of sugar–PVP and water–PVP interactions has been investigated by Raman spectroscopy, and these interactions all seem to occur *via* hydrogen bonding. In addition, results from ^{13}C NMR analysis of hydrated glassy dispersions of PVP–sucrose demonstrated that sugar–PVP interactions, as measured by the reduced mobility of sucrose, had no significant effect on the hydration of the glassy phase.[27]

It is clear that a mechanistic interpretation of water absorption isotherms for amorphous solids must take into account both molecular interactions and effects of water on the structure of the solid. Although site-binding (*e.g.* BET) and solution models provide two different interpretations of interactions of water with amorphous solids, there are commonalities between the two approaches. Recently, an attempt has been made to correlate the parameters obtained from BET and Vrentas analyses of the water sorption isotherms of amorphous solids.[20] The two BET parameters with physical significance are the amount of water needed to achieve monolayer coverage or the hydration limit, W_m, and the energy term, C_B, which is related to overall energy of absorption. Conversely, the

Vrentas solution model is described by two parameters, the Flory–Huggins affinity parameter, χ, which is a measure of water–solid interactions, and a parameter that accounts for the free energy change due to plasticization, f. It was found that W_m from BET correlates with the Flory–Huggins affinity parameter, and that the C_B (BET) parameter correlates with the energy change due to water–induced structural relaxation, f. Amorphous solids that contain the same monomer unit have similar W_m, independent of the state of the system (*i.e.* glassy *vs* rubbery). This is one indication that W_m is strongly dependent on the chemical nature of an amorphous solid, and that it depends on the number of available hydration sites. A comparison of W_m values for different classes of glassy solids (*e.g.* proteins, sugars and PVP) is shown in Tables 1 and 2.

2.3 Property Changes at the Hydration Limit

A number of physicochemical properties change at the hydration limit. Consider, for example, the change in physicochemical properties of lysozyme at the hydration limit, which is at about 0.06 g water per g dry solid (see Figure 3). Changes are observed in diamagnetic susceptibility, heat capacity, and protein conformational state (as illustrated by the changes in carboxylate absorbance at $1580 \, \text{cm}^{-1}$ and amide-1 shift at $1660 \, \text{cm}^{-1}$). In addition, a second discontinuity is observed in some properties at about 0.12 g water per g dry solid.

Analysis of lysozyme's hydration by time domain dielectric spectroscopy indicated the existence of a population of 32 water molecules per lysozyme molecule, which, during their lifetime at the lysozyme surface, are irrotationally bound to the lysozyme structure (see Figure 4).[32] These water molecules appear to constitute the hydration limit. Above the hydration limit, the number of water molecules irrotationally bound to the protein seems to be conserved (8 MHz frequency domain). However, the proportion of high-mobility water molecules (1 GHz frequency domain) increases rapidly for hydration levels above the

Table 1 *Hydration limit,* W_m, *for selected proteins*

Substance	W_m % w/w dry basis	MW kD	Reference
Isolated Soy	3.5	NA	28
rhIFN-γ	4.5	16.4	29
rhIGF-I	4.7	7.65	29
rRhGH	5.0	22.3	29
Lipase	5.1	40–60	28
β-Glucoronidase	5.4	68	28
rhuMAbE25	5.7	163	29
α-Amilase	6.0	46–52	28
Urease	6.0	480	28
rtPA	6.6	63.6	29
rhDnase	8.0	32.7	29

W_m values from Costantino *et al.*[29] were calculated from protein stoichiometry. W_m values from Teng *et al.*[28] were experimentally determined from water sorption isotherms at 25 °C.

Table 2 *Hydration limit, W_m, for selected carbohydrates and PVP*

Substance	W_m, % w/w dry basis	C_B	MW kD	T_g, °C	Reference
PVA 103	3.5	<1	20	73	30
PVA 125	3.8	<1	125	74	30
Glucose	5.4	0.3	0.180	30	20
Dextran	6.2	13.5	40	200	20
Trehalose	6.4	5.0	0.342	115	20
Starch	6.6	17.9	100	225	20
PVP 17	10.1	2–4	10	156	30
PVP 90	11.0	6.2	1000–2000	100	20
PVP 12	11.1	2.1	4	148	30
PVP 30	12.0	4.8	40	150	20
PVP 12	14.0	2.1	4	175	20
PVP LMW	>14	<2	<3.5	175	31

Data from Lai *et al.*[30] were at 50 °C; other data at 30 °C.

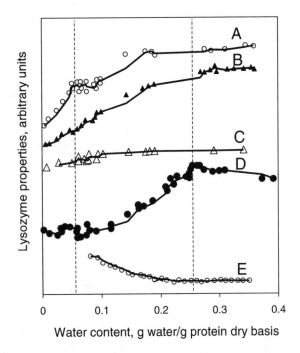

Figure 3 *Measured dependence of some lysozyme properties on water content at 25 °C: (A) carboxylate absorbance at 1580 cm^{-1}; (B) amide-1 shift at 1660 cm^{-1}; (C) OD stretching frequency at 2570 cm^{-1}; (D) apparent specific heat capacity; and (E) diamagnetic susceptibility. From Rupley* et al.[33] *in Hageman[5]*

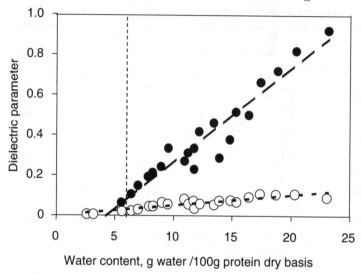

Figure 4 *Dependence of the dielectric parameter, F(ε), on water content for samples stored under various RH conditions at 25°C: low frequency, 8 MHz (open circles), and high frequency, 1 GHz (closed circles) dispersion data. A glassy sample is subjected to an electric field of various frequencies, in an experiment to probe the mobility of water within the hydrated sample. The mobility of water molecules is proportional to the frequency imposed. Two distinct populations of water with different mobilities are observed. See Bone[32] for additional experimental information*

hydration limit. In that case, a larger population of water molecules is relatively free to respond to the electric field and exhibits a dipole moment close to that of vapor phase water molecules. Discontinuities in dielectric characteristics, in the range 0.1 to 0.14 g water per g dry solid (see Figure 4), correlate with IR spectroscopic changes corresponding with the carboxylate and amide-1 groups (Figure 3), and could be due to hydration-induced conformational changes.

2.3.1 Isothermal Heat of Hydration.   As discussed above, analysis of gravimetric water sorption isotherms provides some insight into interactions between water and amorphous solids. A less common approach involves isothermal calorimetric measurement of the energy of interaction between water molecules and a glassy solid as a function of RH (see Figure 5). The isothermal heat of hydration displays a distinct change in the energy of interaction between water molecules and a glassy solid at the storage RH corresponding to the hydration limit, RH_m. At this point, the heat of interaction approaches the heat of condensation of water.[34] This is consistent with a high energy of interaction at low water contents, due to the preferred interaction between water molecules and the glassy solid. At higher water contents, there is a decrease in the hydrogen bonding capacity of the glassy solid, which is eventually exceeded at the hydration limit. That is, at hydration levels above the hydration limit, water molecules can no

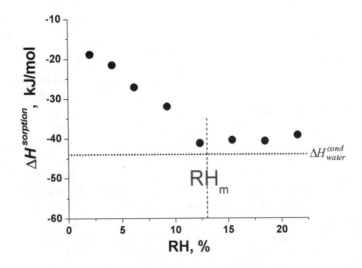

Figure 5 *Molar enthalpy of hydration of sucrose at 25°C. Reported values are calculated from isothermal sorption–desorption experiments on the same batch of sample as that shown in Figure 6. The dotted line indicates the latent heat of condensation of water at 25°C. At low RH (or low residual water), interactions between water molecules and glassy sucrose are more energetic than water–water interactions. At RH_m, the water content reaches the hydration limit, and water–water interactions dominate, as suggested by the heat of sorption, which approaches the heat of condensation of pure water. See Lechuga-Ballesteros et al.[35] for experimental details*

longer hydrogen-bond with the glassy solid, and water–water interactions become predominant.

Another important feature of the heat of interaction of water with an amorphous solid is that the interactions between water vapor and glassy solid are reversible below the hydration limit.[35] The energy hysteresis for sorption–desorption cycles, shown in Figure 6, is only observed above RH_m. This suggests that exceeding the hydration limit induces irreversible structural changes.

2.3.2 The Effect of Temperature on the Hydration Limit. The amount of water needed to reach the hydration limit in a lysozyme sample increases as temperature is decreased[32] (see Figure 7). This behavior resembles the dependence of solubility of gases into a condensed phase, which also decreases as temperature increases. Dissolution of gases in liquids involves changes in heat of solvation and entropy. The entropy of the gas is greater than the entropy of a solvated gas, so an increase in temperature should favor the free gas over the solvated gas.

2.3.3 Is W_m a Critical Property? An important outcome of a mechanistic study of hydration isotherms for glassy solids is that the hydration limit, W_m, has an important physical meaning. Both the heat of hydration, approaching that of water condensation at W_m, as well as its temperature dependence, support the analogy of the hydration limit to the solubility of water vapor in glassy solids. Solubility, of course, denotes thermodynamic equilibrium, and the hydration limit is, at best, a *quasi*-equilibrium state with long relaxation times. For hydrated samples, the storage RH at which the hydration limit is reached achieved is denoted as RH_m. This can be thought of as the amorphous state analog of RH_o, the deliquescence point of a crystalline material. For a crystalline solid, RH_o denotes the water activity of a saturated solution in equilibrium with the crystalline solid. Below RH_o, crystalline solids are physically stable; above RH_o, the solid dissolves. It has been proposed by Lechuga-Ballesteros and collaborators[35] that RH_m has a similar effect on an amorphous solid in equilibrium with water vapor. Below RH_m, a glassy organic material is physically stable; above RH_m, water vapor condenses into liquid water, and a supersaturated (with respect to the solubility of the amorphous solid) solution is formed, which is physically unstable. This is consistent with the fact that water vapor interactions with a glassy solid are reversible below RH_m,[35] as shown in Figure 6.

3 The State of Water Within Amorphous Solids

The state of water refers to its molecular mobility in a hydrated glassy phase, which depends on the degree of interaction with the glassy solid. Water interacts in similar ways with all water-soluble glassy solids. For the sake of simplicity, we have subdivided this section by type of glassy solid (*i.e.* protein, polymer and small molecule solids).

3.1 Proteins

The state of water in hydrated glassy proteins has been studied for many years.[1,4,11,36] It is now generally accepted that water interacts with glassy proteins in stoichiometric proportions *via* nonspecific interactions. Pauling recognized the relationship between the amount of water bound to a protein and the protein's chemical structure. He suggested that the number of moles of water in a monolayer roughly corresponds to the number of moles of polar amino acid side chains in a protein, minus the number of polar amino acid side chains that are able to form hydrogen bonds with imido groups.[11] Thus, water sorption is a consequence of polar amino acid side chains that are unable to satisfy their intramolecular hydrogen bonding requirements.

The amount of water sorbed by globular proteins has been calculated with a reasonable degree of accuracy.[37] The calculation is based on protein sequence and structure, in combination with experimentally determined hydrogen-bonding affinity coefficients for the different functional groups, as a function of the RH above the glassy protein.

In general, empirical and semi-empirical models to describe water sorption properties of proteins need to be refined to account for water-induced conforma-

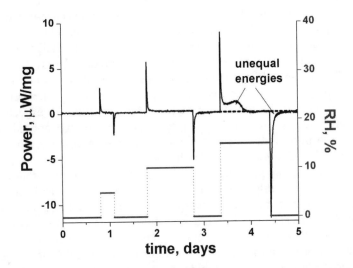

Figure 6 *Heat flow for hydration–dehydration of sucrose at 25°C as a function of relative humidity (exothermic = upward). Above RH_m, the RH corresponding to the hydration limit, energy hysteresis due to irreversible structural changes below the calorimetric T_g is observed (sample $T_g \approx 55°C$ at RH_m, i.e. experiment is performed 30°C below T_g). The area under the curve for each peak is the total heat of hydration, which, divided by the corresponding total amount of water sorbed, yields the heat of hydration shown in Figure 5. See Lechuga-Ballesteros et al.[35] for experimental details*

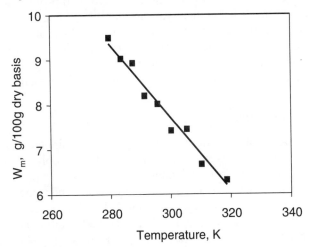

Figure 7 *Temperature dependence of the hydration limit calculated from BET analysis of experimentally determined water vapor sorption isotherms of lysozyme at various temperatures. From Bone[32]*

tional changes that in turn affect a protein's absorption properties.[38] The problem of water sorption-induced protein folding-unfolding is age-old and has not yet been fully resolved. An amino acid sequence alone is insufficient to completely specify a local conformation, because non-specific long-range forces tend to impart a protein structure different from that predicted only from covalent bonds. The magnitude of the energy of absorption of water (and other polar molecules) by proteins is comparable to that needed to stabilize a native configuration, so that the solvation resulting from absorption is expected to progressively modify protein conformation. Models have been proposed to account for water sorption-induced conformational changes in proteins, which result in denaturation and formation of a 'molten globule'.[39]

The stoichiometric approach to predicting water sorption isotherms for proteins works reasonably well at low water contents, and has been shown to correlate well with the BET adsorption model. Data in Table 1 show the hydration limits, calculated from a BET analysis of experimentally determined water sorption isotherms, for several proteins. The estimated W_m values are in good agreement with theoretical predictions based on amino acid composition of the *accessible* surface of a globular protein and on its relative propensity to bind water molecules.[29]

Detailed spectroscopic studies (NMR, dielectric resonance spectroscopy (DRS)) have helped to identify the nature of interactions of water molecules with glassy proteins. The time scale for molecular mobility is a convenient measure of the strength of association of water with a glassy solid. For example, water molecules bound to polar groups on a protein surface exchange with bulk water on time scales ranging from nanoseconds in highly hydrated proteins to tens of seconds at hydration levels below the hydration limit.[40] Water molecules that form hydrogen bonds with accessible hydrophilic groups of a protein are collectively referred to as a hydration shell, which has been suggested to correspond to the BET monolayer. Because of its low molecular mobility, water in the hydration shell remains unfrozen upon cooling to sub-zero (°C) temperatures.[11,36]

3.2 Amorphous Polymers

The interaction of water vapor with amorphous polymers is similar in nature to its interaction with proteins. Water molecules absorbed into a polymer structure form hydrogen bonds with specific functional groups of the polymer. The amount of water sorbed per unit mass of dry polymer is independent of the polymer's molecular weight,[20] as shown in Table 2. This is also reflected in the similar plasticization patterns observed for PVPs of different molecular weights. That is, although PVP K12 (MW 4 kD) and PVP K90 (MW 1000–2000 kD) have different intrinsic T_g values (*i.e.* T_g of the dry amorphous solid), both polymers exhibit comparable water sorption-induced T_g depressions.[41]

FT-Raman spectroscopy has been used to show that the strength of hydrogen bonds formed between water and a given polymer depends on the chemical nature of the polymer. For example, comparison of the hydration of PVP and poly(vinyl pyrrolidone-*co*-vinyl acetate) (PVP/VA) shows that water molecules interact more strongly with the pyrrolidone group than the acetate group. Although both polymers have similar intrinsic T_g values, PVP/VA has a consistently lower T_g at comparable water contents.[26] Thus, since water interacts more strongly with the pyrrolidone group, it is inferred that water has lower mobility and is therefore less able to plasticize PVP.

Detailed solid-state NMR analyses on hydrated samples of PVP demonstrated that there were at least two populations of water with different mobilities.[41] Furthermore, the translational and rotational mobility of water decreased in the hydrated glassy polymer below the hydration limit. Note, however, that even below the hydration limit, water mobility does not cease completely.[41] Aldous *et al.*[42] showed, for example, that water can be removed from an amorphous matrix by drying at temperatures below T_g.

3.3 Low Molecular Weight Glassy Solids

In contrast to proteins or polymers, some low molecular weight organic glassy solids have a high propensity to crystallize following water sorption. Such systems are widely studied, since crystallization can be used as an intrinsic marker of molecular mobility. Although crystallization is a convenient indicator of physical instability, other physical stability issues such as stickiness or structural collapse usually precede crystallization.[1] In recent years, it has been reported that crystallization can occur at temperatures well below T_g.[43] The effects of temperature and water content on collapse and crystallization of oligosaccharides have been extensively studied,[1,15,44-47] and it is generally accepted that, for crystallization to be observed on practical time scales, an amorphous sample needs to be stored near or above its T_g. Nucleation and crystallization are diffusion-controlled processes and correlate with the overall molecular mobility of a glassy solid.[1] As the storage temperature decreases below T_g, the onset of crystallization takes longer to be observed. It has been suggested that molecular mobility below the hydration limit is low enough to prevent crystallization from occurring for years.[35]

3.4 Water Sorption by Molecularly Mixed Organic Glassy Solids

Many important pharmaceutical formulations consist of molecularly mixed glassy solids. In many cases, this is by design, since glassy substances have been found to protect therapeutic molecules (*e.g.* proteins). A debate about the protective mechanism of glassy solids centers around two prevailing hypotheses: the vitrification and water replacement hypotheses.

The vitrification hypothesis posits that the ability to form an amorphous solid is important for long-term storage of biologicals.[1] In this view, a glass restricts the molecular motions that lead to loss of structure and function. In support of this idea, Green *et al.*[48] noted a correlation between the T_g values for several mono- and disaccharides and their ability to protect Ca^{2+}-transporting microsomes during lyophilization.

An alternative mechanism of preservation during dehydration was first suggested in 1971 by Crowe[49] and has come to be known as the 'water replacement hypothesis'. To account for the ability of certain organisms to survive during desiccation, Crowe proposed that certain physiological solutes replace the lost water around polar residues of biological macromolecules. Carpenter and Crowe[50] suggested that this might be the same mechanism by which certain solutes maintain native protein structure during dehydration. According to those authors, some of the hydroxyl groups of a saccharide form hydrogen bonds with polar residues of proteins.

As discussed above, the hydration of a glassy solid depends on its chemical structure. Generalizations can be made across different materials; however, the chemical nature of individual constituents cannot be ignored. For a mixture of two or more solids (either amorphous or crystalline) in which there are no interactions among components, the amount of water sorbed is simply given by the weighted sum of the amount sorbed by each individual component.[51,52] See, for example, Figure 8.

Even though the mixtures shown in Figure 8 were freeze-dried from a solution containing both protein and mannitol, crystallization of mannitol promoted the formation of a product with two distinct phases, glassy protein and crystalline mannitol.[52] This is made evident by the additivity of their weighted water sorption isotherms. Conversely, when a solution containing a glass-forming sugar and protein is freeze-dried, a single amorphous phase is formed. In that case, water sorption additivity of the glassy formulation is no longer observed, as shown in Figure 9.[52] Figure 9 represents indirect evidence of hydrogen bonding interactions between sugar and protein, which prevent water from binding to the glassy solid, as indicated by the negative deviations. Thus, the sugar satisfies some of the hydrogen bonding sites on the protein and *vice versa*. In more than one instance, however, water sorption additivity is observed, as sugar concentration is increased in the pre-freeze-dried solution. This suggests the formation of an amorphous solid that comprises two separate glassy phases. The nature of such phase separation has been suggested to originate in the solution state, where, at high concentrations, sugar molecules associate with themselves more

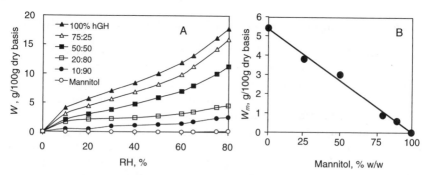

Figure 8 *Water sorption behavior of lyophilized rhGH:mannitol: (A) water vapor sorption isotherms for powders consisting of the indicated rhGH:mannitol compositions (w:w, dry basis). Isotherms were measured gravimetrically at 25°C; (B) hydration limit (W_m) obtained from BET analysis of the isotherms. After Costantino et al.[52]*

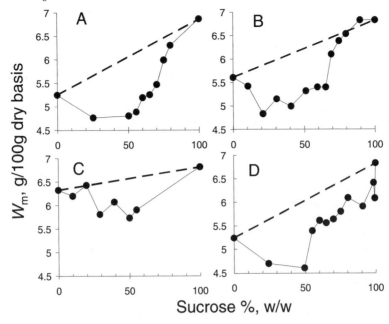

Figure 9 *Hydration limit, W_m, calculated from a BET analysis of sorption isotherms at 25 °C for various co-lyophilized protein:sugar systems: (A) rhGH:sucrose; (B) rhIGF-I:sucrose; (C) rhuMAb:sucrose; and (D) rhGH:trehalose. The dashed line is expected W_m value from linear addition of the weighted isotherms. Deviations from expected values suggest protein–sugar interactions. Interactions (i.e., deviations) decrease at high sugar content, which suggests phase separation between protein and sugar. After Costantino et al.[52]*

strongly than with protein.[29,52–54] An important corollary from these studies is that stabilization of proteins by amorphous sugars requires formation of a uniform glassy phase, where protein–sugar interactions are favored over protein–water and sugar–water interactions. Such phase separation has been shown to have implications for the stability of such mixtures, and should be considered when optimizing a particular formulation.[55] It also implies that the stabilizing effects of glass-forming saccharides may depend not only on their ability to form a glass with a high T_g but also on their ability to interact with protein.

It has also been shown that the protective efficacy of glass-forming saccharides appears to follow the order of heats of solution of the amorphous compounds.[56] This heat represents the sum of: (a) the enthalpy of breaking inter- and intramolecular hydrogen bonds in the amorphous state, (b) the enthalpy of breaking water–water hydrogen bonds, (c) the enthalpy of forming hydrogen bonds with water in the solvated, aqueous state, plus (d) the enthalpy due to the change in a saccharide's molecular conformation induced by solvation in water.

If it is assumed that, among these compounds, molecular interactions in the amorphous solid state are comparable, then the heat of solution is largely a measure of hydrogen bonding of saccharide and water and, presumably, proteins and other biological macromolecules. This is consistent with a two-step stabilization mechanism, where first, the ideal excipient readily forms hydrogen bonds with proteins in solution, in order to prevent their unfolding during dehydration, and second, the excipient has a high T_g, in order to stabilize them during storage.[57]

Interestingly, Miller and de Pablo[56] found that T_g correlates with the heat of solution of an amorphous saccharide (Figure 10A). This correlation provides an alternative explanation for the T_g-based hierarchy of protective efficacy proposed by Green *et al.*[48] Miller and de Pablo[56] used structural arguments to explain this correlation between thermodynamic and dynamic quantities. Disaccharides that contain a furanose ring (sucrose, lactulose, leucrose, turanose and palatinose) have lower conformational flexibilities than those composed of two pyranose residues. Not only does this result in a reduced ability to form hydrogen bonds with water (*i.e.* a lower absolute heat of solution), but it could also lead to less efficient packing in the solid state.

Furthermore, the trend in the data for dry disaccharides shown in Figure 10A is consistent with that for partially hydrated disaccharides shown in Figure 10B. That is, sorption of small amounts of residual water by either amorphous trehalose or amorphous sucrose results in a decrease in T_g and absolute heat of solution. Figure 10B shows that the 'depressed' values of T_g and heat of solution for these plasticized sugars still satisfy the correlation illustrated in Figure 10A. In this sense, the hydrated disaccharides behave similarly to the dry compounds. One interpretation is that, upon exposure to water, some of the hydrogen-bonding sites on a saccharide become occupied by sorbed water (at the expense of saccharide–saccharide hydrogen bonds). Because some of a saccharide's hydrogen-bonding capacity is already fulfilled, further interactions with water will be reduced upon complete solvation. On the basis of these observations, it is tempting to infer that the T_g values of these systems are proportional to the extent of saccharide–saccharide hydrogen bonding.

The use of water sorption isotherms to prove or disprove molecular interactions between two or more components in a glassy dispersion should be done

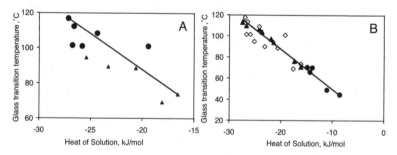

Figure 10 (A) *Relationship between* T_g *and heat of solution for various disaccharides. Symbols denote the types of ring moieties present in each compound: (circles) hexose–hexose and (triangles) pentose–hexose. Linear regression (solid line):* $R^2 = 0.7$. (B) (diamonds) dry disaccharides, (triangles) humidified trehalose, and (circles) humidified sucrose. Solid line is from the linear regression of data in (A). From Miller and de Pablo[56]

with care. It has been shown that a water vapor absorption isotherm is not a sensitive indicator of possible sugar–PVP interactions in an amorphous molecular dispersion.[24,51] Water vapor absorption isotherms for co-lyophilized PVP–sugar mixtures can be predicted from the weighted sum of the isotherms for the individual components, whereas sugar crystallization is significantly retarded in the molecular dispersions.[24,51] Crystallization inhibition could be attributed to dilution effects, with no significant hydrogen bonding between sucrose and PVP. However, specific hydrogen bonding between sucrose and PVP has been identified by Raman spectroscopy.[58] In addition, results from ^{13}C-NMR analysis of hydrated glassy dispersions of sucrose–PVP suggest that the residual water (at storage RH < 25%) had no significant effect on sugar–PVP interactions, as measured by the reduced mobility of sucrose carbon atoms.[27]

4 Residual Water and Stability of Hydrated Glassy Solids

There are many examples of the influence of residual water on the rates of chemical reactions in the dry state[59] (see Figure 11). The effect of residual water on the solid-state stability of glassy systems is complex, since water can play many different roles in a degradation process. The roles of water in degradative reactions that occur in hydrated glassy solids have been described as: (1) reactant, (2) solvent and (3) plasticizer.[2] The nature of a specific reaction will determine whether water acts as a reactant or product, and whether water, as a

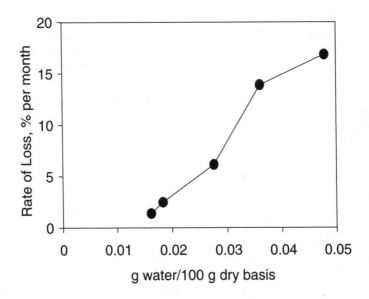

Figure 11 *Increase in the initial zero-order rates of degradation for recombinant bovine somatropin (rbST) in a lyophilized formulation stored at 47°C, measured by reverse-phase HPLC. From Hageman et al.*[13]

polar medium, will increase, decrease or have no effect on the rate of a reaction. Depending on the particular role(s) water plays in a reaction, the overall rate of a typical degradation process could be increased by a factor of 10, when water acts as either a reactant or solvent, or by a factor of up to 10^6, when it acts as a plasticizer.[2] However, generalization is difficult, since water may play more than one role in the degradation of a given system, and may play different roles under different storage conditions.

In all cases, water is expected to influence the overall mobility of a glassy solid and, thus, affect reactions that require molecular motion as a rate-limiting factor. This would certainly be critical for any reaction involving translational diffusion of two or more reactants, but it may also play a role, when certain rotational or intramolecular motions are critical for reactivity.

The stability of glassy solids has been studied extensively. Theoretical and semi-empirical models have been developed to relate the stability of a glassy solid to its intrinsic properties, such as T_g, relaxation time constants, and molecular mobility. However, the effects of water have not been explicitly accounted for in most cases. In the following sections, we summarize the major approaches used to relate glassy state properties to stability. We also propose how the effects of residual water can be accounted for in current predictive models. This is of great practical importance, since the water content of commercial glassy products continuously changes during storage.

4.1 Relationship Among T_g, Relaxation Time, and Molecular Mobility

A glass can be considered as a highly viscous liquid. The dynamics of structural changes in a glassy phase are often referred to as 'relaxation behavior', which is associated with molecular mobility and macroscopic fluidity (*i.e.* viscosity). This determines many of the important properties of amorphous materials. Viscosity is observed to increase by 10 to 12 orders of magnitude as a liquid undergoes a vitrification process. At and slightly below T_g, the viscosity of a glassy phase is in the range of 10^{12} to 10^{14} Pa s^{-1}, and decreases exponentially above T_g.[1] The extent and nature of this temperature dependence is material-specific. The viscosity of 'strong' glass formers, which can form networks in the liquid state, obeys an Arrhenius relationship over a wide temperature range; 'fragile' glass formers show strong deviations from such behavior. As the constituent molecules of a glass undergo a glass transition, molecular motions decrease and are said to be correlated with, or coupled to, the total viscosity of a system.[60] For fragile glass-formers, such as many carbohydrates, the temperatures of physical state changes (*e.g.* glass softening or stickiness, structural collapse, water and solute crystallization) can be predicted with reasonable accuracy, based on knowledge of T_g. Parametric equations have been used to relate viscosity changes to the difference between the temperature of a system and its T_g, (*i.e.* $T - T_g$).[61] For amorphous carbohydrate solutions, the bulk viscosities at which structural collapse, glass softening, and solute crystallization take place are predicted to be in the order of 10^8 to 10^{11}, 10^7 to 10^{10}, and 10^5 to 10^7 Pa s^{-1}, respectively.[62]

The glass transition may also be considered in terms of relaxation processes that occur as a liquid is cooled. T_g is defined as the temperature at which a relaxation time is equal to the time scale of an experiment. At a calorimetric T_g, the relaxation time is approximately 100 seconds.[60] Relaxation times, or the time scales of molecular motion, are temperature-dependent, with longer times being observed as a storage temperature drops below T_g. If the 'structural' relaxation time for a material is short with respect to the time of observation (which would be the case at temperatures above T_g), then macroscopically, the material will appear 'liquid-like'. Below T_g, however, molecular motions are considerably reduced, and the relaxation time will be longer than the time scale of a cooling process, and therefore, the material will appear 'solid-like'.

In the region immediately above T_g, the relaxation time of glassy materials follows a Vogel–Tammann–Fulcher (VTF) relationship, which relates relaxation time to 'fragility' (a parameter that accounts for the temperature dependence of viscosity above T_g for fragile and strong liquids) and the 'zero mobility' temperature, T_o. Hancock *et al.* have suggested that T_o can be roughly estimated as $T_g - 50$ K.[63]

Kauzmann proposed the existence of a 'critical' temperature, T_K (the Kauzmann temperature[64]), where the configurational entropy of a supercooled liquid, or *ideal glass*, becomes equal to that of a crystal. The ideal glass experiences total loss of structural mobility at T_K. Shamblin *et al.*[65] have estimated the T_K for several organic glassy solids from heat capacity and enthalpy of fusion measurements. They demonstrated that, at T_K, relaxation times were in the order of three years, indicating low (but not zero) mobility.[65] Shamblin *et al.* have also proposed a more refined approach to predict the temperatures at which relaxation times of glassy solids are in the order of years. This was done using a form of the VTF equation derived from the Adam–Gibbs[66] formalism for non-equilibrium systems below T_g. This treatment introduced the concept of 'fictive temperatures' to account for deviations from an ideal glass below T_g.[65]

The concepts behind the definitions of fragile and strong glasses have been interpreted on a thermodynamic basis,[67] through the Adam–Gibbs model.[66] In this model, the rate of structural relaxation is linked to the excess entropy[68] in the amorphous state. A key feature of this model is the identity between T_o (from kinetic measurements) and the Kauzmann temperature (from thermodynamic measurements). The reader is directed to recent reviews on the subject for more details.[69,70]

4.1.1 Relaxation Time and Stability. Since the average relaxation time at the calorimetric T_g is approximately 100 seconds,[60] molecular mobility is high enough to allow physical changes such as structural collapse and crystallization to be observed in real time. For example, chemical instability, structural collapse, and crystallization can be observed to occur within hours or even minutes, when glassy solids are stored at temperatures well above T_g.[15]

Molecular motion in the glassy state persists at temperatures well below T_g.[65] Crystallization in glassy systems can occur, even when they are stored well below their T_g values.[43] There have also been several reports of *chemical* instability in glassy formulations stored below their T_g.[30,71,72] These observations have important practical implications for the physical and chemical stability of commercial glassy products. For this reason, some recent research has focused on characterization of the time scales of molecular motion in organic glassy solids below T_g. This was done to identify environmental conditions where molecular processes that lead to unwanted changes in amorphous systems (*e.g.* chemical reactivity, crystallization, structural collapse) are improbable.

It has been postulated that long-term physical and chemical stability can be achieved, only if an organic glass is stored below its T_o, whose value was suggested to be about 50 K below its calorimetric T_g.[63] This hypothesis was based on empirical evaluation of available kinetic models such as the VTF model. Shamblin *et al.*[65] also reported that 'real glasses' near their Kauzmann temperatures experience relaxation times on the order of years. For a series of organic glasses, they reported Kauzmann temperatures that ranged from 40 to 190 K below the corresponding T_g values.[65] In practical terms, this means that there is no universal rule to define a storage temperature that ensures a shelf-life on the order of years for a glassy product. Thus, T_o values need to be determined on an *ad hoc* basis.

The concepts of relaxation time and molecular mobility outlined above have been successfully applied to predict the stability of pharmaceutical formulations, based on molecular mobility measurements.[73-75] For example, Duddu *et al.*[73] have correlated a short average relaxation time with aggregation of a monoclonal antibody in lyophilized formulations. Van den Mooter *et al.*[74] have used relaxation time constants to predict the stability of amorphous benzodiazepines. Guo *et al.*[75] have used relaxation time constants to predict the degradation rate of amorphous quinapril hydrochloride. Using this concept, they recognized that physical collapse can also affect observed degradation rates.[75] Physical changes in amorphous solids may occur concomitantly with chemical degradation processes, and, if overlooked, may lead to inaccurate conclusions.[75]

The aforementioned models have been developed assuming a dry state as the reference state. In this context, the term 'intrinsic' glass transition temperature, T_g^o, is useful to describe the T_g of a dry material, and to distinguish it from the T_g of a glassy phase at a given water content. These models account for the effect of residual water on relaxation time by considering its plasticizing effect on T_g. This tacitly assumes that the dependence of relaxation time on water content is independent of $T - T_g$. However, until now, no straightforward interpretation has been proposed to explicitly account for the combined effects of residual water and temperature on the molecular mobility of glassy materials.

4.1.2 Mobility of Water Within an Amorphous Solid and Chemical Stability. Water molecules are always mobile within a hydrated glassy solid. In fact, it has been shown that the diffusivity of water molecules within a glassy matrix is finite and substantial below T_g.[42] However, as discussed above, water mobility decreases at W_m, for small molecule glasses, amorphous polymers and proteins alike.[26,40,41,77] As also discussed, the mobility of water is restricted below W_m; for example, water molecules are irrotationally bound to a protein surface at such hydration levels.[32] Moreover, at constant temperature, the water mobility above W_m and below W_g (the amount of water required to depress T_g to the temperature of the surroundings) is higher, but less than the mobility in liquid water. It has also been shown, for polymers and proteins, that the molecular mobility of a glassy matrix itself increases, as a consequence of water-induced plasticization.[1,27,41,77–81]

The molecular mobility of water in hydrated amorphous formulations of pharmaceutical interest has been experimentally determined by NMR spectroscopy.[77,79,81,82] A composition-dependent temperature, below which the molecular mobility of water is limited, has been identified. Such a temperature is referred to as a 'critical mobility temperature', T_{mc}. Like T_o, T_{mc} may be analogous to T_g, in that it represents a temperature below which molecular mobility is decreased within a glassy matrix. Similarly, as T_g represents a state change from glassy solid to rubbery liquid, T_{mc} or T_o represent a change in the state of water mobility in a hydrated glassy system. In addition, the change in the state of water mobility is related to the stability of a hydrated glass, since T_{mc} has been shown to correlate with the chemical stability of labile reactants incorporated in hydrated glassy solids below T_g,[77] and T_o has been shown to correlate with the physical stability of saccharide glasses.[35] It is expected that, as concepts of *mobility transition* temperatures (T_{mc}, T_o, T_K) become more prevalent in the literature, and as methods for their determination become more accessible, a system-specific T_o will replace the empirical $T_g - 50$ rule for determining storage conditions. Furthermore, a harmonization or convention may be made among the various (sub-T_g) transition temperatures used throughout the literature.

4.2 Chemical Stability Below W_m at Constant Temperature

Many reactions in amorphous systems have been shown to exhibit a bell-shaped relationship (see Figure 12) between water content and reaction rate at constant temperature. Such a relationship has been attributed to the effect of water on molecular mobility and reactant concentration.[83] For a glassy formulation stored below its T_g, reaction rate is lower at low water contents, because the diffusion and mobility of reactants are limited. As water content increases, to a level sufficient to depress the formulation's T_g to a temperature below the storage temperature (*i.e.* the system becomes rubbery), diffusion-limited reactions are accelerated by the increased molecular mobility of reactants. At even greater water contents, reaction rates decrease, due to dilution of reactants. Generally, the minimum in the rate of bimolecular reactions is found near the hydration limit.[5,78,83] As illustrated in Figure 5, it is at or just above W_m that water–water interactions increase, thereby favoring the formation of microscopic regions of condensed water, such that chemical species can dissolve, diffuse and react.

Oxidation reactions are, in many instances, an exception to the case discussed

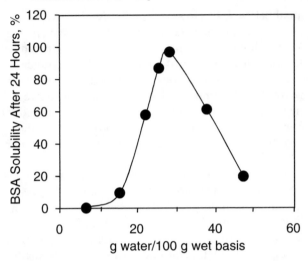

Figure 12 *Bell-shaped relationship between residual water and solid-state aggregation (evidenced by a reduction in solubility) of bovine serum albumin. Solubility was measured following a 24-hour incubation at 37°C. From Liu et al.[84]*

above. The rate of oxidation is observed to have a minimum at the monolayer hydration level, and to increase at lower or higher water contents.[5,78,83,85] This 'antioxidant' effect of water has been ascribed to its interaction with functional groups, which blocks these reaction sites, thereby preventing them from interacting with oxygen.

Improved chemical stability below W_m, for foodstuffs[83] and therapeutic proteins,[13,29,78] has been extensively reported. In general, freeze-dried protein formulations are more stable at lower water contents. Examples include 'pure' proteins and formulated ones with sugars and amino acids. It is generally implied that the mobility of water and the glassy matrix at hydration levels above W_m should be considered as the main factor affecting overall stability. In at least one instance, stability with respect to W_m of a glass was process-independent, since the stability of a sample adjusted to a given water content by desorption (during freeze-drying) was identical to the stability of a sample prepared by sorption of water on a previously highly dried sample.[78]

In many instances, T_g of formulations cannot be determined because of inherent limitations of traditional experimental methods. However, it is safe to assume that T_g of a protein-rich formulation, even at its hydration limit, is much higher than typical storage temperatures (*i.e.* $T_{storage} < 40°C$ or $T_g \gg 40°C$). This is an important consideration, since it suggests that molecular motions implicated in decomposition processes of protein-rich formulations at low water contents are independent of T_g, possibly due to a decoupling of water mobility and protein mobility. Thus, local changes in proteins can result in loss of overall structure and function.

4.3 Physical Stability Below W_m at Constant Temperature

Low molecular weight organic glassy solids are known for their propensity to crystallize when exposed to temperatures above their T_g. This can occur as a result of a temperature change at constant water content or *vice versa*. Although crystallization can be used as an intrinsic marker of molecular mobility, chemical instability and major physical stability issues such as stickiness and structural collapse often precede crystallization.[1,15,86]

In recent years, it has been reported that isothermal crystallization can occur, even when $T_{storage} \gg T_g$.[43] It has also been suggested that long-term physical stability can be achieved, only if an organic glass is stored below its T_o. More recently, specific values of T_o have been calorimetrically determined for sucrose (55 °C[65]) and raffinose (65 °C[87]). This means that, in order to keep sucrose and raffinose amorphous, they must be stored 19 °C and 40 °C below their corresponding $T_g°$ values of 74 °C and 105 °C, respectively.

The effect of hydration on the physical stability of sucrose is illustrated in Figure 13. Physical changes induced by water sorption – stickiness, collapse and crystallization – are similar to those induced by increasing temperature of a dry sample. This is due to plasticization. As water content increases, T_g decreases and approaches the storage temperature. Interestingly, the time scale for the occurrence of physical changes increases exponentially as water content decreases. Glassy sucrose stored at 12% RH and 25 °C has been shown to remain amorphous for more than two years.[88] The equilibrium water content at 12% RH, 2.0 wt% H_2O, is just below the hydration limit at 25 °C. Similarly, it has been shown that, like sucrose, other small saccharides (lactose, raffinose and trehalose) can

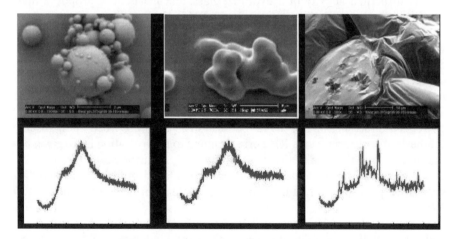

Figure 13 *Water-induced physical changes in amorphous sucrose stored at 25 °C at <5% RH for years ($T_g \gg 25$ °C), left; 33% RH for hours ($T_g \approx 8$ °C), center; and 75% RH ($T_g \ll 25$ °C) for a few minutes, right. The lower panels show the corresponding X-ray powder diffraction patterns; x-axis is 2θ, and y-axis is intensity in arbitrary units. Axes are not labeled, since the patterns are only used qualitatively to demonstrate state changes. From Lechuga-Ballesteros et al.[35]*

remain amorphous for long periods of time (*i.e.* years), as long as their water content is kept below the monolayer hydration level during storage at constant temperature.[35] Lechuga-Ballesteros *et al.*[35] have suggested that $T_g(W_m)$ is equivalent to T_o at a given water content. T_g of sucrose is 44 °C* at W_m ($\approx 2\%$ at 25 °C), which indicates that $T_g - T_{storage}$ is about 19 °C, in agreement with the *dry* T_o prediction of 55 °C from Shamblin *et al.*,[65] or $T_g^o - T_o = 19$ °C.

The proposed mechanism for reaching the onset of physical instability at the hydration limit is consistent with the mechanism proposed for the observed onset of chemical instability. As illustrated in Figure 5, the predominance of water–water interactions above W_m favors the formation of microscopic regions of condensed water, such that chemical species can dissolve. The formation of a supersaturated solution provides the driving force for nucleation and crystal growth.

4.4 Combined Effect of Temperature and Relative Humidity on Physical Stability

It is clear that W_m is an important stability indicator for glassy solids. In addition, the above observations suggest that T_g at the water content corresponding to the hydration limit, $T_g(W_m)$, also has an important significance. As discussed earlier, the viscosity of a dry glass changes drastically, when it is stored at a temperature at or above its T_g. Such a measure of molecular mobility has been used as the operational definition of T_g^o, the T_g of a dry amorphous solid. Similarly, T_g of a hydrated amorphous solid at its hydration limit, $T_g(W_m)$, serves as a convenient reference temperature for describing the observed sudden increase in thermal activity of a hydrated glass, and it has been proposed that $T_g(W_m)$ is equivalent to T_o at a given storage RH_m.[35]

Isothermal microcalorimetry has been used to measure the reactivity of water with amorphous sucrose.[35] By measuring the heat evolved or absorbed while continuously increasing the RH at constant temperature, one can determine a characteristic calorimetric trace. This trace has been referred to as a moisture-induced thermal activity trace (MITAT)[35] (see Figure 14). Analysis of a MITAT for each of several amorphous solids reveals general behavior. Since calorimetric power is a direct measure of the rate of heat evolved or absorbed, these results can be used to determine the RH corresponding to the hydration limit, given by the onset of the increase in thermal activity (RH_m). Furthermore, the onset of collapse and sucrose crystallization can also be readily identified. Experiments at several temperatures can be used to determine the temperature dependence of RH_m and RH_c.

The temperature dependence of RH_m (storage RH to obtain W_m) for sucrose and other small saccharides has been experimentally determined[34] (see Figure 15). For sucrose, when the experimentally determined RH_m, as a function of storage temperature, is extrapolated to 'dry conditions', the extrapolated tem-

---

*T_g at W_m was estimated by assuming a T_g^o of 74 °C (347 K) and a fitted Gordon–Taylor parameter of 0.13675.[14]

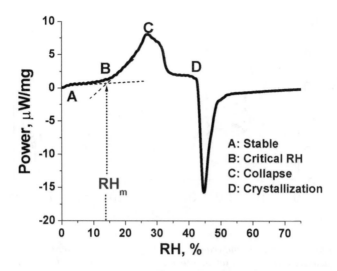

Figure 14 *Moisture-induced thermal activity trace (MITAT) of sucrose at 25°C, showing a region of low thermal activity, A (0 < RH < 13%), 'take off' point B (RH ≈ 13%), which corresponds to W$_m$, exothermic region C (13% < RH < 27%), which represents the onset of collapse, and endothermic region D (RH > 27%), resulting from sucrose crystallization. See Lechuga-Ballesteros* et al.[35] *for experimental details*

perature, 57°C, is similar to the calorimetrically measured T_o for sucrose under dry conditions, 55°C.[65] The extrapolated T_o (RH=0) for raffinose, 87°C, is comparable to the T_o value reported by Kajiwara *et al.*,[87] 65°C. The agreement found is remarkable, considering the completely different experimental methodologies used.

As mentioned above, the T_g associated with a hydrated glassy solid at its hydration limit corresponds to a low-mobility temperature, *i.e.* $T_g(W_m) \approx T_o$. In addition, the experimentally determined dependence of RH_m on temperature is similar to that found for T_g as a function of RH, as shown in Figure 16.[35] We postulate that, in the area below the line described by $T_g(W_m)$ in this RH *vs* T diagram (W_m is converted to RH_m using a sorption isotherm, or RH_m is determined from a MITAT), the relaxation times are sufficiently long (*i.e.* years) to achieve long-term stability.[35] In addition, we suggest that the dependence of RH_m on temperature (or T_o(RH) contour) represents an iso-relaxation time contour[1] sensitive to changes in water mobility. This is analogous to a calorimetric T_g measured at different water contents, in that the resulting T_g contour represents the locus of relaxation times of 100 seconds, whereas the RH_m contour represents relaxation times on the order of years.

Since amorphous solids are thermodynamically unstable, stability is only meaningful when time is noted. In this context, the stability or iso-relaxation diagram is a useful tool for determining the temperature and RH conditions that ensure long relaxation times. A conceptually similar diagram has been developed by modeling the dependence of T_g on water content, using the Gordon–Taylor

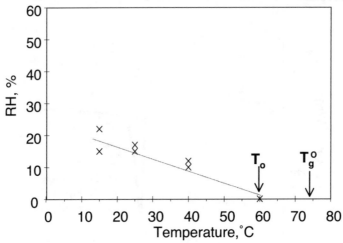

Figure 15 *Effect of storage temperature on* RH_m *of sucrose, determined from MITATs measured at various temperatures. Temperature intercept of extrapolation to RH = 0 is proposed to correspond to a 'mobility transition temperature', T_o, for long-term stability. See Lechuga-Ballesteros et al.[34] for experimental details*

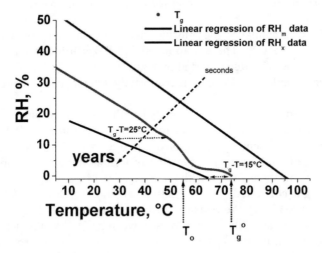

Figure 16 *Iso-relaxation time diagram for amorphous sucrose, constructed from the experimental dependence of T_g on water content. Average relaxation time increases exponentially from the $T_g(RH)$ contour toward the $T_o(RH)$ contour. The stability of an amorphous solid is expected to be similar at (T, RH) storage conditions equidistant from the iso-relaxation time contours. The dependence of T_g on RH is obtained using sucrose's water sorption isotherm and knowledge of $T_g(W)$. The $T_g(RH)$ contour converges at T_g^o for RH = 0. The $T_o(RH)$ contour is experimentally determined by analysis of MITATs (for RH_m) at various temperatures. The $T_o(RH = 0)$ value has been shown to agree with one determined by Shamblin et al.[65] See Lechuga-Ballesteros et al.[34] for additional experimental details*

model, and drawing the curve for the expected water content at $T_g - 50$ K.[89]

The iso-relaxation time diagram (Figure 16) for sucrose is qualitatively supported by experimental observations. Freeze-dried sucrose remained amorphous for more than two years,[88] when stored below its hydration limit at 25 °C. In our laboratory, glassy sucrose (and other small saccharide) samples have remained amorphous for over a year, when stored at ambient temperature and an RH below RH_m.

4.5 Combined Effect of Temperature and Relative Humidity on Chemical Stability

The combined effect of temperature and RH on the chemical stability of labile reactants in glassy solids has been the subject of many recent studies.[30,31,71,72,90] The consensus thus far is that chemical stability is a complex matter, since the effects of water content, storage RH*, storage temperature, and T_g on the chemical stability of reactants embedded in a hydrated glassy solid cannot be easily separated.[30]

Likewise, recent studies have challenged the simplistic interpretation that the relevant variable is $T_{storage} - T_g$, regardless of water content.[91] Heat inactivation of lactase and invertase enzymes in dried, amorphous matrices of sugars (trehalose, maltose, lactose, sucrose, raffinose) and some other selected systems (casein, PVP, milk) was observed in glassy matrices (*i.e.* below T_g) and appeared to depend on $T_{storage}$ *per se* rather than $T_{storage} - T_g$.

Many recent studies attempting to discriminate between the effects of storage RH and T_g have been performed at a constant temperature, $T_{storage}$, but at different $T - T_g$ differentials. This is accomplished by using water to change T_g, thereby also changing $T - T_g$. Therefore, the dependence of T_g on water content is an obvious complication to interpreting the stability of glassy systems,[1] as is also the bell-shaped relationship (see Figure 12) between reaction rate and storage RH.

To decouple the effects of water content from those due to T_g, PVP is a preferred system because its T_g increases with molecular weight, yet its equilibrium water sorption isotherm is relatively independent of molecular weight. Thus, at a given RH, two PVP samples of different average molecular weights will have similar water contents, but different T_g values.[31] These systems of similar composition can then be used to study the chemical decomposition of co-lyophilized model reactants. In addition, glycerol has been used as a plasticizer to decrease PVP's T_g at constant water content.[72] This general approach has been used to study the effect of storage RH and T_g on the stability of various model reactants, such as sucrose and invertase,[90] aspartame,[31] and Asn-hexapeptide (Val-Tyr-Pro-Asn-Gly-Ala).[30,72]

An extensive study of the isothermal degradation of Asn-hexapeptide in PVA

---

*Note that water activity is often equated with storage RH/100. This is only true when the vapor phase is ideal. For consistency, we have chosen to use the term 'storage RH',[1] even when 'water activity' was used in the cited articles.

and PVP glassy matrices[30] offers additional insight on the effects of water content, formulation T_g, and storage RH. As would be expected, the rate of Asn-hexapeptide deamidation in PVA and PVP matrices increased with increasing water content and storage RH and decreasing T_g. However, the degradation mechanisms in the two polymers differed, so that chemical reactivity could not be predicted from water content, storage RH, or formulation T_g alone (see Figure 17A–C). Asn-hexapeptide was more stable in low molecular weight PVP at all water contents (or storage RH) (see Figure 17A and B). Degradation of Asn-hexapeptide in both PVA and PVP appeared to correlate well with storage RH, as shown in Figure 17B. Values of k_{obs} expressed as a function of $T_{storage} - T_g$, see Figure 17C, show that PVA and PVP offered quite different levels of protection to Asn-hexapeptide below T_g. However, degradation of Asn-hexapeptide in the rubbery state was similar, regardless of the polymer matrix. The decomposition of Asn-hexapeptide in PVP below T_g was persistent, suggesting that the decomposition reactions were decoupled from polymer relaxation (or mobility or T_g).

As discussed above, water mobility below the glass transition depends on the chemical nature of a glassy matrix. In addition, the molecular mobility of a glassy matrix itself plays an important role in degradation kinetics. For instance, at low water content, see Figure 17A, the estimated half-life for deamidation of Asn-hexapeptide at 50°C in a co-lyophilized matrix with PVA (formulation $T_g° = 75°C$, *i.e.* $T_g° - T_{storage} = 25°C$) is about 20 years, which is much longer than the 2 year half-life for a freeze-dried mixture with PVP (formulation $T_g° = 150°C$, *i.e.* $T_g° - T_{storage} = 100°C$).[30] This difference, however, becomes less important as water content increases. At higher water contents, the trend reverses, and Asn-hexapeptide becomes stable in PVP (Figure 17A). These findings offer some insight with respect to decomposition well below T_g, and suggest that the $T_g - 50$ rule, widely used to ensure chemical or physical stability, needs to be applied with care, since its utility will depend on the chemical nature of a glassy phase and on water content.

In contrast, and to emphasize the importance of the specificity of reactant–water–glassy solid interactions, consider the following example. For an insulin formulation, below its T_g, the rate of protein aggregation was found to be faster in freeze-dried formulations with PVA than in those with dextran, even though the water content of the PVA formulation was lower than that of the dextran formulation (0.10 *vs* 0.18 wt% H_2O). The lower stability in PVA was found not to correlate with water content, but rather with water mobility within the glassy matrix, since the T_{mc} determined by NMR was lower for PVA than for dextran.[82]

Sucrose hydrolysis in amorphous PVP (see Figure 18) was reported to be more dependent on the physical state of the matrix (*i.e.* glassy *vs* rubbery), and less dependent on water mobility.[90] In contrast, degradation of aspartame, *via* intramolecular arrangement, in amorphous PVP (see Figure 19) was reported (by the same authors) to be more dependent on water mobility, and less dependent on the molecular mobility of the glassy matrix, as indicated by $T - T_g$.[31]

Based on the data in Figure 18, it is difficult to determine whether reaction rate was impacted more by storage RH than by T_g. Indeed, what stands out is that

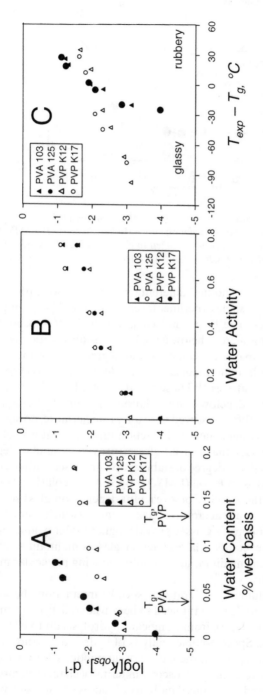

Figure 17 *Chemical stability of Asn-hexapeptide, co-lyophilized with either PVP or PVA, as represented by the observed rate constant for deamidation (k_{obs}) at 50 °C: (A) effect of water content; (B) effect of storage RH; (C) effect of formulation T_g. From Lai et al.[30]*

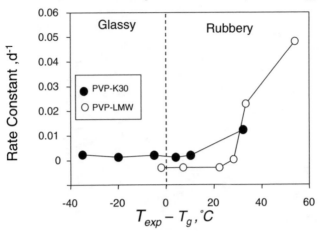

Figure 18 *Sucrose hydrolysis rate constant at 30°C, as a function of* $T_{storage} - T_g$. *Amorphous sucrose was co-lyophilized with either PVP-K30 or PVP-LMW. Storage RH, for each point of each series from left to right, was 32, 44, 53, 62, 70 or 75% RH. See source reference (Chen et al.[90]) for details*

hydrolysis occurred below T_g, and that different reaction rates were observed for sucrose hydrolysis under the same experimental conditions (of $T_{storage}$ and RH) in different polymer matrices. The lower molecular weight PVP appeared to be more effective in protecting sucrose below 62% RH. This observation is consistent with results from comparative studies of the degradation of the Asp-Pro dipeptide and the unimolecular dissociation of 2-(4-nitrophenoxy) tetrahydropyran in several glassy matrices.[71] Those results showed that both reactions (studied separately) occurred below a formulation's T_g, albeit at significantly reduced rates compared to those observed above T_g. Furthermore, no significant increase in the temperature dependence of reaction rate constant was observed near T_g. In addition, the fact that not all glasses provide a similar degree of stabilization was confirmed; the degree of stabilization increased with decreasing molecular weight (*i.e.* sucrose > Ficoll (LMW) > Byco A ≈ Ficoll (HMW) > dextran). This suggests that the stability of a reactant depends on glass relaxation and on the molecular 'fit' of a reactant in a glassy matrix (which contributes to reactant mobility). Lower density glasses provide higher mobility and therefore lower stability to small reactants, whereas denser glasses maintain small reactants in a lower free volume medium, substantially reducing molecular mobility and hence chemical reactivity.[71,92]

In addition to free volume considerations, specific interactions between an amorphous matrix and a labile reactant need to be considered. For example, the fact that PVP can protect sucrose from significant hydrolysis up to 62% RH at 30°C is quite remarkable. Spectroscopic analysis of hydrated mixtures of PVP and sucrose offers insight regarding the nature of interactions between sucrose and PVP, as measured by Raman spectroscopy in terms of changes in molecular chemical environment[58] and the reduced mobility of sucrose.[27] Sucrose hydrolysis involves attack of a glucose–fructose bond, which might be obstructed by

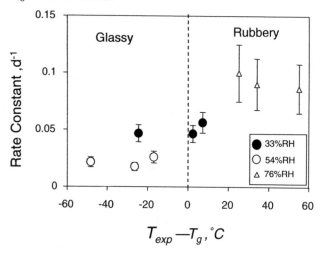

Figure 19 *Rate constants for aspartame degradation at 25°C in lyophilized PVP matrices. Each point in each triad corresponds (from left to right) to PVP LMW, PVP 15 and PVP 30. Water contents, for each sample in the corresponding triads from left to right, were: (open circles) 10.6, 10.1, 11.1; (closed circles) 20.3, 19.5, 20.0; and (triangles) 38.4, 37.1, 36.3 wt% H_2O (dry basis). See source reference (Bell and Hageman[31]) for details*

PVP.

A similar analysis can be described for the degradation of aspartame, *via* intramolecular arrangement, in glassy and rubbery PVP matrices. Based on results shown in Figure 19,[31] this reaction was reported to be more dependent on water mobility than on the molecular mobility of a glassy matrix. Aspartame degradation was observed to occur below T_g. However, the molecular weight of PVP did not affect decomposition rate. Rather, decomposition rate decreased with decreasing water content (or storage RH). Samples stored at 76, 54 and 33% RH contained about 37, 20 and 11% (dry basis) water, respectively.

In contrast, in a more recent study by Liu *et al.*[76] the degradation of aspartame in glassy matrices of sucrose or trehalose did show that degradation correlated with molecular mobility of the glassy matrix. Such apparently contradictory results may suggest that the intramolecular rearrangement of aspartame in glassy matrices is limited by the free volume within a glassy system. In the case of PVP, the free volume evidently exceeds that necessary for aspartame to rearrange. In contrast, in the cases of sucrose or trehalose, aspartame rearrangement is enabled by water-induced free volume changes resulting from plasticization of the glassy system.

The observed increases in the degradation rates for aspartame or sucrose in PVP matrices stored above T_g were not significant, compared to the corresponding viscosity changes. Even the increase in degradation rates experienced at $T - T_g > 20°C$ was not large enough to correspond to the expected viscosity changes. This suggests that the mobilities of reactant molecules and of water did not correlate with volume changes at T_g, and that the rate increase at

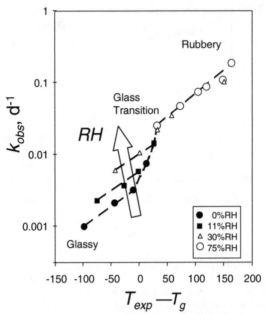

Figure 20 *Observed rate constant, k$_{obs}$, for Asn-hexapeptide degradation, as a function of T$_g$ at T$_{exp}$=50°C (T$_{exp}$ is T$_{storage}$), at different RH conditions. Water content (in wt%) at 0% RH was <0.4, at 11% RH was 2.0, at 30% RH was 6.0, and at 75% RH was 19.0. At each RH, formulations with increasing (T$_{storage}$−T$_g$) (i.e. with lower T$_g$) were produced by adding increasing amounts of glycerol. The dashed lines denote trends in the data and do not represent fits to any equation. Adapted from Lai et al.[72]*

$T-T_g>20°C$ may have been due to concentration effects, as has been reported for other systems.

As mentioned earlier, the degradation of Asn-hexapeptide has also been studied in glassy matrices of PVP/glycerol.[72] Glycerol was used in an attempt to separate the plasticization effects of water from its role as a solvent, since glycerol can plasticize a PVP matrix without significantly changing its water content. Results in Figure 20 show that degradation of Asn-hexapeptide occurred below formulation T_g (i.e. $T_{storage}-T_g<0$). An increase in rate of peptide deamidation was observed with increasing water content (i.e. as a response to increasing storage RH) and decreasing $T_{storage}-T_g$ (i.e. increasing glycerol content). These results suggest that matrix mobility increased, perhaps due to an increase in free volume caused by the presence of glycerol. Such an increase in free volume would enable increased mobility of water and Asn-hexapeptide, which would facilitate deamidation. In rubbery systems ($T_{storage}-T_g>0$), deamidation rates appeared to be independent of water content (or storage RH) in formulations with similar T_g values. An increase in water content also affected the distribution of degradation products, decreasing the formation of cyclic imide intermediate and favoring that of hydrolytic products, isoAsp- and Asp-hexapeptides. Thus, residual water appears to facilitate deamidation in these solid PVP formulations by enhancing

molecular mobility and acting as a solvent/medium. In addition, water participates as a chemical reactant in the subsequent breakdown of cyclic imide.

From the studies outlined above, it seems clear that the degradation of labile
chemicals in hydrated amorphous matrices is affected by the specific composition of a formulation. Interactions of glassy matrix with water, as well as
potential interactions with reactant, need to be considered. The overall mobility
of a glassy matrix may or may not dictate the availability of water to act as a
solvent or reactant. Frequently, reactions are observed to occur below a formulation's T_g, and reaction rate increases as storage RH and temperature increase. In
addition, the extent of a reaction below a formulation's T_g depends on the
'molecular fit' between reactant and glassy solid, with denser glasses (*i.e.* with
lower free volume) being more protective of labile chemicals.[92]

5 Iso-relaxation Time Diagrams

Thus far, we have shown that the effect of residual water on chemical stability of
labile chemicals can be related to storage RH through a water sorption isotherm.
In addition, the effect of residual water on overall mobility of an amorphous
system has been determined from the effect of plasticization on T_g of a formulation. Finally, residual water and storage temperature have been linked by
measuring a water sorption isotherm at several temperatures. The representation
of a reaction rate as a function of $T_{storage} - T_g$, a measure of molecular mobility,
attempts to combine the effects of T_g, water content, storage RH and temperature. However, it has also been shown that the contribution of residual water to
overall molecular mobility can be independent of $T_{storage} - T_g$. That is, reaction
rate can vary as a function of storage RH at constant $T_{storage} - T_g$, as can be
verified by simply looking at the RH arrow in Figure 20.

Therefore, the use of $T_{storage} - T_g$ as the unifying factor to explain the overall
molecular mobility of a glassy system seems incomplete, mainly because it is not
predictive of the mobility of water below T_g. This has been further reinforced by
studies of the isothermal induction time for crystallization of sucrose.[34] The
same $T_{storage} - T_g$ can be attained either by keeping a sample at constant water
content (or constant storage RH) and varying the storage temperature or by
holding storage temperature constant and varying T_g by varying the water
content (*e.g.* increasing the storage RH and relating that to water content by
means of a water sorption isotherm). Results shown in Figure 21 clearly indicate
that interpretation of molecular mobility on the basis of $T_{storage} - T_g$ can be
overly simplistic. The contribution of residual water to overall molecular mobility of an amorphous phase, especially below T_g, needs to be considered. As
shown in Figure 21, the induction time for crystallization at constant water
content increases exponentially at $T_{storage} - T_g = 0$, whereas the induction time
for crystallization of a hydrated sample is measurable even below
$T_{storage} - T_g = 0$. The induction time for crystallization of hydrated amorphous
sucrose is no longer measurable on a reasonable time scale (days) at
$T_{storage} - T_g^\circ \approx -20\,^\circ C$. This experiment was performed at $25\,^\circ C$; thus, T_g was

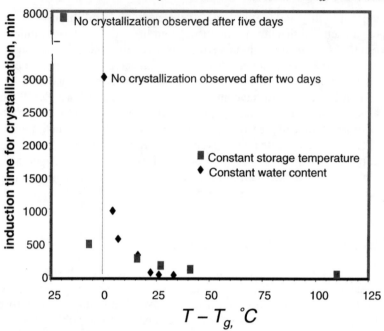

Figure 21 *Isothermal induction time for sucrose crystallization at constant storage tempera-*
ture (squares) and constant storage RH (diamonds). Data at constant water
content were obtained from individual samples of spray-dried sucrose heated to a
given temperature, until the onset of crystallization was observed (by DSC). Data
at constant temperature were obtained from individual samples of spray-dried
sucrose exposed to a constant RH at 25 °C (TAM using RH steps). Correspond-
ing T_g *values were calculated from the water sorption isotherm at 25 °C and a*
plasticization curve ($T_g(W)$*). For further experimental details, see Lechuga-*
Ballesteros et al.[34]

$\approx 50 °C$, which corresponds to T_g at the hydration limit or $T_0(RH = RH_m)$ and
compares to the 'dry T_0' or $T_0(RH = 0)$ of 57 °C.[65]

We propose that the contribution of residual water to overall mobility and
chemical stability may be described using the concepts of W_m and T_0, just as W_g
and T_g are used as a measure of mobility of a glassy matrix. In addition, because
W_m and T_0 are formulation-specific, they will account for the specific composition of a formulation. These parameters can be related to the molecular mobility
of a glassy system *via* iso-relaxation time diagrams. This was exemplified earlier
for the physical stability of sucrose, in the section describing the combined effect
of RH and temperature on physical stability.

In order to describe the combined effect of storage RH and temperature on the
chemical stability of labile chemicals in an amorphous system, we have constructed a rudimentary stability diagram for aspartame or Asn-hexapeptide in an
amorphous PVP matrix (see Figure 22). In the absence of an experimental value
for *dry* T_0, it has been estimated as $T_g° - 50$ (in K or °C); W_g and W_m have been
estimated from water sorption isotherms. Average relaxation time constants, τ,

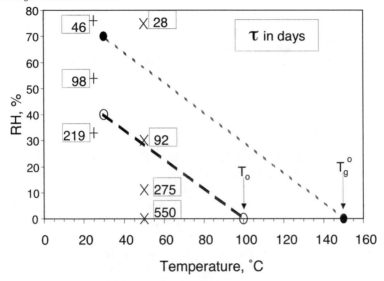

Figure 22 *Average relaxation times, calculated from observed degradation rate constants, for aspartame (×) or Asn-hexapeptide (+) in a PVP matrix, as a function of storage RH and temperature. Storage conditions (i.e. T, RH) are labeled with the corresponding relaxation times (numbers in boxes) for aspartame (from Bell and Hageman[31]) and Asn-hexapeptide (from Lai et al.[72]). Iso-relaxation time lines were calculated using experimentally determined values. RH_m values at 30 and 50°C (open circles, dashed line) were estimated from the corresponding W_m from BET analysis of PVP's water sorption isotherm at 30°C[20] and from GAB analysis of PVP's water sorption isotherm at 50°C.[72] RH_g (i.e. storage RH corresponding to W_g, the water content required to depress T_g to $T_{storage}$) for PVP at 30°C;[20] $T_g^o = 150°C$;[20] and assuming dry $T_o = T_g^o - 50°C$ (closed circles, dotted line)*

were calculated assuming that[60] $\tau_g = t\tau(T_g) \approx 100$ s and that $k_g/k = \tau/\tau_g$, where k is the observed degradation rate, k_{obs}, under given conditions, and k_g is the observed degradation rate at $T_{storage} = T_g$. Relaxation times calculated in this manner are similar to first-order half-lives (*i.e.* $t_{1/2} = \ln(2)/k_{obs}$).

The fact that the relaxation times for aspartame and Asn-hexapeptide are similar, with respect to their positions relative to the iso-relaxation time contours, is remarkable and suggests the utility of this approach to help unify the concepts of 'zero mobility' temperature and hydration limit with molecular mobility.

The qualitative impact of the combined effect of storage RH and temperature on chemical stability can be readily assessed, and it can be related, at a glance, to molecular mobility *via* iso-relaxation time contours. This is particularly useful, since it can provide an idea of how excursions from a given set of storage conditions could potentially affect the quality of a glassy product. Obviously, excursions of 10% RH or 10°C from given storage conditions will have different impacts on molecular mobility, depending on relative position in an iso-relaxation contour. In addition, iso-relaxation time contours enable the prediction of storage conditions (T, RH) where the relaxation time constants for a glassy formulation exceed practical storage times. Finally, such a stability diagram also points out the limitation of performing experiments at a single constant temperature, since results are limited to a small region of storage RH and temperature space.

6 Conclusions

Certainly, our mechanistic understanding of the effect of residual water on the stability of hydrated amorphous solids has increased considerably in the last decade. Spectroscopic and calorimetric analyses are now routinely used to identify molecular-level interactions and estimate molecular mobility of water within a hydrated glassy system and of the glassy system itself. The science has also been advanced by attempts to provide thermodynamically correct interpretations of sorption isotherms, as well as by the ingenuity of experimenters using novel model systems to challenge current state-of-the-art assumptions.

The refinement of kinetic and thermodynamic, semi-empirical models has also served to improve stability predictions for amorphous formulations. Models based on relaxation time or molecular mobility theories have been refined to predict stability below T_g. However, such models do not explicitly account for the effect of residual water.

The concept of *mobility transition* temperatures (T_{mc}, T_o, T_K), at which a sudden change in the molecular mobility of water within a glassy matrix occurs, seems to be a promising approach to link the hydration limit, W_m, with current relaxation models. It is expected that the use of mobility transition temperatures, as predictors of stability below T_g, will become more prevalent in the pharmaceutical literature. As methods for their determination become more accessible, a system-specific mobility transition temperature will likely replace the empirical $T_g - 50$ rule for determining storage conditions. In addition, we need to further explore the use of W_m and T_o to 'map' the combined effect of storage RH and temperature in terms of iso-relaxation time contours, to describe the shelf-life of a glassy product. In this regard, use of the proposed stability diagrams appears to be a reasonable methodology.

A general conclusion that can be drawn, from the many studies reviewed here, is that glassy solids of the same chemical nature provide different degrees of protection to a given model reactant. The stability of a reactant depends on the correlation between glass relaxation and reactant mobility or the molecular 'fit' of a reactant in a glassy matrix. Lower density glasses provide higher mobility and, therefore, lower stability to small reactants, whereas denser glasses can maintain small reactants in a lower free volume environment, substantially reducing molecular mobility and hence chemical reactivity. However, exceptions are found, due to more specific interactions between reactant and glassy solid. The dependence of stability on the chemical nature of a glassy solid is also observed for proteins formulated in different glassy matrices.

As a final remark, it can be said that, in spite of improved understanding, theoretical predictions of stability will be hindered by the irrefutable fact that the chemical nature of a formulation's constituents, in the end, determines the stability of a glassy product. However, this very fact provides the basis to predict, with a high degree of confidence, that the study of residual water in hydrated glassy systems will continue to provide fertile ground for work by academics, and will constitute an element of job security for many industrial scientists in the years to come.

Nomenclature

Symbol	Meaning
C_B	BET energy parameter constant
F	Vrentas plasticization parameter
$F(\varepsilon)$	Dielectric parameter
k_g	observed degradation rate at the glass transition temperature
k_{obs}	observed degradation rate
RH	relative humidity
RH_m	storage RH necessary for a glass to attain its hydration limit
T	Temperature
T_g^o	intrinsic glass transition temperature of dry solid
T_g	glass transition temperature
$T_g(W)$	glass transition temperature of a sample as a function of water content
$T_g(W_m)$	glass transition temperature of a sample with a water content equivalent to its hydration limit
T_K	Kauzmann temperature
T_{mc}	critical mobility temperature
T_o	zero mobility temperature
$T_{storage}$	storage temperature
W	Water content
W_m	hydration limit, BET monolayer

Greek Symbol	Meaning
χ	Flory–Huggins affinity parameter
τ	relaxation time constant
τ_g	relaxation time constant at the glass transition temperature

Acknowledgements

We thank Dr Felix Franks (BioUpdate Foundation, London, UK) for his advice and encouragement to write this manuscript. Many thanks to Dr Harry Levine (Nabisco, East Hanover, New Jersey) for his excellent editorial comments, which have greatly enhanced the quality of this paper. Our most sincere gratitude goes to Professor Aziz Bakri (University Joseph Fourier, Grenoble, France) for being a constant source of ideas, and for sharing his knowledge of and experience with isothermal calorimetry. DLB dedicates this work to Cecilia Espadas and thanks her for her unfailing love, faithful support and for taking care of the boys (David Alberto and Luis Daniel).

References

1. L. Slade and H. Levine, *Crit. Rev. Food Sci. Nutr.*, 1991, **30**, 115.
2. E.Y. Shalaev and G. Zografi, *J. Pharm. Sci.*, 1996, **85**, 1137.
3. Y.H. Roos, M. Karel and J.L. Kokini, *Food Technol.*, 1996, **50(11)**, 95.
4. G. Zografi, *Drug Dev. Ind. Pharm.*, 1988, **14**, 1905.
5. M.J. Hageman, *Drug Dev. Ind. Pharm.*, 1988, **14**, 2047.
6. Y.H. Roos, 'Phase Transitions in Foods', Academic Press, San Diego, 1995.
7. R.B. Anderson, *J. Am. Chem. Soc.*, 1946, **68**, 686.
8. E.A. Guggenheim, 'Applications of Statistical Mechanics', Clarendon Press, Oxford, 1966.
9. J.H. De Boer, 'The Dynamical Character of Adsorption', 2nd edn., Clarendon Press, Oxford, 1968.
10. S. Brunauer, P.H. Emmett and E. Teller, *J. Am. Chem. Soc.*, 1938, **60**, 309.
11. L. Pauling, *J. Am. Chem. Soc.*, 1945, **67**, 555.
12. G. Zografi, M.J. Kontny, A.Y.S. Yang and G.S. Brenner, *Int. J. Pharm.*, 1984, **18**, 99.
13. M.J. Hageman, P.L. Possert and J.M. Bauer, *J. Agric. Food Chem.*, 1992, **40**, 342.
14. B.C. Hancock and G. Zografi, *Pharm. Res.*, 1994, **2**, 471.
15. Y.H. Roos and M. Karel, *J. Food Sci.*, 1991, **56**, 38.
16. M. Gordon and J.S. Taylor, *J. Appl. Chem.*, 1952, 493.
17. F. Franks, in 'Chemistry and Technology of Water Soluble Polymers', C.A. Finch (ed.), Plenum Press, New York, 1981, 157.
18. B.C. Hancock and G. Zografi, *Pharm. Res.*, 1993, **10**, 1262.
19. P.J. Flory, *J. Chem. Phys.*, 1942, **10**, 51.
20. J. Zhang and G. Zografi, *J. Pharm. Sci.*, 2000, **89**, 1063.
21. S. Rosenbaum, *J. Polym. Sci.*, 1970, **31(C)**, 45.
22. J.S. Vrentas and C.M. Vrentas, *J. Polym. Sci.*, 1990, **28**, 241.
23. J.S. Vrentas and C.M. Vrentas, *Macromolecules*, 1991, **24**, 2404.
24. J. Zhang and G. Zografi, *J. Pharm. Sci.*, 2001, **90**, 1375.
25. L.S. Taylor and G. Zografi, *J. Pharm. Sci.*, 1998, **87**, 1615.
26. L.S. Taylor, F.W. Langkilde and G. Zografi, *J. Pharm. Sci.*, 2001, **90**, 888.
27. Y. Aso, S. Yoshioka, J. Zhang and G. Zografi, *J. Pharm. Sci.*, 2002, submitted.
28. C.D. Teng, M.H. Zarrintan and M.J. Groves, *Pharm. Res.*, 1991, **8**, 191.
29. H.R. Costantino, J.G. Curley and C.C. Hsu, *J. Pharm. Sci.*, 1997, **86**, 1390.
30. M.C. Lai, M.J. Hageman, R.L. Schowen, R.T. Borchardt, B.B. Laird and E.M. Topp, *J. Pharm. Sci.*, 1999, **88**, 1073.
31. L.N. Bell and M.J. Hageman, *J. Agric. Food Chem.*, 1994, **42**, 2398.
32. S. Bone, *Phys. Med. Biol.*, 1996, **41**, 1265.
33. J.A. Rupley, E. Gratton and G. Careri, *Trends Biochem. Sci.*, 1983, **8**, 18.
34. D. Lechuga-Ballesteros, D.P. Miller and A. Bakri, *Pharm. Res.*, 2002, in preparation.
35. D. Lechuga-Ballesteros, D.P. Miller and A. Bakri, *Pharm. Res.*, 2002, submitted.
36. M. Karel, *CRC Crit. Rev. Food Technol.*, 1973, **3**, 329.
37. J.D. Leeder and I.C. Watt, *J. Colloid Interf. Sci.*, 1974, **48**, 339.
38. S.L. Shamblin, B.C. Hancock and G. Zografi, *Eur. J. Pharm. Biopharm.*, 1998, **45**, 239.
39. G.F. Cerofolini, *Langmuir*, 1997, **13**, 995.
40. R. Pethig, *Ann. Rev. Phys. Chem.*, 1992, **43**, 177.

41. C.A. Oksanen and G. Zografi, *Pharm. Res.*, 1993, **10**, 791.
42. B.J. Aldous, F. Franks and A.L. Greer, *J. Mater. Sci.*, 1997, **32**, 301.
43. M. Yoshioka, B.C. Hancock and G. Zografi, *J. Pharm. Sci.*, 1994, **83**, 1700.
44. S.L. Shamblin, E.Y. Huang and G. Zografi, *J. Thermal Anal.*, 1996, **47**, 1567.
45. G. Buckton and P. Darcy, *Int. J. Pharm.*, 1996, **136**, 141.
46. P. Bonelli, C. Schebor, A.L. Cukierman, M.P. Buera and J. Chirife, *J. Food Sci.*, 1997, **62**, 693.
47. E.A. Schmitt, D. Law and G.Z. Zhang, *J. Pharm. Sci.*, 1999, **88**, 291.
48. J.L. Green, J. Fan and C.A. Angell, *J. Phys. Chem.*, 1994, **98**, 13780.
49. J.H. Crowe, *Am. Nat.*, 1971, **105**, 563.
50. J.F. Carpenter and J.H. Crowe, *Cryobiol.*, 1988, **25**, 459.
51. S.L. Shamblin and G. Zografi, *Pharm. Res.*, 1998, **16**, 1186.
52. H.R. Costantino, J.G. Curley, S. Wu and C.C. Hsu, *Int. J. Pharm.*, 1998, **166**, 211.
53. T. Arakawa and S.N. Timasheff, *Biochem.*, 1982, **27**, 8338.
54. S.T. Tzannis and S.J. Prestrelski, *J. Pharm. Sci.*, 1999, **88**, 360.
55. S.T. Tzannis and S.J. Prestrelski, *J. Pharm. Sci.*, 1999, **88**, 351.
56. D.P. Miller and J.J. de Pablo, *J. Phys. Chem. B*, 2000, **104**, 8876.
57. S.J. Prestrelski, K.A. Pikal and T. Arakawa, *Pharm. Res.*, 1995, **12**, 1250.
58. L.S. Taylor and G. Zografi, *J. Pharm. Sci.*, 1998, **87**, 1615.
59. M.J. Hageman, in 'Chemical and Physical Pathways of Protein Degradation', T.J. Ahern and M.C. Manning (eds.), Plenum Press, New York, 1992, 273.
60. C.A. Angell, *Science*, 1995, **267**, 1924.
61. M.L. Williams, R.F. Landel and J.D. Ferry, *J. Am. Chem. Soc.*, 1955, **77**, 3701.
62. W.Q. Sun, A.C. Leopold, L.M. Crowe and J.H. Crowe, *Biophys. J.*, 1996, **70**, 1769.
63. B.C. Hancock, S.L. Shamblin and G. Zografi, *Pharm. Res.*, 1995, **12**, 799.
64. W. Kauzmann, *Chem. Rev.*, 1948, **43**, 219.
65. S.L. Shamblin, X. Tang, L. Chang, B.C. Hancock and M.J. Pikal, *J. Phys. Chem. B*, 1999, **103**, 4113.
66. G. Adam and J.H. Gibbs, *J. Chem. Phys.*, 1965, **43**, 139.
67. C.A. Angell, *J. Non-Cryst. Solids*, 1991, **13**, 131.
68. K. Ito, C.T. Moynihan and C.A. Angell, *Nature*, 1999, **398**, 492.
69. D.Q.M. Craig, P.G. Royall, V.L. Kett and M.L. Hopton, *Int. J. Pharm.*, 1999, **179**, 179.
70. L. Yu, *Adv. Drug Del. Rev.*, 2001, **48**, 27.
71. L. Streefland, A.D. Auffret and F. Franks, *Pharm. Res.*, 1998, **15**, 843.
72. M.C. Lai, M.J. Hageman, R.L. Schowen, R.T. Borchardt, B.B. Laird and E.M. Topp, *J. Pharm. Sci.*, 1999, **88**, 1081.
73. S.P. Duddu, G. Zhang and P.R.D. Monte, *Pharm. Res.*, 1997, **14**, 596.
74. G. Van den Mooter, P. Augustijns and R. Kinget, *Eur. J. Pharm. Biopharm.*, 1999, **48**, 43.
75. Y. Guo, S. R. Byrn and G. Zografi, *J. Pharm. Sci.*, 2000, **89**, 128.
76. M.J. Pikal, J. Liu and D. Lechuga-Ballesteros, unpublished results.
77. S. Yoshioka, Y. Aso and S. Kojima, *Pharm. Res.*, 1999, **16**, 135.
78. M.J. Pikal, K. Dellerman and M.L. Roy, *Pharm. Res.*, 1991, **8**, 427.
79. S. Yoshioka, Y. Aso and S. Kojima, *Pharm. Res.*, 1996, **13**, 926.
80. F. Separovic, Y.H. Lam, X. Ke and H.K. Chan, *Pharm. Res.*, 1998, **15**, 1816.
81. S. Yoshioka, Y. Aso, S. Kojima, S. Sakurai, T. Fujiwara and H. Akutsu, *Pharm. Res.*, 1999, **16**, 1621.
82. S. Yoshioka, Y. Aso, U. Nakai and S. Kojima, *J. Pharm. Sci.*, 1998, **87**, 147.
83. T.P. Labuza, *Food Technol.*, 1980, **34**, 36.

84. W.R. Liu, R. Langer and A.M. Klibanov, *Biotechnol. Bioeng.*, 1991, **37**, 177.

85. M.C. Lai and E.M. Topp, *J. Pharm. Sci.*, 1999, **88**, 489.

86. W.Q. Sun, *Cryo-Letters*, 1997, **18**, 99.

87. K. Kajiwara, F. Franks, P. Echlin and A.L. Greer, *Pharm. Res.*, 1999, **16**, 1441.

88. B. Makower and W.B. Dye, *J. Agric. Food Chem.*, 1956, **4**, 72.

89. P.G. Royal, D.Q.M. Craig and C. Doherty, *Int. J. Pharm.*, 1999, **192**, 39.

90. Y.-H. Chen, L.N. Bell and M.J. Hageman, *J. Agric. Food Chem.*, 1999, **47**, 504.

91. C. Schebor, L. Burin, M.P. Buera, J.M. Aguilera and J. Chirife, *Biotechnol. Prog.*, 1997, **13**, 857.

92. L. Slade and H. Levine, *Adv. Food Nutr. Res.*, 1995, **38**, 103.

A Decrease in Water Adsorption Ability of Amorphous Starch Subjected to Prolonged Ball-milling is Accompanied by Enthalpy Relaxation

Toru Suzuki, Yu Jin Kim, Chidphong Pradistsuwana* and Rikuo Takai

DEPARTMENT OF FOOD SCIENCE AND TECHNOLOGY, TOKYO UNIVERSITY OF FISHERIES, JAPAN
*DEPARTMENT OF FOOD TECHNOLOGY, CHULALONGKORN UNIVERSITY, THAILAND

1 Introduction

Prolonged ball-milling, even at room temperature and under low-moisture conditions, converts partially crystalline native starch into a completely amorphous state.[1] We have reported that the amorphous starch prepared by ball-milling shows a clear glass transition in a DSC heating scan, and that prolonging the milling procedure beyond a few hours, when almost complete amorphisation was achieved, caused an acceleration of enthalpy relaxation, as shown in Figure 1.[1] Enthalpy relaxation in amorphous materials is usually accompanied by a structural relaxation, such as a decrease in free volume. Such a change in the amorphous state should affect physical properties of amorphous materials. It is ordinarily assumed that well-ball-milled starch has an increased water adsorption capacity, because of its less crystalline character.[2,3] However, according to our previous finding[1] on enthalpy relaxation induced by ball-milling, it would be predicted that prolonged ball-milling might cause a decrease in water adsorption, accompanied by decreased free volume. In the present study, aimed at confirming the above assumption, water uptake during prolonged ball-milling of amorphous starch was investigated.

2 Materials and Methods

Potato starch (Wako Chemical Ind.) was used as received, without further treatment; initial moisture content, before and after ball-milling, was constant at about 16%. Ball-milling was carried out using an Irie Shokai V-I mill. Stainless-

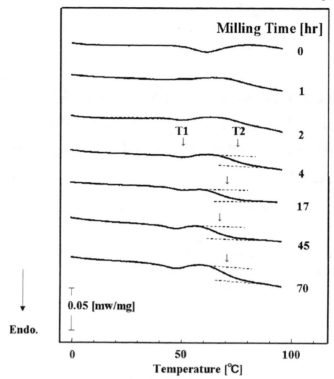

Figure 1 *DSC curves for potato starch samples, measured after ball-milling for different amounts of time, as indicated. T1 represents a recovery peak for enthalpy relaxation, and T2 represents a glass-to-rubber transition*

steel balls, 1 cm in diameter (total weight, 297 g), were added to 10 g of potato starch in a stainless-steel cylindrical container of 9 cm diameter and height. The container was tumbled, at 90 rpm, at room temperature for 1, 2, 4, 17, 45, 70 and 140 h. This milling condition was the same as that used previously for enthalpy relaxation experiments.[1] A water adsorption isotherm was measured, for each milled sample, by storing samples in humidity-controlled desiccators containing various saturated salt solutions. Details have been described elsewhere.[4]

3 Results and Discussion

Water adsorption isotherms for potato starch shifted down with increasing milling time, as shown in Figure 2. From isotherms for samples milled for different amounts of time, monolayer adsorption values were calculated using the BET equation. This BET monolayer value showed a decreasing trend with prolonged milling time, as illustrated in Figure 3. This result lent support to our assumption that enthalpically relaxed amorphous starch might exhibit reduced water absorption. For a more detailed analysis, we attempted to use a dual-mode sorption model,[5,6] which combined Henry's Law and the Langmuir adsorption

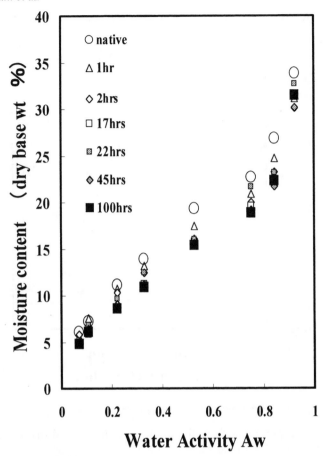

Figure 2 *Water adsorption isotherms for potato starch samples, measured after ball-milling for different amounts of time, as indicated*

equation. This dual-mode model is often used for analysis of sorption of penetrants into glassy polymers, which is expressed by Equation 1,

$$C = k_d p + \frac{C'_H b p}{1 + b p} \tag{1}$$

where C and p are the sorbed amount and relative vapor pressure of penetrant, respectively, k_d is a Henry's Law constant, and b is an affinity constant. C'_H is an adjustable parameter, called a capacity constant in the Langmuir equation, which has been taken to relate to the amount of free volume in a given glassy polymer.[5-8] In our study, water was used as the penetrant, so the sorption amount, C, and relative vapor pressure, p, were replaceable by water content and 'water activity' (A_w), respectively. In Figure 4, a curve fit to experimental sorption data, using Equation 1, for a sample ball-milled for 22 h, is shown as an example.

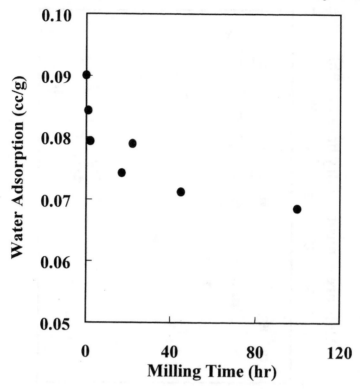

Figure 3 *Variation in BET monolayer value with ball-milling time*

Equation 1 could represent fairly well the equilibrium water adsorption up to 0.8 A_w, so that C'_H could be determined. The C'_H value clearly decreased with increasing milling time, as shown in Figure 5, which is similar in appearance to Figure 3, which showed the corresponding trend in the BET monolayer value. During the early phase of ball-milling, the decrease in C'_H was steep, but subsequently, C'_H appeared to level-off to a constant value after 20 h of milling.

We have also reported that this phenomenon of decreasing water sorption ability for starch during ball-milling shows a trend analogous to that for the relationship between enthalpy relaxation and aging time.[4] These results suggest that free volume in amorphous starch would decrease rapidly with increasing ball-milling time.

4 Conclusion

It was found that the water adsorption ability of starch is reduced by ball-milling. This phenomenon cannot be explained in terms of decreasing crystallinity in starch, but it is thought to be affected by the degree of enthalpic and/or structural relaxation in amorphous starch molecules.

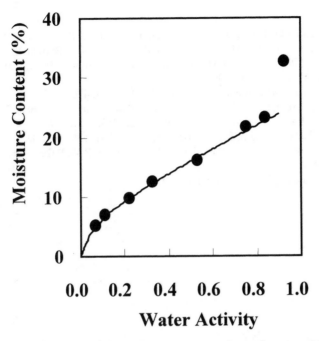

Figure 4 *Water adsorption isotherm for a potato starch sample, measured after 22 h of ball-milling, and a curve (solid line) fit to the data using Equation 1*

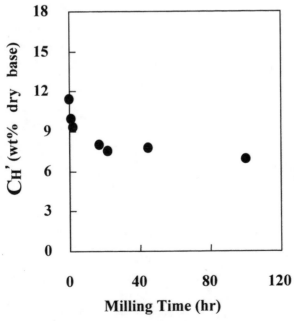

Figure 5 *Variation of capacity constant in Langmuir adsorption equation with ball-milling time*

References

1. Y.J. Kim, T. Suzuki, T. Hagiwara and R. Takai, *Carbohydr. Polym.*, 2001, **46**, 1.
2. T. Yamada, S. Tamaki, M. Hisamatsu and K. Teranishi, 'Starch, Structure and Functionality', P.J. Frazier, A.M. Donald and P. Richmond (eds.), Royal Society of Chemistry, Cambridge, UK, 1997, 59.
3. P. Chinachoti and M.P. Steinberg, *J. Food Sci.*, 1986, **51**, 456.
4. Y.J. Kim, T. Suzuki, C. Pradistsuwana and R. Takai, *Japan J. Food Eng.*, 2001, **2**, 121.
5. R.M. Barrer, J.A. Barrie and J. Slater, *J. Polym. Sci.*, 1958, **27**, 177.
6. A.S. Michaels, W.R. Wieth, and J.A. Barrie, *J. Appl. Phys.*, 1963, **34**, 1.
7. V.T. Stannett, W.J. Koros, D.R. Paul, H.K. Lonsdale and R.W. Baker, *Adv. Polym. Sci.*, 1979, **32**, 69.
8. J.H. Petropoulos, *Adv. Polym. Sci.*, 1985, **64**, 93.

Novel Experimental Approaches to Studies of Amorphous Aqueous Systems

Use, Misuse and Abuse of Experimental Approaches to Studies of Amorphous Aqueous Systems

D.S. Reid

DEPARTMENT OF FOOD SCIENCE AND TECHNOLOGY,
UNIVERSITY OF CALIFORNIA, DAVIS, CA 95616, USA

1 Introduction

In recent years, there has been considerable interest shown in characterizing the behavior of amorphous aqueous systems, particularly under conditions where there are transformations between what might be described as essentially solid-like behavior and behaviors that appear to reflect increased molecular mobility of the constituents.[1] The exact behavior patterns that are observed are mediated by both temperature and moisture content. The kinetic nature of the processes underlying the changes is made evident by observations that demonstrate that both temperature history and compositional history can greatly influence the observed properties of systems that are nominally of the same composition, and also are at the same temperature at the exact moment of measurement.

In order to probe the patterns of change experienced by experimental systems under a range of conditions, a variety of techniques can be, and have been, employed.[2,3] These techniques are employed to monitor the changes that take place in selected properties of the systems as a function of temperature and/or composition and/or time. The challenge to the experimentalist is that of relating the observed properties of a system to the microscopic and molecular processes that are occurring in the system under study, as the measurements are taking place. An additional, non-trivial challenge is to understand the inter-relationships that exist between observations of the same, or similar, systems made using different experimental techniques. Among the techniques employed in the study of amorphous aqueous systems are: differential scanning calorimetry (including modulation methods and stepwise methods),[4–7] dielectric relaxation (including dielectric thermal analysis techniques),[8] nuclear magnetic resonance relaxation and electron spin resonance relaxation measurements,[9,10] and dynamic mechanical thermal analysis (and other related mechanical characterization methods).[11–13]

325

While each of these techniques yields data that exhibit changes in the monitored property (or properties) at identifiable temperatures, or over identifiable temperature ranges, often ascribed to transformations from the solid amorphous state to a more mobile state, the interpretation of the changes can be controversial. The processes under study are often time-dependent, and additionally, the study methodologies have their own time scales for meaningful measurement. Furthermore, the coupling between the experimental sample and the measuring system may be difficult to describe in any quantitative manner. How then do we reconcile the observations made by the different methods, and how do we combine the information to reach a better understanding of the behavioral characteristics of the system?

It is the purpose of this paper to discuss the different perspectives provided by the application of a variety of techniques to the study of amorphous aqueous systems, and to examine the many potential sources of artifact present, and to try to help in discriminating between facts and artifacts generated in such studies.

2 Discussion of Issues

We are a visual people. Charts and diagrams are more easily understood than long strings of numbers in tables. Graphical presentations of data are easier to discuss and characterize. It is appropriate, therefore, for many instrumental techniques to yield some form of graphical display as an end result, for example, the magnitude of a measured property plotted against temperature, or against time, or some other significant variable. These graphical displays can be considered as pictures, or as maps that outline the apparent dependence of the property upon the selected variable. Examination of data sets presented in such a fashion can often lead to the identification of an 'event', where there is a discernable change in the pattern of the record. The locations of such events can be catalogued by specifying the coordinates of the event on the graphical representation of the data. The next step towards better understanding is to characterize the manner in which the locations of these events are influenced by the moisture content of the experimental sample, or by its temperature history. In some cases, it may also be necessary to consider the size of the experimental sample. Studying the patterns of change which result from these procedures can lead to important insights into the factors that influence the system's behavior.

It can be useful to consider the map analogy further. Imagine that you are being given directions to a house. This house is some distance from any main road. It is easier to find the destination, if the directions can identify some significant landmark at the point of turn-off from the main road. To have an instruction that merely specifies 'turn off around 1/3 of the way around the long curve you reach some time after town' will lead to confusion. Turn off to the right, on the road that exits the main road just after the 'direction sign', is much easier to follow. So it is with the description of graphical data presentations. It is much easier to specify a location, if it is associated with some distinct feature of the trace, such as a gradient change, or a clear change in curvature. It is much more difficult to specify an easily located point on a smooth, featureless curve.

For this reason, our data sets for analysis tend to be lists of the graphical coordinates of readily identifiable points on the instrumental data plots. These we presume are directly associated with identifiable causes. This then poses the questions:

- In what way is the position of an identifiable point on the instrument plot related to the occurrence of identifiable change within the experimental sample?
- In what ways are changes within the sample coupled to changes in the instrumental plots?
- What factors influence what we see, and how we perceive it?

In order to answer these questions, it is necessary to look much more closely at the operational characteristics of the instruments that we use to collect our data.

The data set we examine is often in the form of a plot, relating some set of properties to a variable that we manipulate. For example, we might, in a calorimeter, plot heat flow against time, or heat flow against temperature. This raises several more questions; for example:

- How well is the sample coupled to the sensor?
- How homogeneous is the sample?
- How homogeneous is the temperature within the sample?
- Is the temperature we plot the actual temperature of the sample, or the temperature of some specific location that may, or may not, track the temperature of the sample?
- How long does it take for the instrument to detect a change in the sample?
- What is the response time of the sensor?
- Does the size of the sample influence the result obtained from the instrument?
- Does the rate of change of temperature influence the result obtained from the instrument?
- Is the sample composition changing during the measurement?

Some of these questions we will examine more closely, as we consider the situation further. We will examine issues relevant to techniques identified earlier as ones frequently employed in studies of amorphous aqueous systems. The detailed relationships between the changes in the system properties under investigation, and the motions of, and within, the molecules that comprise the system, are the topic of other chapters in this volume, and will not be discussed further here.

It is important at this point to note that the true specific values of properties under kinetic control might well be influenced by rate of change of temperature, and by sample size, whilst the true specific values of thermodynamic properties should be independent of sample size, and of rate of change of temperature. Since the temperature of a sample in an instrument is a result of the instrumental program, and instrumental operational factors, it is quite likely that the actual sample temperature will be influenced by the design of the instrument, by the characteristics of the 'furnace' or other controlled-temperature environment within which the sample resides, by the heat transfer characteristics of both instrument and sample, and by the ability of the instrument to respond rapidly to program calls for temperature change, as well as by the uniformity of temperature fields within the instrument. The existence or otherwise of a sealed environment will also influence sample composition, in that the tendency for a sample to gain or lose moisture will be influenced both by the sample's immediate environment, and by the existence of mass-transfer paths between the sample and the environment beyond the immediate vicinity of the sample. We must ask the question, is the sample isolated from the external environment, or are there mass-transfer paths that will allow for changes in the composition of the sample?

Some system properties are expected to be temperature-dependent. These include thermodynamic properties and rates of processes, and molecular mobility changes observed as a consequence of the alteration of the temperature of a sample may reflect contributions from any or all of these. System properties are frequently found to exhibit time-dependence. Time-dependent properties include the extent of a process, and absolute rate. When considering experimental data, therefore, we must consider the whole temperature–time history, including the history imparted by the utilization of the experimental apparatus. If we were somehow able to instantaneously monitor at all times the actual properties of the sample, often what we would see would differ significantly from the instrumental display.

Differential scanning calorimetry (DSC) is a technique frequently used to help characterize amorphous aqueous systems. In DSC, a sample and a reference are subjected to a steady change in temperature, and the differential heat flow into sample and reference is recorded. Both sample and reference may be contained in hermetically sealed pans or crucibles, each placed in an instrumental location well-coupled to a sensitive temperature measurement capability. Two principal types of instrument exist. In one type, the property measured is the difference in temperature seen between sample and reference, in an instrument in which the heat flow system is passive. The temperature difference is proportional to the different rates of heat flow required by sample and reference, as the DSC follows a preset instrument temperature program. In the other type of DSC, the temperature difference between sample and reference is used to drive a power supply system that seeks to actively maintain sample and reference at the same temperature. Thus, though the principles of operation are different, one type being passive and the other active, in both cases, the temperature difference between sample and reference is the key measurement.

It may be instructive to give some consideration to typical data sets obtainable from the use of DSC. Since we are often interested in systems in which some of the water may form ice at lower temperatures, we will consider such systems. Figure 1a illustrates the form of idealized enthalpy–temperature plots for a variety of aqueous systems. In DSC, what we expect to see is the heat flow associated with the changing heat content of the sample, as the temperature is changed at some specified rate. To see what one might expect, we need to look at the derivative plots from Figure 1a, on the assumption that the rate of temperature change with time is constant, such that dH/dt is proportional to dH/dT.

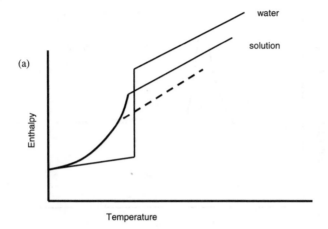

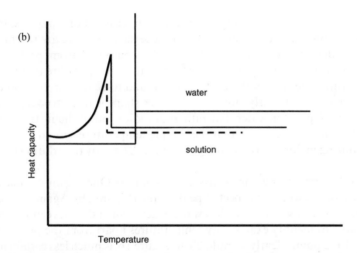

Figure 1 (a) *Enthalpy–temperature plots for water and a binary aqueous solution. The dashed line represents a higher concentration solution.* (b) *Heat capacity–temperature plots for the systems in* (a)

These derivative plots are shown in Figure 1b. Figure 2 shows a typical instrument plot resulting from the melting of ice. Heat flow appears to be plotted against temperature. However, is this truly so? Figure 1b shows what we should expect, given that ice melts at 273 K. Why the difference? What can we learn from the distorted view provided by the DSC? Why is the view distorted?

First, we notice that the DSC melting curve starts at 273 K. This truly is the melting point. However, since the differential heat flow into the sample is driven by a temperature difference, the rate of heat supply is limited. At the same time, the instrument is programmed to maintain a linear rate of temperature increase. Though the ice melts at a constant temperature, the program temperature is

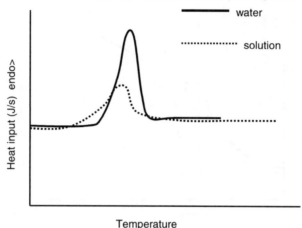

Figure 2 *Schematic DSC plots for water and a binary aqueous solution. The dashed line represents the solution*

increasing, and we are plotting the heat flow (measured or computed from ΔT) against the program temperature. This is necessary, since we are plotting dH/dt, and in order for the total heat flow to be computed from peak areas, the temperature axis for the plots must be signifying a temperature that is changing linearly with time. The peak area is indeed found to be the correct heat of fusion for the ice sample. Clearly, however, the peak shape is not a true representation of the dH/dT plot for water, but rather an indication of the instrumental lags. Since we know, in this case, that the sample temperature should be constant at 273 K during melting, it is possible to estimate ΔT at any time during the melting step.

Consider now the melting behavior of a solution. Once again, compare a DSC trace (Figure 2) with the expected peak shape (Figure 1b). Again we see distortion, but in this case, unless we know the exact form of the curve in Figure 1b, we are unable to identify exactly any temperatures apart from the peak start. The final melting point, clearly identifiable in Figure 1b, is much less certain in Figure 2, as we know that the existence of a temperature differential is accompanied by an uncertainty in the actual sample temperature. We can correct for this uncertainty, but the accuracy of the correction can be problematic.

The best indication of the difficulty we face is the realization that it is not possible to construct the melting curve for a phase diagram from just one DSC melting scan for a dilute solution, even though, as Figure 1b shows, the true derivative curve for a dilute solution overlays that for a more concentrated solution, up until the temperature of final melting of the more concentrated solution.

In Figure 3, the enthalpy–temperature and heat capacity–temperature plots are extended to a lower temperature, below the eutectic temperature for a binary system. These diagrams show what is to be expected for a binary system in which the second component also crystallizes out. In passing, it should be noted that the test of acquiring the true melting curve for a solution from data obtained

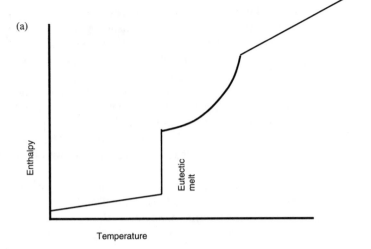

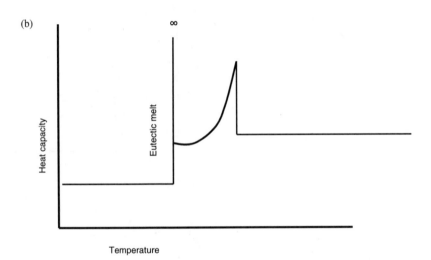

Figure 3 (a) *Enthalpy–temperature plot for a binary aqueous solution with eutectic.* (b) *Heat capacity–temperature plot for the system in* (a)

using a particular calorimetric procedure can be a very powerful tool to validate proposed procedures.

As yet, we have not considered the effect of sample size, or of temperature scanning rate. For the curves of Figure 1, sample size and rate of change of temperature do not influence the curve shapes in any way. These curves can be plotted in units of J g^{-1} sample, and no matter what the sample size, the plots for the same material will be identical. This, however, is not the case for DSC plots. After normalizing by dividing the measured heat flow by the appropriate sample mass, one sees that the plots for different sample sizes of the same material do not superpose. This is a result of the finite time taken for heat to transfer into the sample, and of the internal temperature gradients established within the measuring cell. It is also found that scans at different rates do not exhibit the same temperature profiles, again as a result of the finite time taken for heat transfer to be accomplished.

The above suggests that an important test of the value of a datum is that it be independent of sample size, and of temperature scanning rate. If this does not prove to be the case, there exists an uncertainty as to whether the discrepancy is a result of a property of the sample, or a result of instrumental effects. The first may provide important insights; the second may lead to misinterpretations of system properties.

In Figure 4, the expected enthalpy change and corresponding change in dH/dT, which accompany the transformation between a liquid and an amorphous solid, are indicated. While the melting behavior in the region of Figures 1–3 has been assumed to be controlled by thermodynamic factors, the behavior in the liquid–amorphous solid region results from kinetic factors, and so the

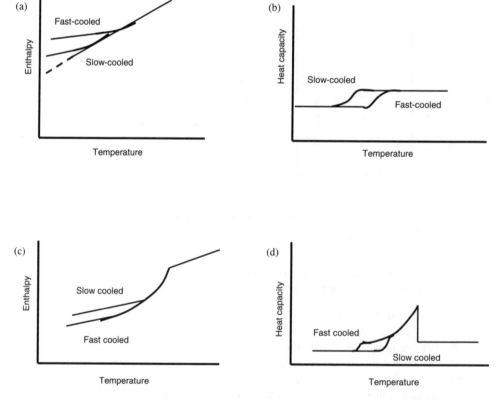

Figure 4 (a) *Enthalpy–temperature plot for a liquid/glass transformation in a homogeneous system.* (b) *Heat capacity–temperature plot for the system in (a).* (c) *Enthalpy–temperature plot for a binary aqueous system forming ice, followed by a concentrated aqueous glass.* (d) *Heat capacity–temperature plot for the system in (c)*

history of a sample may influence the exact pathways seen. The plots shown may be assumed to represent the boundaries of the enthalpy–temperature plot. Figure 4 shows schematically the pattern of change expected for samples cooled at different rates. Figures 4a and 4b show what to expect, if the system is homogeneous. Note that more rapid cooling results in a higher temperature of transition. Figures 4c and 4d show what to expect, if the system, on cooling, separates out ice, but, unlike in Figure 3, does not form a eutectic solid. Ice will continue to separate, and the unfrozen phase to increase in concentration, as cooling proceeds. At some point, the unfrozen component will undergo a liquid–amorphous solid phase transition. Note that in this case, due to the finite rate of ice formation at lower temperatures, more rapid cooling results in a lower temperature of transition, because the amorphous phase is of lower concentration than that formed with slower cooling.

In addition to standard DSC, a variety of approaches have been suggested to help eliminate some of the shortcomings of the technique. For power-compensated DSC instruments, one approach has been to program the temperature change in a series of jumps, with isothermal interludes (a stepwise approach); the total differential heat flow of each step can be determined, and the instrumental contribution subtracted out by having an appropriate control experiment. This eliminates many of the problems encountered in establishing the true thermodynamic behavior, and indeed, this type of procedure allows for the construction of the true melting curve from the melting behavior of one dilute sample. This procedure, in that it requires an isothermal equilibration period after each temperature step, is more limited, when kinetic factors are under investigation. If the response time of the calorimeter is sufficiently fast, the signal decay, after the initial instrument transient, may provide information about the kinetics of the approach to equilibrium after the step change. A related approach, in which step time and hold time are of equivalent duration, can be subjected to a variety of analyses claimed to yield kinetically appropriate data.

In heat flow instruments, an approach in which the steady change in instrument temperature is modulated by the superposition of a smaller cycling temperature fluctuation is used. Here, the analysis also attempts to separate kinetic and thermodynamic contributions. Here, it is helpful to validate results by using different sample sizes, different average scan rates, and different modulation amplitudes and frequencies. One concern with both approaches that separate out kinetic and thermodynamic factors is that neither seems able to produce a true melting curve for a frozen dilute system.

The stepwise approach, which can yield the true melting curve, has a hold time that is assumed to be long enough to allow the process at each temperature to reach equilibrium. Much more needs to be done to define the limitations of the techniques that seek to probe the kinetically controlled changes. It is not entirely clear to what extent these approaches can truly separate out kinetic factors characteristic of the sample from kinetic factors characteristic of the instrument. Many more tests are required, studying systems whose behaviors are already well characterized, to quantify the limitations of these potentially valuable techniques. An instructive experiment might be to look at the melting of a binary aqueous eutectic system by these techniques.

One further question that must be considered, when interpreting data provided in DSC traces, is that of the nature of the event to be characterized. First-order phase transitions, such as ice melting, have well-defined temperature relationships. In a binary system, the eutectic temperature and the final melting point for any composition are well-defined. However, if in a binary system, the second component fails to crystallize, such that the liquid phase, as its concentration in this component increases during cooling, transforms into an amorphous solid phase, then only the final melting temperature for the system is well-defined. Unlike the eutectic temperature of the first system, which identifies the start of melting, the start of melting in the system with an amorphous component is not a well-defined temperature, but rather a temperature range, with the exact temperature being dependent upon the exact conditions of formation of the amorphous phase, and also the exact conditions of heating. If one examines a melting curve for such a system, the point at which the curve departs from the baseline is difficult to specify, and may depend to some extent upon the instrumental characteristics. It also depends to some extent upon the sample history.

In the phase/state diagram of Figure 5, this point is made by plotting well-defined temperatures as sharp lines, and temperature ranges as wider bands. Sharp lines indicate a property under thermodynamic control, while broad bands indicate a temperature influenced by kinetic and historic factors.

One approach to defining a more easily observed temperature, for characterization of amorphous state transitions, than the onset temperature is to define the mid-point temperature. This is indeed more easily characterized, but it is still difficult to define the breadth of the transitional event for which the reported temperature is the mid-point. Also, since the mid-point exhibits a larger differential temperature change than that at the onset, errors due to instrumental deficiencies are likely to be greater. It is therefore important to check whether the observed mid-point temperature is dependent upon the rate of temperature scanning, or the size of the experimental sample. Any differences could be a result

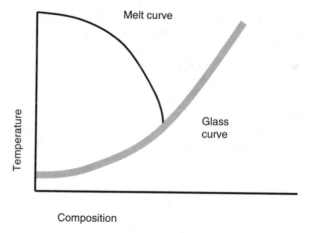

Figure 5 *Schematic state diagram for a glass-forming binary aqueous system*

either of kinetic factors within the sample, or of instrumental factors.

In a similar vein to our discussion of DSC, let us look at measurement of mechanical properties. Several mechanical parameters might be of interest, as a system transforms between amorphous solid, rubbery, and liquid states. Viscosity, viscoelasticity, elasticity, volume, and expansibility are all parameters that would be expected to reflect the changing state. Measurement of each property brings with it a particular set of challenges. One challenge is that of containment. A sealed sample container, appropriate for calorimetry, is seldom feasible for mechanical measurement. The nature of the sample also has an influence on the approach. A liquid sample can be contained in some form of cup, and a probe immersed in the sample can measure viscoelastic properties. A solid sample in the same cup may have the probe in contact with the surface, and expansion can be followed. When the solid begins to transform into a more mobile state, the probe will begin to sink below the surface. The rate of sinking will depend upon the viscosity. A solid may also be formed into some appropriate shape, and clamped into a measuring system. A particular challenge of many systems is how to achieve thermal coupling to the sample, for temperature control, without interfering with the capabilities of the mechanical sensor.

When we measure mechanical properties of amorphous aqueous systems, as their temperatures are raised, once again we have to ask the question, is what we see a representation of reality, or a distorted representation of reality. Also, we must ask to what extent we can make allowances for the distortion, and to what extent the distortion will prevent us from learning what we seek to know. As temperature is raised, the system becomes more mobile, and the mechanical properties reflect this.

As mentioned, unlike the DSC situation, it is difficult, and often impossible, to provide a sealed environment for the sample, without significantly compromising either the coupling of sample to sensor, or of sample to temperature regulating system. There may exist vapor transfer paths. The coupling of mechanical sensors to a sample can be a major instrumental challenge. Couplings that are suited to liquid, viscous systems are unsuited to solid, elastic systems. For example, while the coupling of a solid sample to a mechanical sensor can be achieved by a variety of clamp designs, and a variety of sample shapes can be utilized to simplify the coupling, the coupling of the same mechanical sensor to a liquid is a much greater challenge. One solution has been to use an inert, flexible and porous substrate, infused with the sample, and to follow the change in the properties of the sample/substrate complex as the temperature is changed. These properties can be compared to the properties of the substrate alone, to deduce the contribution made to the combined properties by the experimental sample. An alternative approach, following the liquid flow properties of the system by flow rheology, becomes problematic, once the system becomes solid.

Given the complexity of the sensor–sample coupling, there exist many opportunities for ambiguity as to the source of observed change. In addition, both the size of the sample and the potential for moisture transfer result in a sample whose temperature and composition are difficult to specify exactly, as there may be both internal temperature gradients and internal moisture content gradients within the sample. Sample size and the time–temperature parameters of the experimental protocol may both influence the observed behavior.

Even absent the complications of sample interaction with the environment, and sample–sensor coupling, the mechanical behavior is expected to show a dependence upon the frequency of any applied mechanical perturbation. The higher the frequency of the applied stress, the higher the temperature at which the amorphous transition is seen to influence the modulus. This is illustrated schematically in Figure 6. The shape of the curves suggests that it is difficult to specify an onset temperature, and that it may be easier to identify a mid-point temperature. Once more, the range of temperature from onset to completion of the process is difficult to determine precisely. Once again, too, it is important to look at the effects of sample size and temperature scanning rate on the identified temperatures.

Given that the transformation from the amorphous solid state to a more liquid-like state is a result of increased molecular mobility, both rotational and translational, spectroscopic techniques that identify significant frequencies of motion can provide important information. It is found that dielectric properties detect the increasing motions of the molecules in a system as increases in the absorption of input energy, with the significant temperatures, as is the case with mechanical relaxations, exhibiting frequency dependence. The measurement of dielectric properties is a non-invasive technique, an appropriate measurement being the electrical capacitance in a parallel plate cell. Coupling is simple. The challenge is characterizing the heat transfer, and therefore being able to describe the thermal state of the sample. Another challenge is that of changing temperature sufficiently rapidly to be able to follow kinetic transients. The difficulty of these challenges is very much linked to sample size, and also to the mass and dimensions of the measuring cell. The realities of heat transfer dictate the maximum rate of temperature change possible, and also dictate the maximum rate of temperature change consistent with minimizing temperature gradients

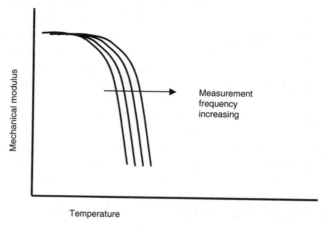

Figure 6 *Effect of temperature on mechanical properties of amorphous systems (schematic). The trace to the left represents the measurement at the lowest frequency, that to the right at the highest frequency*

within the sample.

Both NMR and ESR techniques can be employed to obtain data on molecular mobilities. As with dielectric techniques, the primary challenge in the study of amorphous aqueous systems is the need to change temperature, in order to observe the evolution of the molecular mobilities of the system. Sample sizes tend to be fairly large, and so two challenges exist: that of changing the sample temperature in a well-characterized fashion, and also that of preventing significant temperature gradients within the sample. The important factors are those identified for dielectric measurements.

One question that has arisen frequently, as we consider different study techniques, is that of the uniformity of the sample temperature, and also the relationship between the sample temperature and the temperature indication provided by the experimental apparatus. It can be helpful to have an appreciation of temperature offsets and thermal lags inherent in an apparatus. This is simpler to investigate than one might think. Consider the behavior of a eutectic system on heating. There are two significant events, the onset of melting, at the eutectic temperature, and the end of melting. By placing such a system in the sample receptacle of an apparatus, one can determine the apparent temperatures of onset and completion of melting. The dependence upon amount of eutectic mixture, for the same sample size (achieved by changing the concentration of the system), and upon total sample size (with either the same fraction of eutectic mix or the same amount of eutectic mix) can provide valuable insights about the thermal coupling and thermal lags of the experimental apparatus.

3 Conclusion

In conclusion, when evaluating experimental data on amorphous state transitions, it is important to consider the inevitable distortions of reality resulting from the inherent characteristics of the coupling between sample and measuring system. The size of the sample utilized must be considered. Is a small sample representative of the system? Does a large sample exhibit uniformity of composition and temperature? The coupling between sample and sensor must be understood. In what ways does this coupling distort the information being collected? Are there any interferences present that may cause problems in collecting representative data? Finally, an important factor to consider is the ease with which the data can be interpreted. Not all techniques yield data whose meaning is immediately apparent.

References

1. H. Levine and L. Slade (eds.), 'Water Relationships in Foods', Plenum Press, New York, 1991.
2. H.D. Goff, 'Quality in Frozen Food', M.C. Erickson and Y.C. Hung (eds.), Chapman and Hall, New York, 1997, 29.
3. D. Champion, M. Le Meste and D. Simatos, *Trends Food Sci. Technol.*, 2000, **11**, 41.
4. S. Ablett, M.J. Izzard and P.J. Lillford, *J. Chem. Soc. Faraday Trans.*, 1992, **88**, 789.

5. M.E. Sahagian and H.D. Goff, *Thermochim. Acta*, 1994, **246**, 271.

6. W.L. Kerr and D.S. Reid, *Thermochim. Acta*, 1994, **246**, 299.

7. A.M. Lammert, R.M. Lammert and S.J. Schmidt, *J. Therm. Anal. Calorim.*, 1999, **55**, 949.

8. T.R. Noel, R. Parker and S.G. Ring, *Carbohydr. Res.*, 2000, **329**, 839.

9. M.A. Hemminga, I.J. van den Dries, P.C.M.M. Magusin, D. van Dusschoten and C. van den Berg, 'Water Management in the Design and Distribution of Quality Foods', Y.H. Roos, R.B. Leslie and P.J. Lillford (eds.), Technomic Publ., Lancaster PA, 1999, 255.

10. I.J. van den Dries, D. van Dusschoten, M.A. Hemminga and E. van der Linden, *J. Phys. Chem. B*, 2000, **104**, 10126.

11. W.M. MacInnes, 'The Glassy State in Foods', J.M.V. Blanshard and P.J. Lillford (eds.), Nottingham University Press, Loughborough, UK, 1993, 223.

12. G. Blond, *J. Food Eng.*, 1994, **22**, 253.

13. M.T. Kalichevsky, E.M. Jaroszkiewicz, S. Ablett, J.M.V. Blanshard and P.J. Lillford, *Carbohydr. Polym.*, 1992, **18**, 77.

Glass Transition and Ice Crystallisation of Water in Polymer Gels, Studied by Oscillation DSC, XRD–DSC Simultaneous Measurements, and Raman Spectroscopy

Norio Murase,[1] Masatoshi Ruike,[1] Sumie Yoshioka,[2] Chihiro Katagiri[3] and Hiroshi Takahashi[4]

[1]DEPARTMENT OF BIOTECHNOLOGY, SCHOOL OF SCIENCE AND ENGINEERING, TOKYO DENKI UNIVERSITY, HIKI-GUN, SAITAMA 350-0394,
[2]NATIONAL INSTITUTE OF HEALTH SCIENCES, SETAGAYA-KU, TOKYO 158-8501,
[3]INSTITUTE OF LOW TEMPERATURE SCIENCES, HOKKAIDO UNIVERSITY, SAPPORO 060-0819,
[4]FACULTY OF ENGINEERING, GUNMA UNIVERSITY, MAEBASHI-SHI 371-8510, JAPAN

1 Introduction

Water in polymer gels of a certain crosslink density is considered to remain partially unfrozen during freezing and to transform into a glassy solid state, which then undergoes ice crystallisation during rewarming. This assumption is supported by the fact that the amount of unfrozen water, estimated by differential scanning calorimetry (DSC), is greater in such gels than in other gels differing only in crosslink density.[1] However, a glass transition has not yet been detected in such gels. In this connection, a DSC rewarming trace shows an endothermic trend prior to the exotherm caused by ice crystallisation. The origin of the endothermic event is unclear; is it due to the melting of small ice crystals, where their melting temperature is somewhat depressed by way of the Kelvin effect, or it is due to a glass transition with some enthalpy relaxation?[1–3] A typical DSC rewarming trace illustrating such behaviour is shown in Figure 1 for reference.

The characteristic behaviour shown in a DSC rewarming trace should be rationally explained by corresponding physical events occurring in the gel. For clarification of the mechanism of glass transition, as well as ice crystallisation during rewarming, oscillation DSC and XRD (X-ray diffraction)–DSC simultaneous measurements were performed in this study. Raman spectroscopic studies

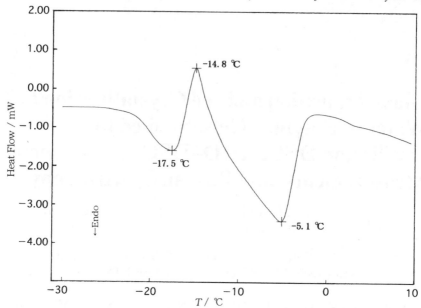

Figure 1 *Typical DSC heating trace obtained for a frozen Sephadex G25 gel containing 50 wt% water. Cooling and heating rates were $10°C$ min^{-1} and $3°C$ min^{-1}, respectively*

were also carried out to obtain further information about the frozen state of the gel.

2 Materials and Methods

2.1 Samples

A type of crosslinked dextran, Sephadex G25, obtained from Amersham Pharmacia Biotech (Uppsala), was used without further purification. The size of the Sephadex G25 bead used was 50–150 μm in diameter. The water content of the gel was adjusted to 50 wt%, as the anomalous freezing behaviour mentioned above is typically observed around that water content.[1] For our Raman spectroscopy study, Sephadex G100, another type of dextran with lower crosslink density than that of Sephadex G25, and which does not exhibit anomalous freezing behaviour, was also used for comparison.

2.2 Oscillation DSC

About 10 mg of Sephadex G25 gel with 50 wt% water content were prepared directly in an aluminium DSC pan, which was then hermetically sealed. Samples were cooled, at 5 °C min⁻¹, from 20 °C down to − 50 °C, and then heated to 20 °C, at 1 °C min⁻¹. The frequency and amplitude of temperature oscillation during heating were 0.01–0.05 Hz and 0.5–3 °C, respectively. The instrument used was a DSC-061 (Seiko Instruments, Japan).

2.3 XRD–DSC Simultaneous Measurements

Several mg of gel were put in an aluminium pan without a lid, promptly cooled to -50 °C at 2–3 °C min^{-1}, then heated at 1 or 0.5 °C min^{-1}. Measurements were performed with an XRD–DSC II (Rigaku Corporation, Japan), equipped with a RINT Ultima$^+$/HP for XRD measurements. X-rays of 50 kV, 40 mA wavelength of CuK$_\alpha$, were used. The scanning rate for diffraction angle (2θ) was 10 or 20° min^{-1} and the scanning range for 2θ was between 20° and 64°.

2.4 Raman Spectroscopy

A cuvette, of outer size $10 \times 6.5 \times 10$ mm (w $\times$ d $\times$ h) and 1.25 mm thickness, was filled with gel of 50 wt% water content, and sealed with a teflon stopper. The cuvette, in contact with a micro-miniature refrigerator (MMR Technologies, USA) on one side of its surface, was placed in a chamber, together with the refrigerator system. The chamber, containing an optical window, was evacuated to prevent water condensation and ice formation on the surface of the cuvette. The sample-containing cuvette was cooled, at about 2 °C min^{-1}, from 27 °C to -40 °C in the cell holder of a Raman spectrometer. The cuvette was subsequently heated to a predetermined temperature with an electric heater. Spectra were obtained with a Raman 960 spectrometer (Thermo Nicolet, USA), in which a 1064 nm YAG laser at 450 mW was used to irradiate the sample through the optical window of the chamber, and Raman scattering at a 180° direction was observed.

3 Results and Discussion

The magnitude of the endothermic event, starting at about -20 °C prior to the exotherm during rewarming, did not vary in oscillation DSC measurements, and was found to be independent of the frequency and amplitude of temperature oscillation used in this study, as shown in Figure 2. Those results suggest that the event was not of kinetic origin, due to enthalpy relaxation, nor was it the change in heat capacity at a glass transition, because the magnitude of the event was too large to be such a change in heat capacity.[2,3]

Five peaks due to ice crystals were observed between $2\theta = 22°$ and 41° by XRD measurement, and they were all ascribed to hexagonal ice, I_h. However, a diffraction peak at 58°, characteristic of cubic ice, I_c, and the presence of which would be indirect evidence of a glass transition, was not observed (as shown in Table 1). The temperature dependence of XRD intensity differed remarkably among the three peaks between 22° and 26°, as shown in Figure 3. Peak 1, representing ice crystal growth along the a axis (in the prism plane), increased most during the exotherm in the DSC rewarming trace, and peak 3 was second-

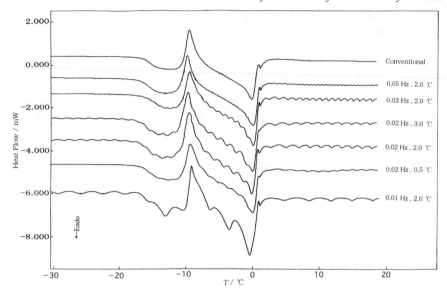

Figure 2 *Heating traces obtained by oscillation DSC, using the indicated oscillation conditions*

Table 1 *Description of XRD peaks*

			I_h (hexagonal ice)		I_c (cubic ice)	
		2θ deg	d (nm)	hkl	d (nm)	hkl
Peak	1	22.1–23.4	0.390	(100)		
	2	23.7–24.6	0.368	(002)	0.368	(111)
	3	25.3–26.4	0.344	(101)		
	4	33.1–33.9	0.267	(102)		
	5	39.3–40.3	0.226	(110)	0.225	(220)

$n\lambda = 2d \sin\theta$; $n = 1$; $d = \lambda/2 \sin\theta$; $\lambda = 0.15418$ nm

most. In contrast, peak 2, representing ice crystal growth along the c axis, did not increase during the exotherm. That result might reflect directional growth of ice crystals in the polymer network.

A significant decrease in XRD intensity, suggesting melting of ice crystals, was not observed for the three peaks during the endothermic event in the DSC scan. Therefore, it could not be concluded from the present study that the endothermic event was due to ice melting. However, it was possible that a decrease in the integrated XRD intensity during the endothermic event was concealed by an integration error for the broad peak representing the formation of small ice crystals, overlapping with the noise level. Another possibility was that a decrease in the integrated XRD intensity was concealed by the remarkable intensity fluctuations caused by gradual melting of ice, which might have induced movement of gel beads and a resultant change in the angle of X-rays irradiating ice

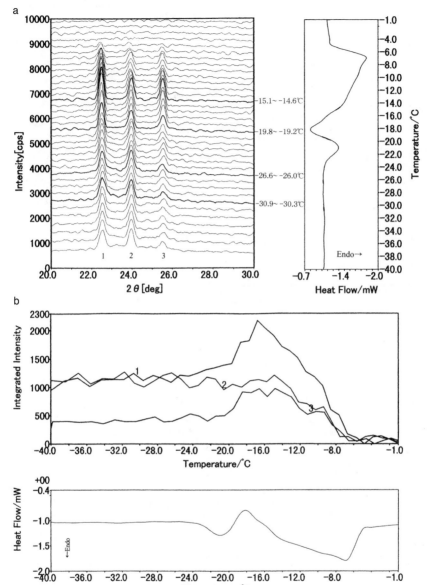

Figure 3 (*a*) *Heating traces obtained by simultaneous XRD–DSC measurements;* (*b*) *Temperature dependence of the integrated XRD intensity calculated from the data shown in* (*a*)

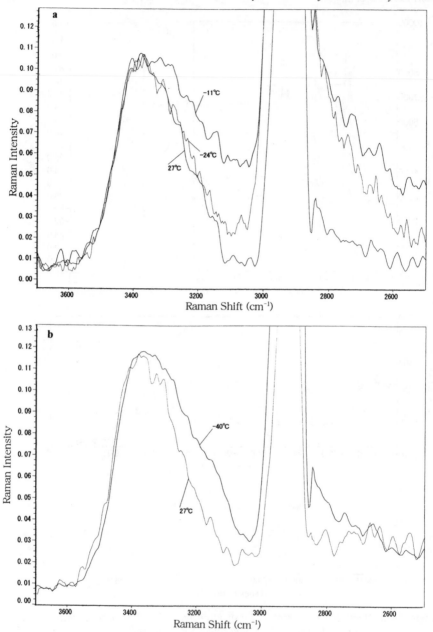

Figure 4 (a) *Raman spectra of OH stretching band, observed for a Sephadex G25 gel during rewarming*; (b) *Raman spectra of OH stretching band, observed for a Sephadex G100 gel*

crystal surfaces. In that case, the size of the ice crystals might not have been so small.

If we assume that the endothermic event was due to the melting of small ice crystals, the size of those ice crystals can be estimated using both the Laplace and Gibbs–Duhem equations, as discussed elsewhere.[3] That size, in terms of radius, was estimated to be several nanometers. For that estimation, however, the interfacial energy between ice and hydrated dextran was assumed to be the same value as that between ice and water, 30 mJ m^{-2}, according to the literature.[4] The actual size of the ice crystals could be larger than the estimated size, because the interfacial energy between ice and hydrated dextran would be expected to be larger than that between ice and water, but that would need to be confirmed by precise XRD analysis.

Raman spectra observed for a Sephadex G25 gel are shown in Figure 4a. The spectrum due to the OH stretching band (~ 3400 cm^{-1}), observed at $-24\,°$C, which is similar to that observed at ambient temperature of $27\,°$C, was compared with the spectrum observed at $-11\,°$C, during rewarming after the occurrence of ice crystallisation. The increase in Raman intensity in the region of wave number lower than 3400 cm^{-1}, when sample was heated to $-11\,°$C, was interpreted as corresponding to a strengthening of the hydrogen bond network resulting from ice crystallisation. Raman spectra observed for a Sephadex G100 gel are shown in Figure 4b. In this case, the spectrum observed for the frozen gel at $-40\,°$C was different from the spectrum at ambient temperature, and was very similar to the spectrum observed for the G25 gel at $-11\,°$C, shown in Figure 4a. Therefore, the Raman spectra observed for the frozen G25 gel at $-24\,°$C, similar to that for liquid water, represents evidence for the presence of glassy water in the frozen gel.

4 Conclusions

(1) The exotherm in the DSC rewarming trace, observed for a Sephadex G25 gel, was confirmed, by XRD measurements, to be due to I_h formation. Moreover, we found that there was a directional growth of ice crystals within the polymer gel.

(2) However, the origin of the endothermic event, *i.e.* whether it was due to enthalpy relaxation at the time of glass transition, or due to the melting of small ice crystals, for which the melting temperature was depressed by the Kelvin effect, still remains unclear.

(3) By means of Raman spectroscopic studies, the presence of glassy water in a Sephadex G25 gel in the frozen state was confirmed.

Acknowledgements

The authors are most grateful to Dr Akira Kishi, Rigaku Corporation, Japan, for support with XRD–DSC simultaneous measurements. They are also grateful to Mr K. Horie for oscillation DSC measurements and to Mr M. Komatsu and Mr

J. Hoshi, Thermo Nicolet Japan, Mr A. Urata, International Servo Data Corp. and Mr D. Terada, Nissei Science Corp. for Raman spectra measurements.

References

1. N. Murase, K. Gonda and T. Watanabe, *J. Phys. Chem.*, 1986, **90**, 5420.
2. N. Murase, *Cryo-Letters*, 1993, 14, 365.
3. N. Murase, T. Inoue and M. Ruike, *Cryo-Letters*, 1997, 18, 157.
4. K.L. Kerr, R.E. Feeney, D.T. Osuga and D.S. Reid, *Cryo-Letters*, 1985, **6**, 371.

Subject Index